ESR and NMR of Paramagnetic Species in Biological and Related Systems

NATO ADVANCED STUDY INSTITUTES SERIES

*Proceedings of the Advanced Study Institute Programme, which aims
at the dissemination of advanced knowledge and
the formation of contacts among scientists from different countries*

The series is published by an international board of publishers in conjunction
with NATO Scientific Affairs Division

A	Life Sciences	**Plenum Publishing Corporation**
B	Physics	**London and New York**
C	Mathematical and Physical Sciences	**D. Reidel Publishing Company** Dordrecht, Boston and London
D	Behavioral and Social Sciences	**Sijthoff International Publishing Company** **Leiden**
E	Applied Sciences	**Noordhoff International Publishing** **Leiden**

Series C – Mathematical and Physical Sciences

Volume 52 – ESR and NMR of Paramagnetic Species in Biological and Related Systems

ESR and NMR of Paramagnetic Species in Biological and Related Systems

Proceedings of the NATO Advanced Study Institute
held at Acquafredda di Maratea, Italy, June 3-15, 1979

edited by

IVANO BERTINI
University of Florence, Italy

and

RUSSELL S. DRAGO
University of Illinois, Urbana, U.S.A.

D. Reidel Publishing Company

Dordrecht : Holland / Boston : U.S.A. / London : England

Published in cooperation with NATO Scientific Affairs Division

Library of Congress Cataloging in Publication Data

NATO Advanced Study Institute, Acquafredda di Maratea, Italy, 1979.
 ESR and NMR of paramagnetic species in biological and related systems.

 (NATO advanced study institute series : Series C, Mathematical and
physical sciences ; v. 52)
 Includes index.
 1. Electron paramagnetic resonance—Congresses. 2. Nuclear magnetic
resonance—Congresses. 3. Biology—Technique—Congresses. I. Bertini,
Ivano. II. Drago, Russell S. III. Title. IV. Series.
QH324.9.E36N37   1979   574'.01'5383   79-24701
ISBN-13: 978-94-009-9526-0   e-ISBN-13: 978-94-009-9524-6
DOI: 10.1007/978-94-009-9524-6

Published by D. Reidel Publishing Company
P.O. Box 17, Dordrecht, Holland

Sold and distributed in the U.S.A., Canada, and Mexico
by D. Reidel Publishing Company, Inc.
Lincoln Building, 160 Old Derby Street, Hingham, Mass. 02043, U.S.A.

CONTENTS

PREFACE

Numerous investigations over the years have illustrated the importance of physical methods, especially magnetic resonance, in the elucidation of the electronic ground states of 3d transition metal ion complexes. The results of these studies coupled with investigations on model compounds are being extended in very meaningful ways to provide electronic structural information in biological systems. In systems not readily amenable to investigation by single crystal X-ray diffraction these methods provide structural information as well. It was felt that a NATO Advanced Study Institute that brought together experts in theoretical aspects of magnetic resonance, experts in biology and experts in both areas would be extremely profitable for all parties attending the school. The enthusiastic response of the participants indicated that our objectives were accomplished to a high degree.

We hope this publication of the proceedings will transmit, to some degree, the stimulating discussion of the Conference. The enormity of the area selected for this Conference can be appreciated by all. We the editors take full responsibility for the many omissions resulting from the imposed limitations of time and resources.

Florence, July 1979

I. Bertini   R. S. Drago

PARTICIPANTS

Andersen, J. P.   Institute of Medical Biochemistry –
                         University of Aarhus DK-8000 Aarhus C
                         Denmark
Andersson, I.   Fachbereich 15.2, Analytische und
                         Biologische Chemie, Universität des
                         Saarlandes – D-6600 Saarbrücken 11
                         F. R. G.
Basosi, R.   Istituto di Chimica Generale, Univer-
                         sità di Siena – Via Pian dei Mantelli-
                         ni, 44 – Siena
                         Italy
Bellomo, P.   Alfa Farmaceutici – Via Ragazzi del
                         '99, 5 – Bologna
                         Italy
Bertini, I.   Istituto di Chimica Generale ed Inor-
                         ganica, Facoltà di Farmacia – Univer-
                         sità di Firenze – Via Gino Capponi, 7
                         Firenze
                         Italy
Bonomo, R.   Istituto Dip. di Chimica e Chimica In-
                         dustriale, Città Universitaria, Viale
                         A. Doria, 18 – Catania
                         Italy
Borremans, F.   Laboratorium voor Organische Chemie
                         Rijksuniversiteit, B-9000 Gent
                         Belgium
Branch, S. K.   Department of Chemistry – University
                         of Sheffield, Sheffield S3 7HF
                         United Kingdom
Burton, D.   Department of Biochemistry, Univ. of
                         Oxford, Oxford OX1 3QR
                         United Kingdom
Cannistraro, S.   Istituto Chimico, Università della
                         Calabria, Arcavacata di Rende, CS
                         Italy
Colligiani, A.   Istituto di Chimica Fisica, Università
                         di Pisa, 56100 Pisa
                         Italy

Cooper, S. R.   Department of Chemistry – University
                                 of California at Berkeley, CA 94720
                                 U. S. A.

Corden, B. B.   School of Chemical Sciences – Univer-
                                 sity of Illinois at Urbana-Champaign,
                                 Urbana Illinois 61801
                                 U. S. A.

Crouigneau, P. M. R.   Laboratoire de Chimie I – 40, Av. du
                                 Recteur Pineau – 86022 Poitiers
                                 France

Cucinotta, V.   Istituto Dip. di Chimica e Chimica In-
                                 dustriale, Città Universitaria, Viale
                                 A. Doria, 18 – Catania
                                 Italy

da Conceiçao Rangel, M.   Instituto de Ciencias Biomedicas
                                 "Abel Salazar" Universidade do Porto,
                                 P-4000 Porto
                                 Portugal

Das Dorez Melo Da Cruz M.   Departamento de Quimica – Faculdade de
                                 Ciencias, Universidade do Porto, P-
                                 4000 Porto
                                 Portugal

de Bolster, M. W. G.   Department of Inorganic Chemistry –
                                 Free University – De Lairessestraat
                                 174, Amsterdam
                                 The Netherlands

Dendooven, M.   Laboratory of Quantum Chemistry –
                                 University of Leuven – Celestijnenlaan
                                 200 F, B-3030 Heverlee
                                 Belgium

Desideri, A.   Dipartimento di Fisica, Università
                                 della Calabria, Arcavacata di Rende, CS
                                 Italy

Devaux, P. F.   Institut de Biologie Physico-Chimique,
                                 Fondation de Rothschild – 13, Rue
                                 Pierre et Marie Curie – 75005 Paris
                                 France

Dias Dos Santos, H.   Centro de Quimica Estrutural – Complexo
                                 Interdisciplinar – Instituto Superior
                                 Técnico – Lisboa 1
                                 Portugal

Dietrich, H.   Fachbereich 15.2, Analytische und
                                 Biologische Chemie, Universität des
                                 Saarlandes – D-6600 Saarbrücken 11
                                 F. R. G.

Do Rosairo Ascenso, J.   Centro de Quimica Estrutural – Complexo
                                 Interdisciplinar – Instituto Superior
                                 Técnico – Lisboa 1
                                 Portugal

Drago, R. S.   School of Chemical Sciences - Univer-
                                sity of Illinois at Urbana-Champaign,
                                Urbana Illinois 61801
                                U. S. A.

Duatti, A.   Istituto Chimico dell'Università di
                                Ferrara, Via Luigi Borsari, 4 - Ferrara
                                Italy

Ehrenberg, A.   Arrhenius Laboratoriet, Biofysiska
                                Institutionen, Stockholms Universitet,
                                S-106 91 Stockholm
                                Sweden

Elias, H.   Anorganische Chemie III, Technische
                                Hochschule Darmstadt - Hochschulstrasse
                                4 - 61 Darmstadt
                                F. R. G.

Ferroni, G.   Laboratoire de Chimie Inorganique -
                                Université de Provence - Place Victor
                                Hugo - 13331 Marseille 3
                                France

Franchini, C.   Istituto di Chimica Farmaceutica -
                                Università di Bari, Via Mendola, 173
                                Bari
                                Italy

Friebel, C.   Fachbereich Chemie der Philipps -
                                Universität Marburg/L. - 3550 Marburg
                                F. R. G.

Ganadu, L.   Centro di Spettroscopia - Facoltà di
                                Agraria, Università di Sassari, Via
                                Enrico De Nicola - Sassari
                                Italy

Gattegno, D.   Laboratorio di Teoria e Struttura Elet-
                                tronica dei Composti di Coordinazione
                                C. N. R. - Via Montorio Romano, 36
                                Roma
                                Italy

Gatteschi, D.   Istituto di Chimica Generale ed Inorga-
                                nica, Università di Firenze, Via J.
                                Nardi, 39 - Firenze
                                Italy

Geraldes, C.   Departamento de Quimica - Faculdade de
                                Ciencias e Tecnologia - Universidade
                                de Coimbra - Coimbra
                                Portugal

Gibson, J. F.   Department of Chemistry, Imperial
                                College of Sciences and Technology,
                                London SW7 2AY
                                United Kingdom

Gueron, M.   Laboratoire Phys. Matière Condensée-
                                Ecole Politechnique, Palaiseau -
                                France

Hadjiliadis, N.   Department of Chemistry, Inorganic
                                Chemistry Laboratory, University of
                                Athens, Navarinon 13A - Athens
                                Greece
Hendriks, H. M. J.   Department of Chemistry - Delft
                                University of Technology - Delft
                                The Netherlands
Hepp, A.   Department of Chemistry - Massachusetts
                                Institute of Technology - Cambridge
                                MA 02139
                                U. S. A.
Horrocks, W. DeW. jr.   Department of Chemistry, The
                                Pennsylvania State University,
                                University Park, Pennsylvania 16802
                                U. S. A.
Jones, P.   Radiation and Biophysical Chemistry
                                Lab., School of Chemistry - University
                                of Newcastle upon Tyne - NE1 7RU
                                United Kingdom
Koenig, S. H.   Thomas J. Watson Research Center,
                                I. B. M. - Yorktown Heights, New York
                                10598
                                U. S. A.
Kokoszka, G. F.   Department of Chemistry, State Univer-
                                sity of New York - College at
                                Plattsburg - New York - NY 12901
                                U. S. A.
Kozłowski, H.   Institute of Chemistry - University
                                of Wrocław - Joliot-Curie 14 - 50383
                                Wrocław
                                Poland
Kudrettin, E.   Department of Chemistry - Middle East
                                Technical University - Ankara
                                Turkey
Lanini, G.   Istituto di Chimica Generale ed Inor-
                                ganica, Facoltà di Farmacia, Università
                                di Firenze, Via Gino Capponi, 7 -
                                Firenze
                                Italy
Lhoste, J. M.   Section de Biologie Lab. 112 - Institut
                                Curie, Centre Universitaire d'Orsay -
                                91405 Orsay
                                France
Libor, S.   Department of Chemistry - University
                                of Surrey - GU2 5XH
                                United Kingdom
Lipscomb, J. D.   Gray Freshwater Biological Institute
                                University of Minnesota, P.O. Box
                                100, Navarre, MN 55392
                                U. S. A.

London, R.   Los Alamos Scientific Laboratory,
                                        Los Alamos - New Mexico 87544
                                        U. S. A.

Luchinat, C.   Istituto di Chimica Generale ed Inor-
                                        ganica, Facoltà di Farmacia, Universi-
                                        tà di Firenze, Via Gino Capponi, 7 -
                                        Firenze
                                        Italy

Maret, W.   Fachbereich 15.2, Analytische und
                                        Biologische Chemie, Universität des
                                        Saarlandes - D-6600 Saarbrücken 11
                                        F. R. G.

Matwiyoff, N. A.   Los Alamos Scientific Laboratory,
                                        Los Alamos - New Mexico 87544
                                        U. S. A.

May, L.   Department of Chemistry - Catholic
                                        University,Washington - DC,20064
                                        U. S. A.

Mc Garvey, B.   Department of Chemistry, University of
                                        Windsor, Windsor Ontario N9B 3P4
                                        Canada

Mc Millin, D.   Department of Chemistry, Purdue Univer-
                                        sity, W. Lafayette - Indiana 47907
                                        U. S. A.

Mc Phail, D. B.   Department of Spectrochemistry - The
                                        Macaulay Institute for Soil Research -
                                        Craigiebuckler Aberdeen AB9 2QJ
                                        United Kingdom

Mendonça Dias, M. H.   Centro de Quimica Estrutural - Complexo
                                        Interdisciplinar - Instituto Superior
                                        Técnico Lisboa 1
                                        Portugal

Mispelter, J.   Section de Biologie Lab. 112 - Institut
                                        Curie, Centre Universitaire d'Orsay -
                                        91405 Orsay
                                        France

Monduzzi, M.   Istituto Chimico Policattedra - Univer-
                                        sità di Cagliari, Via Ospedale, 72
                                        Cagliari
                                        Italy

Monteiro Pirez Basto, M. C.   Departamento de Quimica - Faculdade de
                                        Ciencias, Universidade do Porto, P -
                                        4000 Porto
                                        Portugal

Münck, E.   Gray Freshwater Biological Institute,
                                        University of Minnesota, P. O. BOX 100
                                        Navarre, MN 55392
                                        U. S. A.

Nöthig-Laslo, V.   "Ruder Boskovic" Institute, 41001 Zagreb
                                        Yugoslavia

Nyberg, E. D.   School of Chemical Sciences - Univer-
                          sity of Illinois at Urbana-Champaign,
                          Urbana Illinois 61801
                          U. S. A.

Oldenhof, W.   Laboratory of Quantum Chemistry -
                          University of Leuven - Celestijnenlaan
                          200 F, B-3030 Heverlee
                          Belgium

Onori, S.   Laboratorio delle Radiazioni - Istitu-
                          to Superiore Sanità, Viale Regina Ele-
                          na, 299 - Roma
                          Italy

Owens, C.   Noyes Laboratory - University of
                          Illinois at Urbana- Champaign, Urbana
                          Illinois 61801
                          U. S. A.

Petridis, D.   Department of Physics, Nuclear Research
                          Center "Demokritos", Aghia Paraskevi
                          Attiki
                          Greece

Petrou, A.   Inorganic Chemistry Laboratory,
                          Department of Chemistry, University of
                          Athens - Navarinon 13A-Athens
                          Greece

Raap, A.   Fachgebiet Physicalische Chemie der
                          Proteine, Chemisch-Westfälische
                          Technische Hochschule Aachen, Melatener
                          strasse 213, D 51 Aachen
                          F. R. G.

Reinen, D.   Fachbereich Chemie der Philipps -
                          Universität Marburg/L. - 3550 Marburg
                          F. R. G.

Rossi, C.   Istituto di Chimica Generale, Univer-
                          sità di Siena, Via Pian dei Mantelli-
                          ni, 44 - Siena
                          Italy

Rotilio, G.   Istituto di Chimica Biologica, Città
                          Universitaria, 00185 Roma
                          Italy

Scozzafava, A.   Istituto di Chimica Generale ed Inor-
                          ganica, Facoltà di Farmacia, Università
                          di Firenze, Via Gino Capponi, 7
                          Firenze
                          Italy

Sedlmair, J.   Department of Chemistry, Technical
                          University - München
                          F. R. G.

Shulman, R. G.   Bell Telephone, Lab. Incorp. Murray
                          Hill - New Jersey
                          U. S. A.

Sletten, E.   Department of Chemistry, University
                               of Bergen, N 5014 Bergen
                               Norway

Söylemez, T.   Institut für Strahlenchemie, Max-Planck
                               Institut für Kohlenforschung – Stiftsr.
                               34 D-4330 Mulheim/Ruhr
                               F. R. G.

Stassinopulou, C. I.   Department of Biology, Nuclear Research
                               Center "Demokritos", Aghia Paraskevi
                               Attiki
                               Greece

Sportelli, L.   Laboratorio delle Radiazioni – Istitu-
                               to Superiore Sanità, Viale Regina Ele-
                               na, 299 – Roma
                               Italy

Strouse, J.   Department of Chemistry, University of
                               California , Los Angeles – California
                               90024
                               U. S. A.

Strouse, C. E.   Department of Chemistry, University of
                               California, Los Angeles – California
                               90024
                               U. S. A.

Tanner, S. P.   Faculty of Chemistry, Alpha College,
                               University of West Florida – Pensacola
                               FL 32504
                               U. S. A.

Tomlinson, G.   Laboratorio Teoria e Struttura Elet-
                               tronica dei Composti di Coordinazione,
                               C. N. R., Via Montorio Romano, 36 –
                               Roma
                               Italy

Turner, D. L.   Department of Chemistry, The University
                               of Southampton, SO 9 5NH
                               United Kingdom

van Brakel, G. H.   Department of Molecular Physics,
                               Agricultural University, De Dreijen 6,
                               6700 EP Wageningen
                               The Netherlands

Vogel, G. C.   Department of Chemistry, Ithaca College,
                               Ithaca, NY 14850
                               U. S. A.

Wehrli, F.   Bruker Instruments Inc., Manning
                               Park, Billerica, Massachusetts
                               01821
                               U. S. A.

Wicholas, M.   Department of Chemistry, Western
                               Washington University, Bellingham WA
                               98225
                               U. S. A.

Wüthrich, K.   Institute für Molekularbiologie und
                       Biophysik, ETH, Hönggerberg – CH-8093
                       Zürich
                       Switzerland

Zink, J. I.   Department of Chemistry, University
                       of California, Los Angeles, California
                       90924
                       U. S. A.

PULSE TECHNIQUES AND NUCLEAR SPIN RELAXATION

Felix W. Wehrli

Bruker Instruments, Inc.
Manning Park, Billerica, Mass. 01821, U.S.A.

This chapter is intended to familiarize the reader with the basics of nuclear spin relaxation in liquids from a dynamic, mechanistic and experimental point of view. Besides a treatment of the various mechanisms inducing relaxation, it shall be shown how the latter are separated and interpreted. Some consideration will also be given to the implications of exchange processes on relaxation rates. Additionally, some of the more elaborate pulse techniques related to relaxation, including selective excitation and two-dimensional resolved spectroscopy are presented.

1. SPIN-LATTICE AND SPIN-SPIN RELAXATION MECHANISMS

1.1 Introduction: Phenomenological Definition of T_1 and T_2

The relaxation times are defined as time constants for the return of a component of the macroscopic spin magnetization, to its equilibrium value, following some kind of external perturbation. In the case of T_1, the longitudinal relaxation time, one is concerned with the longitudinal component M_z, which is aligned with the external magnetic field, B_0, and has a finite equilibrium value M_0. The alternative term spin-lattice relaxation implies an energy interaction between the spins and their environment, the lattice. In terms of the Boltzmann populations, saturation may be regarded as a state of infinitely high spin temperature. Hence in order to return to thermal equilibrium the excess heat of the spin system must be dissipated to the lattice.

I. Bertini and R.S. Drago (eds.), ESR and NMR of Paramagnetic Species in Biological and Related Systems, 1–28.

A T_2 process, by contrast, affects the components of M perpendicular to the external field, i.e. M_x and M_y. Following the generation of transverse magnetization, mutual interactions among the magnetic moments lead to a dispersal of their relative phases which results in a decay of the transverse magnetization, M_x and M_y ultimately approaching their equilibrium value of zero. Another distinguishing feature of this relaxation process is conservation of the net spin energy in that the energy is solely exchanged among the spins themselves. In the absence of artificial broadening effects the full width of a line at half height is given by $1/\pi T_2$.

In most situations T_1 is found to be equal to T_2 and most high-resolution relaxation experiments in the past therefore dealt with the more straightforwardly obtainable T_1. However the recent advent of pulse sequence generators and high-precision timing on commercial spectrometers also render amenable the measurement of T_2. The magnitude of the latter may deviate from T_1, notably when slow processes are involved, e.g. chemical exchange or scalar coupling to a rapidly relaxing second nucleus.

1.2 Liquid Dynamics and Relaxation

Spin-lattice relaxation is brought about by transitions between the nuclear Zeeman levels. Requisite therefore is the presence of magnetic fields oscillating at the respective transition frequencies. The time dependence of these fields has its origin in lattice fluctuations which, in fluids, are normally caused by Brownian motion. The nature of the lattice fields themselves is manifold. One frequent source is the presence of a neighbouring magnetic dipole. This constitutes the basis of dipole-dipole (DD) relaxation.

Prior to treating the individual relaxation mechanisms, the properties of above-mentioned lattice motions are briefly reviewed. Local magnetic fields, B_{loc}, may be considered as correlated fluctuations which have a mean value of zero. The so-called autocorrelation function $G(\tau)$ describes the time evolution of an ensemble of moving particles in terms of the local magnetic field associated with them.

$$G(\tau) = \overline{B_{loc}(t) \cdot B_{loc}(t+\tau)} = \overline{B_{loc}^*(t) B_{loc}(t)} \; e^{-|\tau|/\tau_c} \qquad (1)$$

Eqn. 1 implies an exponential decay of the ensemble average $\overline{B_{loc}^*(t) B_{loc}(t)}$ with a time constant τ_c, the rotational correlation time. Physically more relevant is the Fourier inverse of $G(\tau)$, the power spectral density $J(\omega)$, given as

$$J(\omega) = \int_{-\infty}^{\infty} G(\tau) e^{i\omega\tau} d\tau = <B_{loc}^2(0)> \cdot \; 2\tau_c/(1+\omega^2\tau_c^2) \qquad (2)$$

$J(\omega)$ provides the radio frequency power of the local magnetic
field at a given frequency ω. Eqn. 2 indicates maximum power
at $\omega = o$, falling off with increasing frequency as ω becomes
comparable with $1/\tau_C$. The latter may be viewed as an average
rate for the reorientation of the molecules. Fig. 1 shows
three characteristic curves of $J(\omega)$ for reorientation rates,
$1/\tau_C$, large, comparable, and small relative the Larmor resonance
frequency, ω_0. As $\int_o^{\infty} (\omega)\,d\omega$ is constant one realizes that
the power spectral density is largest for an intermediate value
of the correlation time τ_C, which applies to medium-sized molecules, reorienting at a rate comparable to the Larmor frequency.

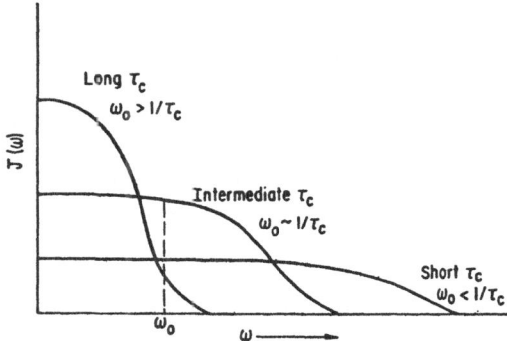

Fig. 1:   Spectral density as a function of frequency
          for three different correlation times

For short correlation times the spectral density is independent
of spectrometer frequency over a very large range of frequencies.
This situation is denoted motional narrowing limit (frequency
invariance of spectral density).

     Clearly the efficiency of relaxation must be closely related
to the spectral density at the transition frequencies involved.
Defining $W_{\alpha\beta}$ as the probability per unit time for a transition to
occur between energy levels labelled α and β and an energy separation $\omega_{\alpha\beta}$, the transition probability is given by

$$W_{\alpha\beta} = \gamma^2 J(\omega_{\alpha\beta}) \tag{3}$$

     For a single isolated spin there are only two energy levels,
hence $1/T_1 = W_{\alpha\beta}$. In coupled spin systems, however, population
changes can ensue through transitions occuring between several
energy levels.

1.3   The Mechanisms (1, 2)

The term mechanism essentially characterizes the source of the local fluctuating magnetic field, B_{loc}, that induces transitions. In the case of spin-lattice relaxation - a similar reasoning also applies to spin-spin relaxation - a specific relaxation process can be described in terms of the type of energy interaction operative between spin system and lattice.

1.3.1   Dipole-dipole relaxation. The simplest case of dipole-dipole relaxation is given by two interacting spins I and S ($I=S=\frac{1}{2}$) which make up four energy levels:

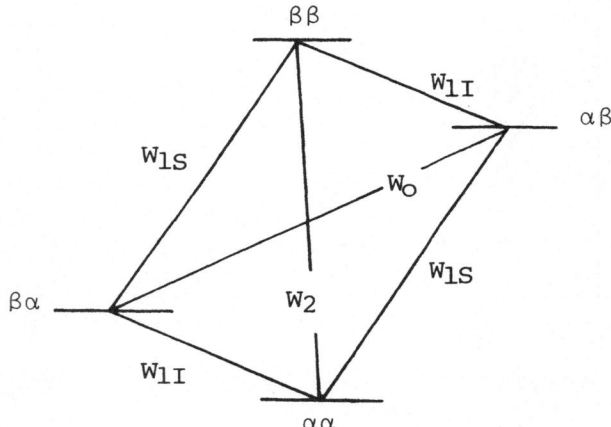

W's represent probabilities per unit time for transitions between the various energy levels, with W_{1I} and W_{1S} representing single quantum transition probabilities. W_0 and W_2 are zero and double quantum transitions, resp., according to the change in total magnetic quantum number.

The equations of motion for the I and S magnetization, $<I_z>$ and $<S_z>$ can be described in terms of two coupled differential equations (3):

$$d<I_z>/dt = -\rho_I(<I_z>-I_z^O) - \sigma_{IS}(<S_z>-S_z^O) \tag{4}$$

$$d<S_z>/dt = -\rho_S(<S_z>-S_z^O) - \sigma_{SI}(<I_z>-I_z^O) \tag{5}$$

with I_z^O and S_z^O representing the respective magnetizations at thermal equilibrium and

$$\rho_I = 2W_{1I} + W_O + W_2 \tag{6a}$$

$$\rho_S = 2W_{1S} + W_O + W_2 \tag{6b}$$

$$\sigma_{IS} = \sigma_{SI} = W_2 - W_o \tag{6c}$$

The general solution for Eqn. 4 may be written as

$$<I_z> = I_z^o + C_1 e^{-\lambda_1 t} + C_2 e^{-\lambda_2 t} \tag{7}$$

$$\lambda_{1,2} = (1/2)\{(\rho_I + \rho_s) \pm (\rho_I - \rho_s)^2 + 4\sigma_{IS}^2\}^{1/2} \tag{7a}$$

From Eqn. 7 one realizes that the relaxation is not governed by a single rate constant (or single longitudinal relaxation time).

If, however, a strong irradiating field acts on the S spins (e.g. decoupling protons while observing ^{13}C) and a saturating 90° pulse affects the I spins, the boundary conditions $<S_z(t)> = S_z^o = 0$ hold and the solution of Eqn. 4 yields

$$<I_z(t)> = I_z^o(1 - e^{-\rho_I t}) \tag{8}$$

indicating that relaxation is exponential with a time constant

$$T_{1I} = 1/\rho_I = (2W_{1I} + W_o + W_2)^{-1} \tag{9}$$

The transition probabilities are proportional to the spectral densities at the respective transition frequencies (3)

$$W_{1I} = (3/20)\gamma_I^2\gamma_s^2\hbar^2 r_{IS}^{-6}\tau_c/(1 + \omega_I^2\tau_c^2) \tag{10a}$$

$$W_o = (1/10)\gamma_I^2\gamma_s^2\hbar^2 r_{IS}^{-6}\tau_c/(1 + (\omega_I - \omega_s)^2\tau_c^2) \tag{10b}$$

$$W_2 = (3/5)\gamma_I^2\gamma_s^2\hbar^2 r_{IS}^{-6}\tau_c/(1 + (\omega_I + \omega_s)^2\tau_c^2) \tag{10c}$$

Explicitly one obtains for the dipolar spin-lattice relaxation rate of a nucleus I interacting with a nucleus S:

$$1/T_{1I}^{DD} = (1/10)\gamma_I^2\gamma_s^2 r_{IS}^{-6} \times$$

$$\{\frac{\tau_c}{1 + (\omega_I - \omega_s)^2\tau_c^2} + \frac{3\tau_c}{1 + \omega_I^2\tau_c^2} + \frac{6\tau_c}{1 + (\omega_I + \omega_s)^2\tau_c^2}\} \tag{11}$$

Hence, the relaxation rate is proportional to the squares of the magnetic moments of the two interacting nuclei and inversely proportional to the sixth power of the internuclear distance r_{IS}. Eqns. 11 and 12 strictly hold only for spherically symmetric systems with no preferred axis for reorientation. If the latter is not the case relaxation can no longer be described in terms of a single correlation time.

Since, among the naturally occurring nuclei the proton possesses the strongest magnetic moment, it is for diamagnetic systems a prime candidate for relaxation of such spin-$\frac{1}{2}$ nuclei as ^{13}C, ^{15}N, ^{29}Si, ^{31}P etc., in particular when these have directly bonded hydrogens. In fact, proton-bearing carbons are generally relaxed by their directly bonded protons (4), unless one is concerned with very small molecules or groups undergoing fast internal motion (5). Since the relaxation rate is predicted to be proportional to the number of directly bonded protons, a methylene carbon is expected to relax twice as fast as a methine carbon, which has generally been observed (4) provided that the C-H bond vector is modulated by the same correlation time. Conversely, NT_1 values (N representing the number of attached hydrogens) may be used to probe internal motion in complex molecules since NT_1 should increase with increasing mobility. In this manner relative sidechain mobility has been studied, which may provide a picture of the conformational properties of such systems. An illustrative example is the cyclic peptide oxytocin whose relaxation map is reproduced in Fig. 2(6):

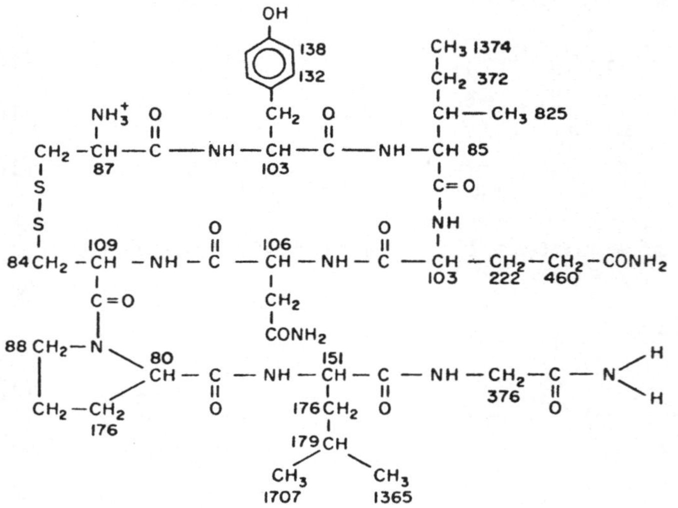

Fig. 2: ^{13}C relaxation map for oxytocin. The relaxation times (NT_1) are in msec and the data were obtained from a 0.1 M solution in D_2O at 30 °C (From Ref. 6 with permission)

Anisotropic overall reorientation is generally predicted for all molecules deviating from spherical symmetry, and description of dipole–dipole relaxation requires treatment in terms of the full rotational diffusion tensor (1).

A side effect and conspicuous manifestation of dipolar relaxation is the nuclear Overhauser effect (NOE) which results from the cross-relaxation term in Eqn. 4. If one supposes again the S spins to be saturated so that $\langle S_z \rangle = 0$ and the I-spin magnetization being monitored after a steady-state has been obtained ($d\langle I_z \rangle/dt=0$), then the solution of Eqn. 4 yields:

$$\langle I_z \rangle = I_o + (\sigma_{IS}/\rho_I) S_o \tag{13}$$

or, since $S_o = I_o \gamma_S/\gamma_I$

$$\langle I_z \rangle/I_o = 1 + (\gamma_S/\gamma_I)(\sigma_{IS}/\rho_I) = 1+\eta \tag{14}$$

If motional narrowing limit conditions apply, the ratio σ_{IS}/ρ_I assumes a value of $1/2$, giving for η

$$\eta = \gamma_S/2\gamma_I \tag{14a}$$

Assuming $S = {}^1H$, maximum enhancement factors for some representative nuclei are:

nucleus	η
${}^6Li$	3.39
${}^{13}C$	1.99
${}^{15}N$	−4.94
${}^{29}Si$	−2.52
${}^{31}P$	1.24
${}^{77}Se$	2.62

Spins with negative magnetogyric ratios (e.g. ${}^{15}N$, ${}^{29}Si$) afford negative signals with the decoupler turned on (on the assumption that the phase had been set for positive absorption in the absence of decoupling). The experimental ramifications of this effect will be discussed at a later stage.

So far, we have only dealt with intramolecular DD relaxation. However, in particular protons, which are located at the periphery of the molecule, are susceptible to intermolecular relaxation.

The relaxation rate of a spin I subjected to intermolecular interaction with a spin S is approximately given by

$$1/T_1{}^{DD}_{inter} = (8/45) N_S \gamma_I{}^2 \gamma_S{}^2 \hbar^2 S(S+1)/D_{IS} a \tag{15}$$

where N_S is the concentration of spins S (per unit volume), a is
the distance of closest approach between the interacting spins,
and D_{IS} represents the mutual translational self-diffusion constant, defined as D_{IS} = 0.5 $(D_I + D_S)$. Eqn. 15 is also applicable
to relaxation by chemically inert paramagnetic relaxation reagents,
in which case γ_S represents the electronic magnetogyric ratio.

1.3.2 Spin Rotation Relaxation (SR). The local magnetic
field responsible for the spin-rotation mechanism is produced
by electronic and nuclear currents associated with molecular
motion. Although currents from symmetric charge distributions
cancel one another, any angular momentum associated with such
charges gives rise to a magnetic field. Molecular interactions
such as collisions will result in changes of both direction and
magnitude of the angular momentum vector. The average period
between such perturbations is denoted spin-rotation or angular
momentum correlation time τ_J. In the simplest case of spherical
molecular symmetry the spin-rotation relaxation rate has the
form

$$1/T_1^{SR} = \frac{2kT}{\hbar^2} I_m C^2 \tau_J \tag{16}$$

with the constants I_m and C designating the momentum of inertia
and the spin-rotation constant, respectively.

The spin-rotation correlation time is in many cases related
in a simple fashion to the rotational correlation time (Hubbard
relation):

$$\tau_J = I_m/6kT\tau_c \tag{17}$$

According to Eqn. 17 a small τ_c implies a large value of τ_J,
hence spin-rotation is the preferred mechanism for relaxation
of spin-½ nuclei in small rapidly reorienting molecules. This
mechanism is rarely ever the sole source of relaxation. Its
identification requires knowledge of the temperature dependence
of the spin-lattice relaxation rate, which is characteristic
because of the reverse temperature dependence of reorientational
and angular momentum correlation time. The former can normally
be written in terms of an Arrhenius-type equation of the form

$$\tau_c = \tau_c^0 \cdot e^{E_a/RT} \tag{18}$$

with E_a indicating an activation energy characteristic of a molecular system. Dipole-dipole (but also chemical shift anisotropy
and quadrupole) relaxation therefore all have the same temperature dependence in that the logarithm of the relaxation rate increases linearly with 1/T, the inverse absolute temperature. By

combining Eqn. 17 and 18 one realizes that SR leads to a straight
line of opposite slope, i.e. $\ln(1/T_1)$ decreases with increasing
values of $1/T$. While examples of pure DD relaxation are plenti-
ful, those with SR being the dominant source of relaxation at all
temperatures are rare.

If, e.g., SR competes to a comparable extent with a mechanism such as DD relaxation, the relaxation rate shows a minimum
at which $T_1 SR = T_1 DD$. ^9Be relaxation in aqueous $Be(NO_3)_2$ (7)
may serve as an illustration of this behaviour. Fig. 3 shows
an Arrhenius plot which exhibits a T_1 maximum at ca. 40°C (here
T_1 rather than $1/T_1$ is plotted). While at high temperatures SR

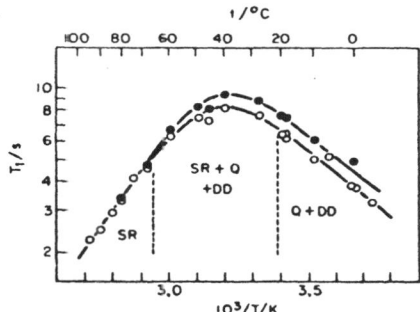

Fig. 3:   Arrhenius plot of $T_1(^9Be)$ in aqueous
          $Be(NO_3)_2$ (from Ref. 7, with permission)

is almost exclusively dominating relaxation, the lower temperatures
favour quadrupolar and some dipolar relaxation. The latter manifests itself in a signal reduction upon irradiation of the water
protons in the hydration sphere of the ion (the magnetic moment
of ^9Be is negative, hence increasing DD relaxation first lowers
the ^9Be signal intensity, nulls it and becomes increasingly
negative).

1.3.3 Chemical Shift Anisotropy Relaxation (CSA). This
type of relaxation may become operative when the chemical
shielding factor σ , varies as the molecular orientation changes
with respect to the applied field direction B_o. If σ is axially
symmetric, the relaxation rate $1/T_1 CSA$ in the motional narrowing
limit is given by

$$1/T_1^{CSA} = \frac{2}{15}\gamma^2 B_o^2 \ (\sigma_{11}-\sigma_{\perp})^2 \tau_c \qquad (19)$$

The characteristic feature of this relaxation mechanism is its
quadratic field dependence. Carbon or nitrogen atoms in linear
X-Y-Z bonding arrangements as they occur for sp hybridization
in nitriles, carbon monoxides, etc., are particularly promising
candidates for chemical shift anisotropy (8-10). This shall be
illustrated with the relaxation behaviour for C-1 and C≡N in
benzonitrile, for which it was possible to extract the chemical
shift anisotropy $|\sigma_{11}-\sigma_{\perp}|$ (10). Fig. 4 shows the ^{13}C signal
recovery at 2.3 and 7T field strength, indicating the much fast-
er decay at the higher field, in particular of the cyano but
also of the quaternary carbon. From the difference of the
relaxation rate T_1^{CSA} could be calculated. At 7T ca. 80% of
the relaxation rate for $^{13}C≡N$ turn out to be CSA-induced.
Further to this a measurement of the dipolar T_1 for the protonated para carbon afforded $\tau_{\perp}$, the correlation time for reorientation of the molecule around its axis perpendicular to
the symmetry axis. This correlation time is responsible also
for the modulation of the CSA tensor and therefore permitted
calculation of $|\sigma_{11}-\sigma_{\perp}|$, which was found to be 330 ppm.

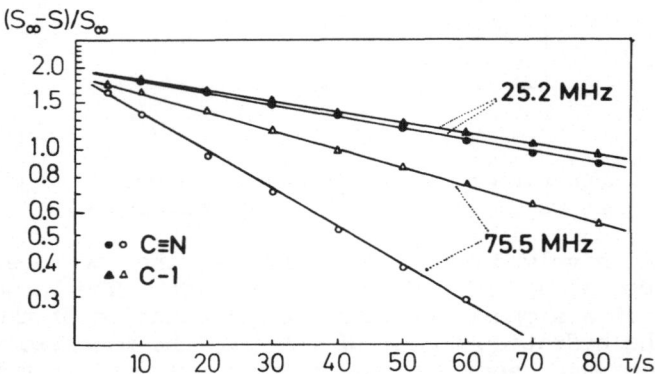

Fig. 4: ^{13}C signal recovery for the cyano and
         C-1 carbon in benzonitrile following a
         180° pulse at two different field
         strengths (from Ref. 10, with permission)

1.3.4 Quadrupole Relaxation (Q). Occurrence of this
mechanism is confined to nuclei possessing ellipsoidal
charge distribution, which is the case for all spins I ≥ 1.
Associated with the spin is a nuclear quadrupole Q which may
interact with an electric field gradient eq caused by the

surrounding valence electrons. Molecular motion modulates this
interaction and leads to transitions between the energy levels
and hence relaxation. The interaction, in spite of being electrical, and not magnetic, still affects nuclear relaxation since the
nuclear quadrupole is coupled to the magnetic quantum number of
the spins. Supposing the electronic charge distribution is
axially symmetric, the expression for quadrupole relaxation may
be formulated as

$$1/T_1^Q = 1/T_2^Q = \frac{3}{40} \frac{2I + 3}{I(2I-1)} \left(\frac{e^2 qQ}{\hbar}\right)^2 \tau_c \qquad (20)$$

The expression e^2qQ/h (in Hz) quantifies the interaction energy.
This quantity is of chemical significance since it describes the
degree of valence electron symmetry. The validity of Eqn. 20 is
confined to the motional norrowing conditions. When these do not
apply, relaxation can no longer be described in terms of a single
exponential time constant. The quadrupolar relaxation mechanism
is usually very efficient and thus prevailing over all other
mechanisms.

    1.3.5 Scalar Relaxation (SC). The fluctuating magnetic
field causing this type of relaxation is engendered by scalar
coupling to a second nucleus or electron. The time dependence
is introduced either by rapid relaxation of the coupled spin S
or chemical exchange at a rate comparable to the difference fre-
quency $|\omega_I - \omega_S|$. Although this mechanism is rarely a source of
longitudinal relaxation, it very often dominates transverse relaxation. The expressions for longitudinal and transverse scalar
relaxation are given below.

$$1/T_1^{SC} = \frac{8}{3}\pi^2 J^2 S(S+1) \frac{\tau_S}{1+ (\omega_I - \omega_S)^2 \tau_S^2} \qquad (21)$$

$$1/T_2 = 1/2T_1^{SC} + \frac{4}{3}\pi^2 J^2 S(S+1)\tau_J \qquad (22)$$

In Eqns. 21 and 22 τ_S denotes the life time of the coupled spin
S which can either be the chemical exchange life time or the
spin-lattice relaxation time of a rapidly relaxing quadrupolar
nucleus. In the latter case measurement of the scalar relaxation
rate permits the determination of a not directly observable scalar
coupling constant.

    An example of scalar relaxation has been found for the
relaxation of ^{71}Ga in the tetrahalogallates (11). In $GaCl_4$ -
^{35}Cl relaxes very rapidly ($T_1 \sim 10^{-6}s$) because the atom is in
an axially symmetric environment. In this case the scalar

coupling constant $(1J(^{71}Ga-^{35}Cl))$ could be determined from a combined measurement of $T_1(^{69}Ga)$ and $T_1(^{71}Ga)$. If quadrupole relaxation prevailed, the ratio of the two relaxation rates would be equal to the square of their quadrupole moments $(Q_{69}^2/Q_{71}^2=2.52)$ In reality the value found was 0.9. By combining the equations for scalar and quadrupolar relaxation and by further taking into account that the observed relaxation rates are the sums of the scalar and quadrupolar contributions, $1J(^{71}Ga-^{35}Cl)\approx 3000$ Hz could be determined.

1.3.6 Paramagnetic Relaxation. Nuclear relaxation by un-paired electron spins can be treated as a special case of the previously discussed dipolar and scalar relaxation by considering that the interacting second spin is electronic. Because of the much larger magnetic moment of the electronic spin, dipolar electron-nuclear relaxation is a considerably more powerful mech-anism than its nucleus-induced counterpart (8):

$$1/T_{1e} = \frac{2S(S+1)\gamma_I^2\gamma_S^2\hbar^2}{15r_{IS}^6} \left(\frac{3\tau_c}{1+\omega_I^2\tau_c^2} + \frac{7\tau_c}{1+\omega_S^2\tau_c^2} \right) \qquad (23)$$

The correlation time τ_c is usually determined by three distinct times as follows:

$$\tau_c^{-1} = \tau_R^{-1} + \tau_S^{-1} + \tau_M^{-1} \qquad (23a)$$

whereby τ_R represents the reorientational correlation time, τ_S is the electron spin relaxation time and τ_M is the residence time of the nucleus at the paramagnetic site, in the case where chemical exchange is operative. It should further be noted that τ_S itself may become frequency-dependent.

1.4 The Impact of Chemical Exchange

Chemical exchange is basically a transverse relaxation pro-cess since it affects the line widths. Derivation of the rate parameters characterizing the exchange process is usually accom-plished by total line shape analysis on the basis of the exchange-modified Bloch equations, which, in matrix formalism for the simple case of exchange between two sites, A and B, may be written as

$$\frac{d}{dt} \begin{pmatrix} M_A \\ M_B \end{pmatrix} = \begin{pmatrix} -1/T_A & 1/\tau_B \\ 1/\tau_A & -1/T_B \end{pmatrix} \begin{pmatrix} M_A \\ M_B \end{pmatrix} \qquad (24)$$

In Eqn. 24 M_A and M_B represent the excess longitudinal or trans-verse magnetizations at sites A and B and T_A, T_B are the average life times at the two sites. T_A and T_B are defined as

$$1/T_A = 1/T_{A2} + 1/\tau_A \tag{25a}$$

$$\text{and } 1/T_B = 1/T_{B2} + 1/\tau_B \tag{25b}$$

Very often one is confronted with exchange of a nucleus between
a "free" and a "complexed" state. Unver favourable circumstances,
NMR allows three types of information to be gained: (a) the
fraction of molecules bound, (b) the rate of exchange and (c) the
relaxation rate of the nucleus in the bound state. In general,
however, only one of these quantities can be derived. Fortunately,
simplifying assumptions are often warrented. For example; the
chemical shift of the nucleus under investigation remains invariant for the two sites. This may be the case for complexation
with a shiftless relaxation reagent such as Gd(III). Above condition is often also fulfilled for quadrupolar ionic nuclei such
as $^{35,37}Cl$ (12) or $^{23}Na^+$ (13). Such a situation may be exemplified with ionophore complexation of sodium ions, the ionophore being a crown ether and the nucleus under investigation
^{23}Na. Since the complexes formed are of very high thermodynamic
stability, it can further be assumed that at excess ion concentration all of the ionophore is complexed. Under these assumptions it has been shown (13) that the life time of the
solvated sodium ion (τ_A) is related to the relaxation times (T_{iA}
and T_{iB} (i = 1, 2) and molar fractions P_A and P_B in the following manner

$$1/\tau_A = \frac{(1/T_{iB} - 1/T_i)\ (1/T_i - 1/T_{iA})\,p_B}{(1/T_{iav} - 1/T_i)} \tag{26}$$

where $1/T_{iav} = p_A/T_{iA} + p_B/T_{iB}$ \tag{26a}

In the high-temperature (fast exchange) limit where $1/\tau_A \gg 1/T_i$, $1/T_{iav}$
becomes equal to $1/T_i$. Under slow-exchange conditions the
^{23}Na resonance in the complexed state is too broad to be detected. The measured relaxation rate at low temperature is there-
fore equal to T_{iA}. Fig. 5 displays an Arrhenius plot of $1/T_{1,2}$ (^{23}Na)
for dibenzo-18-crown-16 (DEC) complexation with Na^+ in DMF, from
which the rate constants of decomplexation could be determined.
Since motional narrowing limit conditions apply, the two curves
for T_1 and T_2 overlap.

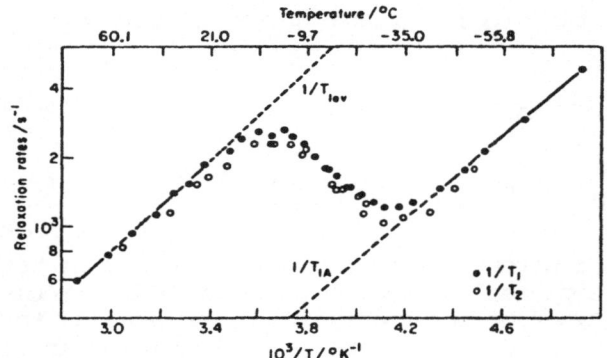

Fig. 5: Semilog plots for $1/T_{1,2}$ (^{23}Na) vs. inverse
absolute temperature for a solution contain-
ing 0.57 M NaSCN and 0.2M DBC in DMF (from
Ref. 14, with permission)

2.   EXPERIMENTAL TECHNIQUES FOR THE MEASUREMENT OF T_1, T_2, $T_{1\rho}$ AND THE NOE

The advent of Fourier Transform methods toward the end of
the last decade has given high-resolution T_1 and T_2 studies an
enormous impetus. The study of relaxation times in multi-line
spectra was initiated by an experiment initially proposed in
1968 (15), which is an extension of pulse methods earlier applied
to single-line spectra.

2.1 Methods Of Measuring T_1

2.1.1 The Inversion-Recovery Method (15). This method is
characterized by inverting the magnetization due to all spins by
a strong nonselective 180° pulse, followed τ seconds later by a
90° detection pulse. The resulting free induction decay (FID)
signal is subsequently sampled and Fourier transformed in the
usual way, thus affording a spectrum whose individual lines reflect a state of partial relaxation. By executing a series of
experiments with varying waiting period between the two pulses,
the line intensities map out the recovery of the longitudinal
magnetization, from which the time constant can be obtained by
fitting the line intensities to the theoretical recovery curve.
This technique turns out to be the most generally applicable one
since it is least prone to systematic errors as they may arise
from finite pulse power and missetting of the pulse flip angle
(16, 17).

2.1.2 The Progressive Saturation Method (16). This method
uses the principle of varying saturation (16). By subjecting the

spins to a train of nonselective 90° pulses at intervals ranging
from 0 to 5 T_1, a steady state is established resulting in an FID
which, after Fourier transformation, provides a spectrum whose
lines correspond to increasing saturation as the pulse interval
is lowered. Mandatory for the applicability of this technique
is a sufficient number of pulses,allowing the system to reach
a steady state, before the data are sampled. The advantage of
this sequence is its efficiency on long runs and its simplicity
(no special pulse program is required). The major limitation on
the other hand is its inapplicability to short T_1's. Moreover
it turns out to be particularly susceptible to systematic errors
arising from pulse flip angle inaccuracies and resonance offset
effects.

   2.1.3 The Saturation-Recovery Method (18). This method is
closely related to the previously discussed one except that it
eliminates its limitations. It is, like the inversion-recovery
method, a two-pulse sequence,differing from the latter in that
the initial perturbing pulse is a saturating 90° pulse. In
order to accomplish elimination of the transverse magnetization
following the saturating pulse, a field gradient or homogeneity
spoiling pulse,or a burst of 90° pulses is applied. Alternatively,
as has recently been shown, the unwanted transverse magnetization
may be cancelled by inverting the phase on successive observing
pulses (19).

   The method is somewhat less susceptible to pulse imperfec-
tions than the progressive saturation technique and is more sensitive for long T_1's.

2.2 Methods For Measuring T_2 ($T_{1\rho}$) (20)

   The measurement of these quantities is much more critical
and puts more stringent demands on the performance of the spectrometer, in particular with regard to timing accuracy. Ideally T_2
is obtained from the line width $(1/\pi T_2 = \Delta \nu_{1/2})$. However, the
natural line width is often obscured by artificial broadening effects arising from the inhomogeneity of the static field, in-
complete decoupling, inadequate digitization, etc. The classical
method for the measurement of T_2 has been introduced by Carr and
Purcell (21) and is based on the idea of refocusing the dispersing
spin isochromats through formation of spin echoes. Following an
initial 90° pulse, the spin isochromats, which make up the transverse magnetization, lose phase coherence due to (i) differing
chemical shifts, (ii) spin-spin coupling, (iii) the spatial inhomogeneity of the external field and (iv) spin-spin relaxation.
While effects (i)-(iii) are reversible, (iv) is not. By applying
a 180° pulse τ seconds after 90° pulse, the spin isochromats are
back on phase τ seconds after the 180° refocusing pulse, the
total magnetization being reduced only by a fraction which is due

to genuine T_2 relaxation. In principle T_2 may be derived from
a series of spectra obtained by varying τ and Fourier transform-
ing the second half echo. Individual lines in these spectra
ideally decay with a time constant T_2.

There are, unfortunately, numerous sources of systematic
errors potentially impeding obtention of reliable data. For example, in order to avoid an irreversible loss of transverse
magnetization by translational diffusion, a large number of
closely spaced refocusing pulses has to be applied before the
last half-echo can be sampled and Fourier transformed. This,
however, means that a minor deviation of the pulse flip angle
from 180° forces the magnetization out of the transverse plane
and thereby results in an apparent T_2 which may be considerably
shorter than the true one. A modification has therefore been
devised which cancels this undesired effect (Meiboom-Gill modification). This consists of shifting the rf phase following
the initial 90° pulse. The effect of this compensation scheme
is diagrammatically shown in a rotating frame representation in
Fig. 6.

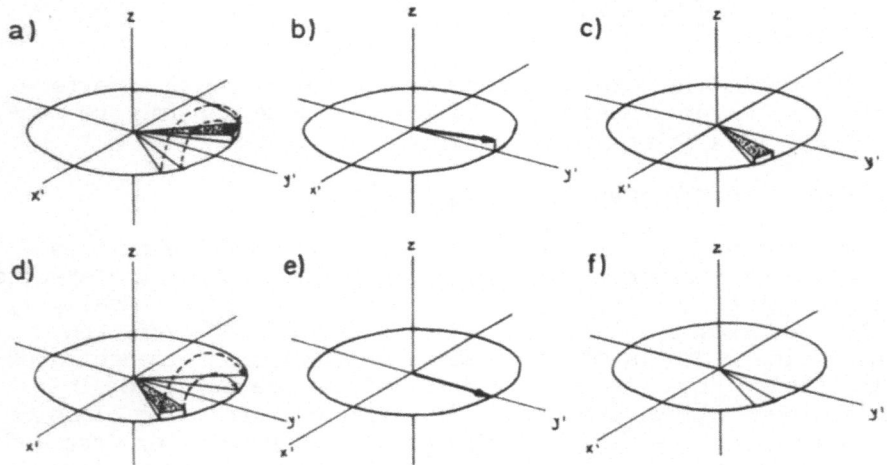

CARR-PURCELL-MEIBOOM-GILL EXPERIMENT

Fig. 6:   Behaviour of the transverse magnetization
          in a CPMG T_2 experiment following two echo
          pulses with $\alpha < 180^\circ$.

A particular problem arises if the spins subjected to a Carr-Purcell spin echo train are coupled to a second species. In the case of a heteronuclear coupling situation (e.g. ^{13}C coupled to protons this may be overcome by broadband irradiation as long as the decoupler is switched to coherent operation during the refocusing period. If, on the other hand spin conpling is of the homonuclear type and the pulses are nonselective, a phenomenon denoted J-modulation occurs which leads to a modulation of the phases of individual multiplet components [20]. The most straightforward remedy in this case consists of recording magnitude rather than phase-sensitive spectra.

If the repetition rate of the refocusing pulses in a CPMG experiment approaches infinity, the isochromats become locked to the rf field B_1. The transverse magnetization decays with a time constant $T_{1\rho}$, the spin-lattice relaxation time in the rotating frame. For liquids this can be assumed to be equal to T_2. One of the experimental difficulties in a spin-lock experiment is to maintain a sufficiently large B_1 field for a period which is on the order of $T_{1\rho}$. A further requirement that is difficult to fulfill concerns the strength of the B_1 field, which has to be large compared with the chemical shift range of the nuclei whose $T_{1\rho}$ is to be measured ($\gamma B_1/2\pi >> |\nu_i - \nu_c|$). Another complication arises when the experiment is conducted in the presence of broadband decoupling of a second spin species S which is coupled in a dipolar fashion to the spins I. Since under noise-decoupling conditions a band of decoupler frequencies is present and at the same time the I-spins are locked to the B_1 field, partial fulfilment of the Hartmann-Hahn condition ($\gamma_I B_{1I} = \gamma_S B_{2S}$) may result in energy transfer between the two spin systems, which accelerates the decay of the I magnetization and therefore yields an apparent T_2 which is too short. It is therefore mandatory that during the spin-locking period the decoupler be switched to coherent and and the B_2 amplitude chosen such that the $\gamma_I B_{1I} \neq \gamma_S B_{2S}$ [20].

2.3   Methods for the Measurement of the NOE.

    The heteronuclear NOE may be obtained from a com-
parison of the integrated signal intensity in the presence and absence of decoupling. This is usually incon-
venient for ^{13}C, ^{29}Si, etc. where scalar coupling often
leads to severe overlap of the spin coupling patterns.
The problem can be overcome by suitably gating the decoupler,so that it is always on during the acquisition
period and alternately on and off during the pulse interval (22). Here the question arises as to the duration of the on and off periods of the decoupler. In order for the NOE to build up to 99% of its full value
the decoupler has to be on for at least 5 T_1 periods.
During the off period, however, the magnetization $<S_z>$
decays more slowly due to the intervention of cross relaxation, thus requiring a longer interval. It has recently been shown that an adequate value is 10 T_1
cycles (23).

2.4   Selective Relaxation, NOE and Related Experiments

    There exist several motivations for selectively
perturbing certain resonances in a spectrum. Most selective T_1 experiments, however, have been concerned
with proton NMR. In dipolar coupled spin systems relaxation is normally non-exponential due to cross relaxation. However, the initial rate of relaxation
turns out to be approximately given by a single time
constant (24). A nonselective 180°-τ-90° experiment
may therefore yield an initial relaxation rate $R_1^A(NS)$
for proton A. A different value $R_1^A(S_A)$ is obtained
from a selective T_1 experiment in which only the A resonance has been perturbed. The ratio $R_1^A(NS)/R_1^A(S_A)$
then gives the extent to which proton A is relaxed
via DD interaction with all other protons. This ratio
is 1.5 if intramolecular DD relaxation is the only acting mechanism. Selective inversion of both A and B
resonances for example allows one to determine the extent of intramolecular DD relaxation of A via dipolar
interaction with B. The fractional contribution of
proton B to the relaxation of A is then given by (24):

$$f^A(B) = \frac{2\left[R_1^A(S_A,S_B)-R_1^A(S_A)\right]}{R_1(S_A)} \qquad (27)$$

Experimentally the problem has been approached in different ways. A single selective inversion is readily accomplished by a weak decoupler pulse of suitable frequency in such a manner that $\gamma B_2/2\pi \sim \Delta\nu$, where $\Delta\nu$ represents the width of the spin multiplet to be affected. Selective T_1's are then obtained by monitoring recovery τ seconds later with a non-selective transmitter pulse. An extension of this method permitting simultaneous selective inversion at two different chemical shifts is achieved by suitably modulating the decoupler frequency. If the latter is modulated with a frequency equal to the chemical shift separation of the two multiplets to be irradiated, the weak decoupler pulse affects both groups of resonances. This is illustrated with the 400 MHz proton spectrum of arabinose tetra (d_3) acetate, obtained by simultaneously inverting the resonances due to H_1 and H_3 (25).

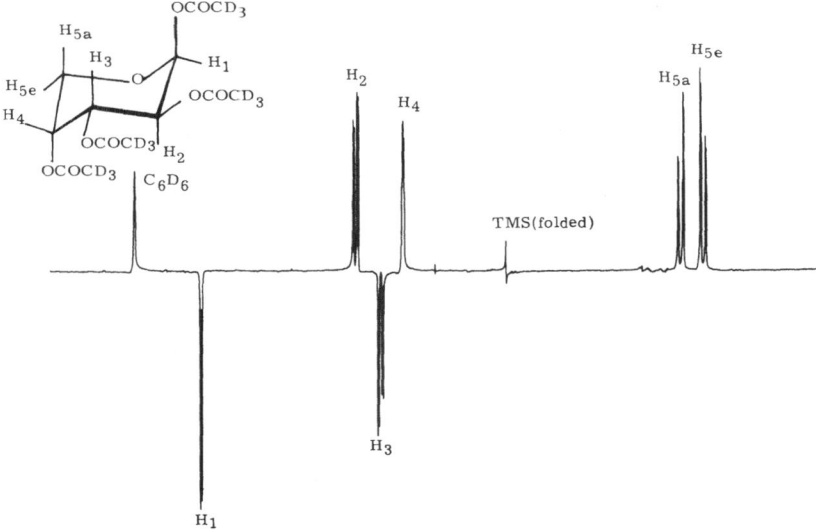

Fig. 7: 400 MHz proton NMR spectrum of arabinose tetra (d_3) acetate obtained by simultaneous selective inversion of resonances H_1 and H_3 using a pulse spacing of 10 msec between 180°(S) and 90°(NS) pulses (courtesy of Dr. L. D. Hall).

The selective excitation method of most universal applicability is based on the elegant idea of applying a pulse sequence generated by Fourier synthesis of a desired frequency spectrum (26, 27). In a multiple selective NOE experiment, for example, in which the experimenter aims at simultaneously irradiating frequencies F_1, F_2, F_3, ..., the required ideal excitation spectrum is essentially uniform except for greatly enhanced amplitudes near the prespecified frequencies. A pulse sequence corresponding to this excitation spectrum can be computed such that the amplitudes or the pulse withs are modulated. Unfortunately neither is instrumentally trivial and the method has therefore not been widely utilized.

A simpler but more limited technique, conceptually derived from the synthesized excitation approach, is the DANTE sequence which consists of a train of n micropulses of flip angle $\alpha(<<\pi/2)$ and interval τ, so that for example $n\cdot\alpha=\pi$ (selective inversion) (28).

If the magnetization is to experience an equal nutation by each micropulse, the pulse interval has to be chosen such that the magnetization carries out an even number of precessions between subsequent pulses. Assuming the line to be excited is displaced f_O Hz relative to the rf carrier, above condition is satisfied if $2\pi f_O\tau=m$ (m = 1, 2, ...). It can readily be shown that all other frequencies are not or only insignificantly affected. In this manner a single resonance line can be excited, since the selectivity condition is fulfilled for a particular line only. Hence the technique is less suited for selective inversion type of experiments where homonuclear coupling is present. However, it lends itself for carbon-13 NMR studies where a combination of selective and nonselective pulses can less easily be realized than in proton NMR. So has the method been tested on saturation - transfer experiments (29) directed toward determining exchange rates for slow conformational interconversion (28).

2.5 Summary of Pulsed Relaxation Experiments

The previously outlined pulse methods related to the measurement of the time evolution of either longitudinal or transverse components of the macroscopic moment may have given the reader an overview on

Table 1: Summary of currently utilised pulse techniques for the measurement of transient phenomena

Pulse sequence[1]	Method[1]	Error sources
$180°_x$ (NS)$-\tau-90°_x$ (NS)	NS IR (T_1)	2)
$180°_x$ (NS)$-\tau-90°_x$ (NS)$-T-180°_x$ (NS)$-\tau-90°_{-x}$ (NS)	phase-inverted NS IR (T_1)	
$180°_x$ (S)$-\tau-90°_x$ (NS) $90°_{-\tau}$	S IR (S T_1)	
	PS (NS T_1)	2) 3) 4)
$90°_x$ (NS)$-$HSP$-\tau-90°_x$ (NS)	SR (NS T_1)	2) 3)
$90°_x$ (NS)$-\tau-90°_x$ (NS)$-90°_x$ (NS)$-\tau-90°_{-x}$ (NS)	phase-inverted SR (NS T_1)	2) 3)
$90°_x$ (S)$-$HSP$-\tau-90°_x$ (NS)	S SR (S T_1)	
$90°_x$ (NS)$-\tau-(180°_y$ (NS)$-\tau)_{2n}$	NS CPMG (NS T_2)	2) 5) 6)
$90°_x$ (S)$-\tau-(180°_y$ (S)$-\tau)_{2n}$	S CPMG (S T_2)	5) 6)
$90°_x$ (NS) B_{1y} (t=0, t=τ)	NS spin lock (NS $T_{1\rho}$)	2) 5) 6)
$90°_x$ (S) B_{1y} (t=0, t=τ)	S spin lock (S $T_{1\rho}$)	5) 6)

1) Acronyms have the following meaning: NS = nonselective, S = selective, HSP = homogeneity spoiling pulse, B_1 = continuous rf field; the superscript following the pulse flip angle refers to the relative transmitter phase in the four quadrants (x, y, -x, -y); IR = inversion - recovery, PS = progressive saturation, SR = saturation recovery, CPMG = Carr-Purcell-Meiboom-Gill

2)-7) Sources of systematic errors

2) resonance offset effects arising from missing compliance with the condition $B_1 >> \Delta B$ (ΔB=resonance offset relative to carrier

3) misadjustment of pulse flip angle

4) failure to reach steady state

5) partial fulfilment of Hartman-Hahn condition in heteronuclear coupling situation (e.g. observing 13C while decoupling protons)

6) spatial inhomogeneity of B_1 field

the current state of art of modern pulse spectrometers. Although
these have reached an unprecedented degree of perfection, there
is no guarantee for the correctness of the derived data and the
newcomer in the field is well advised to test the experiment on
a well defined molecular system where reliable literature data
are available, rather than blindfoldly trusting the experiment.
Particular caution is due in experiments conducted in the transverse plane (T_2, $T_{1\rho}$).

Table 1 which lists the techniques described in sections 2.1 -
2.3 therefore points out the major sources of systematic erros.

3.   TWO-DIMENSIONAL (2D) NMR SPECTROSCOPY

3.1 Principles of 2D NMR

Two-dimensional NMR, as outlined in the fundamental paper by
Aue et al. (30) implies a two-dimensional array of spectral data
with both indices relating to frequencies. In this sense of the
definition stacked plots as they are common for the representation
of spin relaxation spectra (vide infra) cannot be regarded a 2D-
spectrum. The probably most trivial example of a 2D spectrum is
obtained by plotting a series of frequency-domain spectra in the
presence of a second irradiating field such that each spectral
trace corresponds to a different double resonance frequency. This
experiment does not require any particular equipment since it may
entirely be executed in frequency space. By contrast, the category of experiments henceforth denoted "2D" is carried out in
time space and is only subsequently converted to the frequency
domain by two consecutive Fourier transformations:

$$S(\omega_1, \omega_2) = \hat{F}_{t_1, t_2} \ \{s(t_1, t_2)\} \tag{28}$$

In Eqn. 28 $S(\omega_1, \omega_2)$ and $s(t_1, t_2)$ refer to the frequency and time
domain signals, resp., while $\hat{F}$ symbolizes the Fourier transform
operation.

3.2 2D J-resolved Spectroscopy

As of today a host of different 2D experiments have been
characterized, all having in common a two-dimensional array of
time domain data $||S(t_1, t_2)||$ with the time variable t_1 refer-
ring to a time interval between two events (e.g. two pulses) and
t_2 is the data acquisition time.

The basis of 2D J-resolved spectroscopy is the spin echo experiment described in Section 2.2. In this the two time periods

are defined as indicated in Fig. 8. What enables the experiment
is the earlier mentioned J-modulation phenomenon (32) which occurs in a spin echo experiment involving homonuclear (scalar)
spin-spin coupling,provided all spins are equally excited.

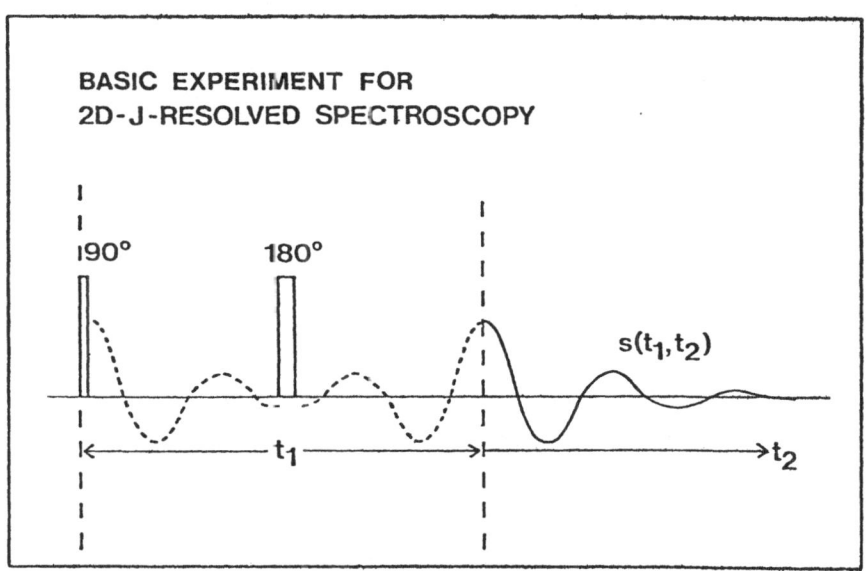

**BASIC EXPERIMENT FOR
2D-J-RESOLVED SPECTROSCOPY**

Fig. 8: Principle of 2D J-resolved spectroscopy

Instead of complete refocusing of the spin isochromats at a time
2τ following the initial 90° pulse, the echo amplitude is modulated. Hence a Fourier transformation according to t_2 affords a
series of spectra in which corresponding multiplet lines rotate
their phases in a cyclic fashion. Since the phases of the spin
multiplets rotate at frequencies related to the coupling constants J_{ij}, a second Fourier transformation of $s(t_1,\omega_2)$ provides
a set of spectra whose projections onto the ω_1 frequency axis correspond to overlays of the various spin multiplets relative to
their center frequencies. Since the echo amplitude decays at a
rate given by $1/T_2$, the individual lines are free of inhomogeneity broadening. By contrast the projection onto the ω_2 axis
affords a spectrum which shows chemical shifts only but no effects due to spin-spin coupling. The technique thus enables obtention of what could be designated "broad-band decoupled" proton NMR spectra.This behaviour is illustrated by the stacked
plots in Fig. 9a displaying the region of the high-field methyl

resonances in the cyclic peptide AW27-400 ($C_{62}H_{111}N_{11}O_2$), recorded at 400 MHz (33). The top trace represents the projection onto the δ axis, revealing separation of twelve of the thirteen magnetically nonequivalent methyls. Cross-sections parallel to the J axis are provided in Fig. 9b, showing individual multiplets with eleven of them being due to isopropyl methyls, stemming either from valine or leucine moieties (doublets), whereas one (number 11) has triplet multiplicity and originates from an ethyl side chain.

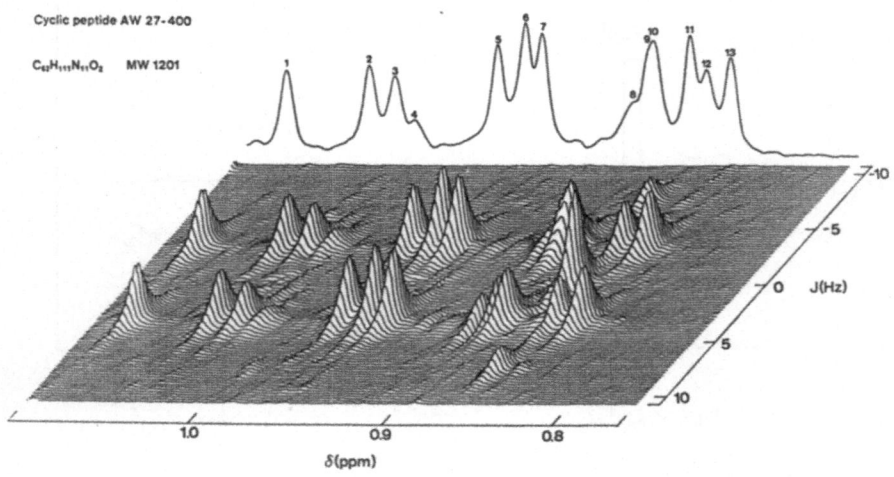

Fig. 9a: 400 MHz 2D J-resolved proton NMR spectrum showing
           high-field methyl region in the cyclic peptide
           AW 27-400. The top trace represents a projection
           onto the δ axis (Courtesy of Dr. E. Bartholdi,
           Spectrospin AG, Zurich)

A prime requisite for the applicability of the method in biology is high magnetic field (34,35), not only because of the chemical shift dispersion required for resolving closely spaced lines, but also because simple splitting patterns are only obtained under weak coupling conditions.

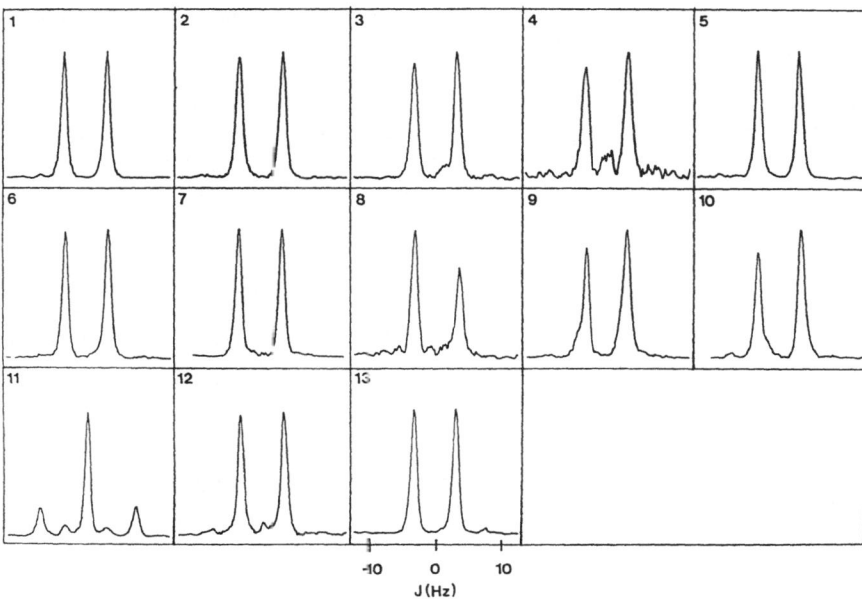

Fig. 9b: Individual cross sections along J-axis obtained
            from the spectrum in Fig. 9a.

References

1. See for example Lyerla, J.R., and Grant, D.M.: in "Interna-
   tional Review of Science,Phys. Chem. Series", Mc Dowell, C.A.,
   Ed., Vol. 4, Chap. 5, Medical and Technical Publ. Comp., 1972;
   Lyerla, J.R. Jr., and Levy G.C., in "Topics in Carbon-13 NMR
   Spectroscopy", Levy, G.C., Ed., Chap. 3, Wiley Interscience,
   1974.

2. Abragam, A. : "The Principles of Nuclear Magnetism" Chap.8,
   Oxford, 1961

3. Noggle, J.H., and Schirmer, R.E. : "The Nuclear Overhauser
   Effect", Chaps. 1 and 6, Academic Press, 1971

4. Allerhand, A., Doddrell, D., and Komoroski, R. : 1971, J.
   Chem. Phys. 55, 189

5. Lyerla, J.R., Jr., Grant, D.M., and Harris, R.K., 1971 : J.
   Phys. Chem. 75, 585; Alger, T.D., Grant, D.M., and Lyerla,
   J.R.,Jr. : J. Phys. Chem. 75, 2539

6. Deslauriers, R., Smith, I.C.P., and Walter, R., 1974 : J.
   Amer. Chem. Soc. 96, 2289

7. Wehrli, F.W. ,1976 : J. Magn. Res. 23, 181

8. Spiess, H.W., Schweitzer, D., Haeberlen, U., and Hausser,
   K.H., 1971 : J. Magn. Res. 5, 101

9. Levy, G.C., Cargioli, J.D., and Anet, F.A.L., 1973 : J. Amer.
   Chem. Soc. 95, 1527

10. Wehrli, F.W., 1979 : J. Magn. Res. 32, 451

11. Tarasov, V.P., Privalov, V.I., and Buslaev, Yu. A., 1977 :
    Dokl. Akad. Nauk SSSR 234, 636

12. See for example Lindman, B., and Forsén, S., "Chlorine, Bro-
    mine and Iodine NMR , Physico-Chemical and Biological Ap-
    plications" in "NMR, Basic Principles and Progress", Diehl P.,
    Fluck, E., and Kosfeld R., Eds., Springer, 1976

13. Mann, B.K., and Harris, R.K., "NMR and the Periodic Table"
    Chap. 6, Academic Press, 1978

14. Shchori, E., Jagur-Grodzinski, J., and Shporer, M., 1973 : J. Amer. Chem. Soc. 95, 3842

15. Vold, R.L., Waugh, J.S., Klein, M.P., and Phelps, D.E., 1968: J. Chem. Phys. 48, 3831

16. Freeman R. & Hill,H.D.W., 1971 : J. Chem. Phys. 54, 3367

17. Canet, D., Levy, G.C., and Peat, I.R., 1975 : J. Magn. Res. 18, 199

18. McDonald, G.C., and Leigh, J.S. Jr., 1973 : J. Magn. Res. 9, 358

19. S. Patt, priv. commun.

20. Freeman, R., and Hill, H.D.W., in "Dynamic Nuclear Magnetic Resonance", Cotton, F.A., and Jackman, L.M., Eds., Chap. 5, Academic Press 1975

21. Carr, H.Y., and Purcell, E.M., 1954 : Phys. Rev. 94, 630

22. Freeman, R., Hill, H.D.W., and Kaptein, R., 1972 : J. Magn. Res. 7, 327

23. Canet, D., 1976 : J. Magn. Res. 23, 361 ; Opella, S.J., Nelson, D.J., and Jardetzky, O., 1976 : J. Chem. Phys. 64, 2533

24. Hall, L.D., and Hill, H.D.W., 1976 : J. Amer. Chem. Soc. 98, 1269

25. Hall, L.D., and Hull, W.E., private commun.

26. Tomlinson, B.L., and Hill, H.D.W., 1973: J. Chem. Phys.59, 1775

27. Freeman, F., Hill, H.D.W., Tomlinson, B.L., and Hall, L.D., 1974 : J. Chem. Phys. 61, 4466

28. Morris, G.A., and Freeman, R., 1978 : J. Magn. Res. 29, 433

29. Forsén, S., and Hoffmann, R.A., 1963 : J. Chem. Phys. 39, 2892

30. Aue, W.P., Bartholdi, E., and Ernst, R.R., 1976 : J. Chem. Phys. 64, 2229

31. Jeener, J., Ampère International Summer School II, Basko Polje, 1971

32. Freeman, R., and Hill, H.D.W., 1971 : J. Chem. Phys. 54, 301 (1971)

33. Bartholdi, E., 1979 : Private communication

34. Nagayama, K., Wüthrich, K., Bachmann, P., and Ernst, R.R., 1977 : Biochem. Biophys. Res. Commun. 78, 99

35. Nagayama, K., Bachmann, P., Wüthrich, K., and Ernst, R.R., 1978 : J. Magn. Res. 31, 133

ARCHITECTURE AND DYNAMICS OF ISOTOPICALLY LABELLED MACROMOLECULES
BY NUCLEAR MAGNETIC RESONANCE SPECTROSCOPY*

N. A. Matwiyoff

Los Alamos Scientific Laboratory
University of California
Los Alamos, New Mexico 87545 U.S.A.

INTRODUCTION

The last two decades were a period of luxuriant growth in
the X-Ray Crystallography of biological macromolecules that has
led to remarkably detailed correlations between their structure
and function (1). A particularly outstanding example is the
elegant work of Perutz and his colleagues on the structure of
hemoglobins which resemble "molecular lungs" (2) in their mode
of O_2/CO_2 binding and release. The family of serine proteases
provide another example and, in this case, x-ray structural
studies have resulted in an elaborate picture of the mechanism
of the action of this class of enzymes (3,4); however, the
detailed picture is not without its controversial elements (5).

As valuable as the x-ray structure of a biological macromolecule is, the picture it provides is a static one (6) of a
molecule whose states of hydration and intermolecular association are much different than those under physiological conditions. The use of NMR spectroscopy in the study of biological
molecules and macromolecular assemblages has grown even more
dramatically than x-ray methods and, in part, this is because
NMR provides a dynamic, conformationally averaged, picture of a
molecule in solution that is complementary to the crystallographic "still photo". The resolution of the dynamic features
of the structure by NMR, of course, is accomplished at the
expense of specificity in the location of the atoms. An additional complication in NMR studies of proteins, for example, is
that the large number of atoms generally prevents resolution of
a single resonance for each nucleus and its assignment to a
specific amino acid residue.

29

*I. Bertini and R.S. Drago (eds.), ESR and NMR of Paramagnetic Species in Biological and
Related Systems, 29–53.*

Of course, as discussed at length in these proceedings, intrinsic and extrinsic paramagnetic shift and relaxation reagents offer great promise in recovering the spatial specificity and in improving the resolution of the NMR experiments. An alternate method for accomplishing these latter objectives without perturbing the experimental system with extrinsic reagents, is to use NMR to monitor a protein specifically labeled with ^{13}C or ^{15}N enriched amino acids. An ancillary benefit is that specific labeling frequently allows an NMR study of the macromolecule in intact cells and tissues. This method is the subject for discussion in this paper which is not a comprehensive treatment but an illustrative survey of what appear to be the most promising areas for future development in the study of the architecture and dynamics of macromolecules and macromolecular assemblages, especially those containing proteins. To illustrate the full power of the technique at the outset, we will first briefly consider the ^{13}C NMR spectra of cell suspensions specifically labeled with ^{13}C.

RED BLOOD CELL SUSPENSIONS

As we all know, the role of the red blood cell in physiological gas exchange is not only to deliver O_2 from the lung to peripheral tissues but also to transport CO_2, the end product of respiration, to the lung. Although the mechanism of the uptake transport, and delivery of O_2 by hemoglobin within the red blood cell is reasonably well understood (2), only in the past few years has the complexity of the physiological transport mechanism for CO_2 begun to be unraveled. As we have learned more about CO_2 transport, we have also come to appreciate that the physiological importance of CO_2 may reside not only in its role of mediating pH and ionic strength but also in the regulation of enzyme activity through formation of carbamino complexes. As outlined in the following, the ^{13}C NMR of ^{13}C labeled systems has played a central role in the development of this understanding.

Matwiyoff and Needham (9) were the first to study the interaction between CO_2 and hemoglobin in the red blood cell by ^{13}C NMR techniques. A summary of their results is contained in Figure 1 and illustrates the ^{13}C NMR spectra of packed human red blood cells in isotonic saline treated with [^{13}C (90 atom %)] CO_2. The resonances identified as a, b, and c (Figure 1b) did not occur in the spectra of red blood cells treated with natural ^{13}C abundance CO_2 and, on the basis of chemical shifts alone, resonances b and c, could be assigned to HCO_3^- ion and dissolved CO_2, respectively. The assignments were confirmed, however, by the addition of carbonic anhydrase which catalyzes the interchange between CO_2 and the HCO_3^- ion, large concentrations of the enzyme leading to chemical exchange broadening (Figure 1c), and

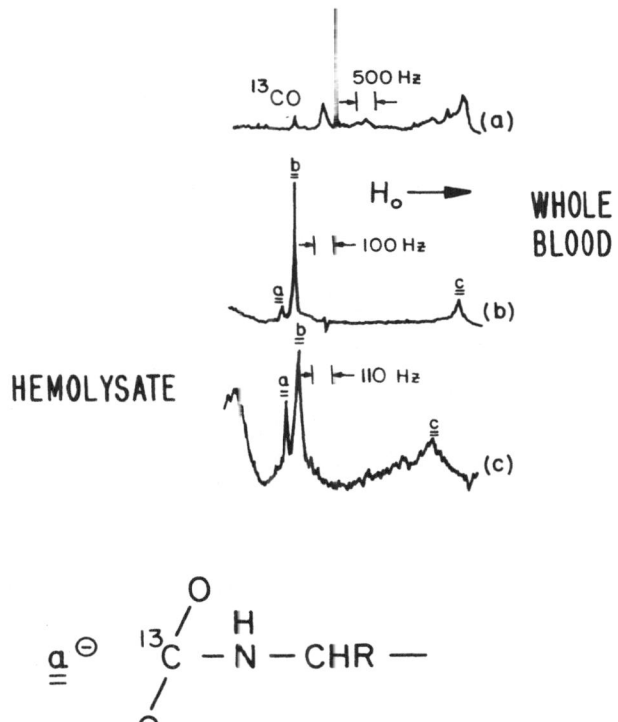

Figure 1. Proton noise decoupled ^{13}C NMR spectra: (a) Dog whole blood + ^{13}CO, 65,000 pulses; (b) Human whole blood + $^{13}CO_2$, 14,000 pulses; (c) Human whole blood + $^{13}CO_2$ + carbonic anhydrase, 12,000 pulses. Signals a, b, and c are assigned to the hemoglobin carbamino complexes, carbonic acid, and CO_2, respectively.

eventual collapse of signals b and c to an exchange narrowed peak at the center of gravity of the original resonances. Signal a was not affected by the carbonic anhydrase treatment.

A clue to the origin of signal a was the early suggestion by Rossi-Bernardi and Roughton (10,11,12) that a significant fraction of the physiological CO_2 transport was in the form of carbamino complexes (structure shown in Figure 1) of the N-terminal valine residues of the hemoglobin chains, a suggestion supported by later studies of Kilmartin and Rossi-Bernardi (13,14,15) who showed that when the terminal α-amino groups of isolated hemoglibin were blocked by cyanate carbonylation, the O_2-linked CO_2 interactions were reduced or obliterated. A confirmation that a was a resonance from a carbamino compound in the red blood cell system was obtained by ^{13}C NMR studies of carbamino derivatives of the common amino acids. Subsequently, Arnone (16) studied the x-ray structure of human deoxyhemoglobin crystals soaked in CO_2/bicarbonate buffer (pH 7.4 and p_{CO_2} = 1 atm) and showed that CO_2 does indeed form carbamino complexes with the valine N-terminal amino groups of the hemoglobin β chains. In more recent work, Gurd and co-workers (17,18) have studied, by ^{13}C NMR techniques, adduct formation of CO_2 with isolated hemoglobins and peptide hormones with the view of elucidating potential regulatory influences of carbamino formation. Because carbamino formation (equation 1) results in the introduction of a charged bulky group on the terminal amino function, one might expect in the general case that regulatory influences could be exerted through the formation of salt bridges and/or changes in the population of rotamers about the N-C bond. In the case of

$$\underset{\text{H}}{\overset{\text{R}}{\underset{|}{\overset{|}{\text{HN-C-C-}}}}} + CO_2 \rightleftarrows \underset{\text{H}}{\overset{\text{R}}{\underset{|}{\overset{H_2|}{O_2C\text{-}N\text{-}C\text{-}C\sim}}}} + H^+ \qquad (1)$$

deoxyhemoglobin itself, Arnone's x-ray studies (16) indicate that CO_2 binding to the terminal valine residues of the β chains merely displaces an ion from an anion binding site and does not produce any marked changes in the conformation of the deoxy β subunits. However, allosterically effective conformational changes may be too small to detect with x-ray methods and, in this context, it is interesting to note that ^{13}C NMR studies have shown (17) that the reduction of carbamino adduct formation accompanying the conversion of deoxy- to oxy-hemoglobin occurs predominantly at the terminal valines of the β subunits. Given the ubiquitousness of CO_2 and HCO_3^- ion in biological systems, it is likely, as Gurd has pointed out, that carbamino adducts may play a role in modifying the activities of enzymes with terminal amino groups having pK's near 7. Indeed, it has been shown that even the peptide hormones, angiotensin and bradykinin, form significant quantities of carbamino adducts at physiological pH and p_{CO_2} (18).

In a related context, there is another interesting feature of the spectra in Figure 1 -- spectrum C is for hemolyzed blood and one notes a pronounced broadening of the HCO_3^- and CO_2 resonances compared to the whole blood spectrum. The red blood cell contains a high concentration of carbonic anhydrase (~ 2 mg) which may be loosely or tightly bound to the membrane. Hemolysis releases it, allowing it to more effectively catalyze the $^{13}CO_2$ and $H^{13}CO_3^-$ exchange which broadens both resonances. The apparent difference in the activity of the enzyme in whole and hemolyzed blood, which we do not understand as yet (but see 19) and which may suggest a function for carbonic anhydrase in the cell other than CO_2 turnover, points to the importance of study-ing whole cell systems where possible.

Whereas the apparent decrease in the activity of the red blood cell carbonic anhydrase may be unusual, I would stress that the carbamino chemistry we are monitoring by looking through the cell with ^{13}C NMR is completely normal. That is, hemoglobin carbamino formation conforms to the laws of mass action, rising with CO_2 pressure and decreasing with decreasing pH in the red blood cell just as it does in homogeneous aqueous solutions as studied extensively by Gurd and co-workers at Indiana University (17,18). This is not too surprising -- physical chemistry is physical chemistry. It is stressed here because there is a folklore or mystique about the properties of the interior fluids of the cell which might lead to unusual chemistry - regions of high ionic strength, high viscosity, abnormal pH, polywater structure, etc. And, of course, it is difficult to get a fix on the properties of the interior of the cell without using a tech-nique that violates its integrity. NMR provides a limited opportunity to probe the environment of intracellular constituent without perturbing the system. One well-known method based on NMR is to probe the intracellular pH by studying the pH dependent chemical shifts of intracellular phosphate derivatives. A less well known, though equally promising, NMR method for probing the intracellular environment is the study of intracellular viscosity by measuring the spin lattice relaxation and rotational correla-tion times of ^{13}C labeled soluble intracellular constituents (21,22). Unfortunately time does not permit elaboration of these investigations but it could be noted that NMR studies have revealed nothing "abnormal" about the pH or viscosities of the intracellular fluids studied to date.

STRUCTURAL CONSTITUENTS OF CELLS

In the preceding, we have outlined briefly a general kind
of NMR approach to the analysis, chemical transformations, and
environment of specific, isotopically labeled intracellular
constituents. In principle, it is possible also to isotopically
label individual <u>structural</u> components of the cells themselves.
This is a promising area of investigation as well and one which
has attracted a great deal of interest recently. Most attention
has been devoted to the lipids of the cell membrane, both
because of the ease with which they are labeled and their
importance in cellular function. The cell membrane lipids,
exemplified by the phospholipids depicted in Figure 2, have been
judged on the basis of a number of physical studies (23) to be
arranged in a bilayer structure and to be involved in a variety
of cellular processes, including the maintenance of the cell
permiability barrier and association with proteins to form
functional entities. With regard to these functions, one
parameter which appears critical is the mobility or fluidity of
the lipid bilayer which, in turn, depends on the stiffness of the
hydrocarbon chains and the way they are packed in the bilayer.
This question has been studied in some detail by the use of ^{13}C
and ^{2}H labeled fatty acids incorporated into natural and artificial membranes; and the nuclear spin lattice relaxation times
of the hydrocarbon chains are not inconsistent with chain flexi-
bilities and more classical measures of membrane permiabilities
and transition temperatures for the transformation from the
"solid" to "liquid crystalline" phases. These questions have
been reviewed previously (24) and will not be considered at length
here. Instead, we focus on the illustrative aspects of a unique
^{13}C labeling experiment which has yielded valuable information
about both the structural and dynamic properties of the polar
phosphoryl, choline head groups of phospho lipids in natural and
artificial membranes.

A central question about the disposition of membrane polar
head groups, which is reciprocal to that asked about hydrocarbon
residues, is whether these polar groups which interface with the
aqueous phases at the extracellular or intracellular boundary
are close packed or loose structures? Specifically, do the
polar - $N(CH_3)_3^+$ groups: (a) form rigid salt bridges with an
adjacent charged phosphate group making a compact structure; or
(b) interact strongly with the aqueous phase forming an extended
mobile structure characterized by rotational freedom favored by
the loose, and dynamic, water structure. We have studied this
question by biosynthetically incorporating ^{13}C-labeled choline
(22,25) into Chinese Hamster Ovary (CHO) cells and studying the
^{13}C spin lattice relaxation times and linewidths. Two types of
labeling experiments were done: (1) the label was inserted into
the N-methyl group alone; and (2) the label was inserted separately

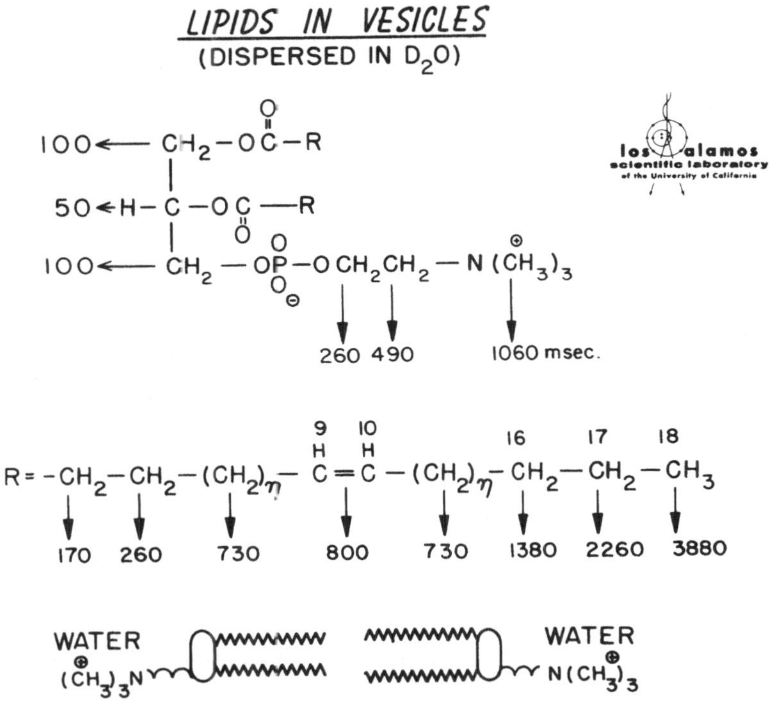

Figure 2.   Spin lattice relaxation times (T_1) in msec for the carbon atoms in lecithin vesicles. The T_1 values observed are consistent with a high degree of mobility or fluidity of the lipid bilayer. The vesicles are a micelle structure with the polar head groups exposed to water and the lipid chains packed together in a hydrophobic environment.

into the N-CH₃ group, the 1-¹³C and the 2-¹³C positions and a
mixture (1:3:3) was incorporated into the polar head group (the
mixture was used rather than a uniformly labeled choline to
avoid the complications of ¹³C-¹³C coupling).

The [¹³C]choline N-trimethyl resonances exhibited by an
aqueous suspension of CHO cells grown in a medium containing
[¹³C-methyl]choline is reproduced in Figure 3. The line is
narrow (14.8 Hz half-width at 21°C) and the T_1 was determined to
be 400 ms at 21°C. Both observations show that the relaxation
of the [¹³C]choline methyl groups in the CHO cells are determined
by the internal motion of the choline methyl carbons. This was
confirmed by showing that the relaxation time increased with
increasing temperature; and exhibited an apparent activation
energy of 4.3 kcal/mole, a value expected for a neopentyl-like
barrier for the motion. The relaxation data could be fit with
the model shown in Figure 4 in which motion of C-1 is restricted,
whereas motion by R_3, R_2, and R_1 occur to a significant extent
($\tau \cong 3.1 \times 10^{-11}$ s). Although a more appropriate model should
incorporate a motional gradient $R_1 > R_2 > R_3$, the point is
established that the polar head groups do not adopt a compact
structure but rather have significant freedom of motions which,
in the aggregate, are isotropic enough that the ¹³C line is
narrow.

The spectra of the [1-¹³C], [2-¹³C], and [methyl-¹³C]
labeled choline in CHO cell suspensions and in suspensions of
vesicles prepared from the CHO cell lipids are even more interesting (Figure 5). In the spectrum obtained under conditions of
simultaneous ¹⁴N and ¹H (¹³C{¹⁴N, ¹H}) decoupling, the two and
three bond ³¹P-¹³C scalar coupling is evident as it is in the
synthetic vesicle. The rationale behind using ¹⁴N decoupling,
though an interesting story in itself, need not concern us here.
The interesting point is that the value of the three bond coupling
constant allows an estimation of the conformation about the C-1-O
bond from the expression derived by Smith and co-workers (26)

$$^3J_{CCOP} = 9.5 \cos^2\theta - 0.6 \theta$$

Trans coupling constants predicted from this equation are
similar to those measured recently for cyclic sugar phosphates
locked into a trans configuration (27). Analysis of the C-2-P
coupling constants using this expression, together with the
assumption that there is a rapid gauche ⇌ trans equilibrium
leads to fractional trans probabilities of 61%, 64%, 65%, and
75% for dioleyl, dipalmitoyl, and egg yolk lecithins, and the
CHO cells, respectively.

It is clear that isotopic labeling in conjunction with NMR
spectroscopy can provide valuable information about intact

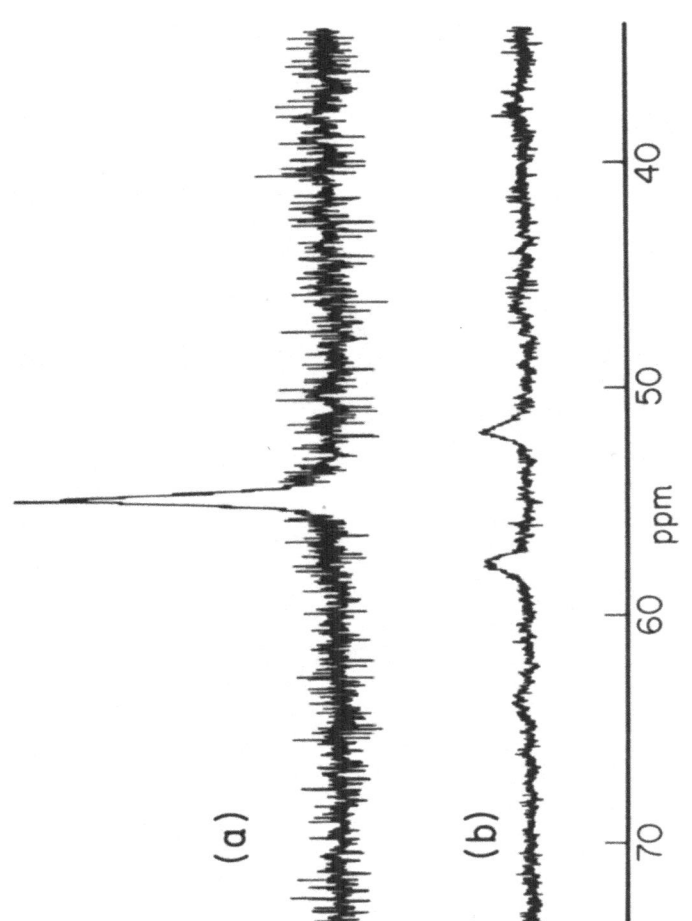

Figure 3. Proton decoupled (a) and coupled (b) ^{13}C FT NMR spectra of the choline resonance of sonicated vesicles prepared from extracted CHO cell lipids. Spectrum a represents 8200 pulses and spectrum b 100 850. Sample temperature was 11°C and a 2-s delay between pulses was used for the accumulation time.

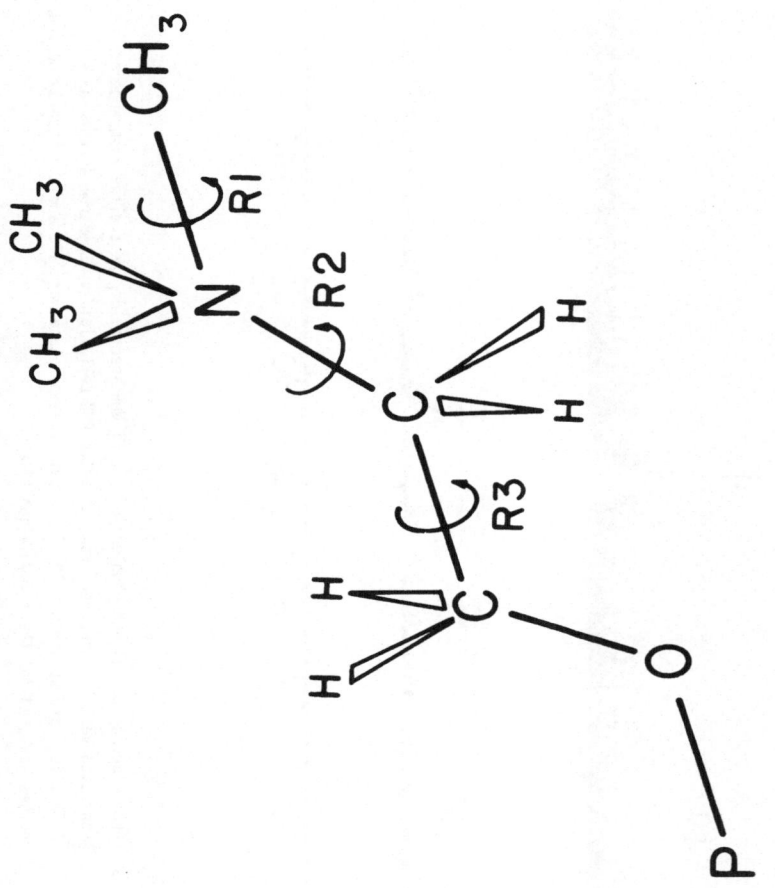

Figure 4. Illustration of the possible motions contributing to the relaxation of the choline methyl groups.

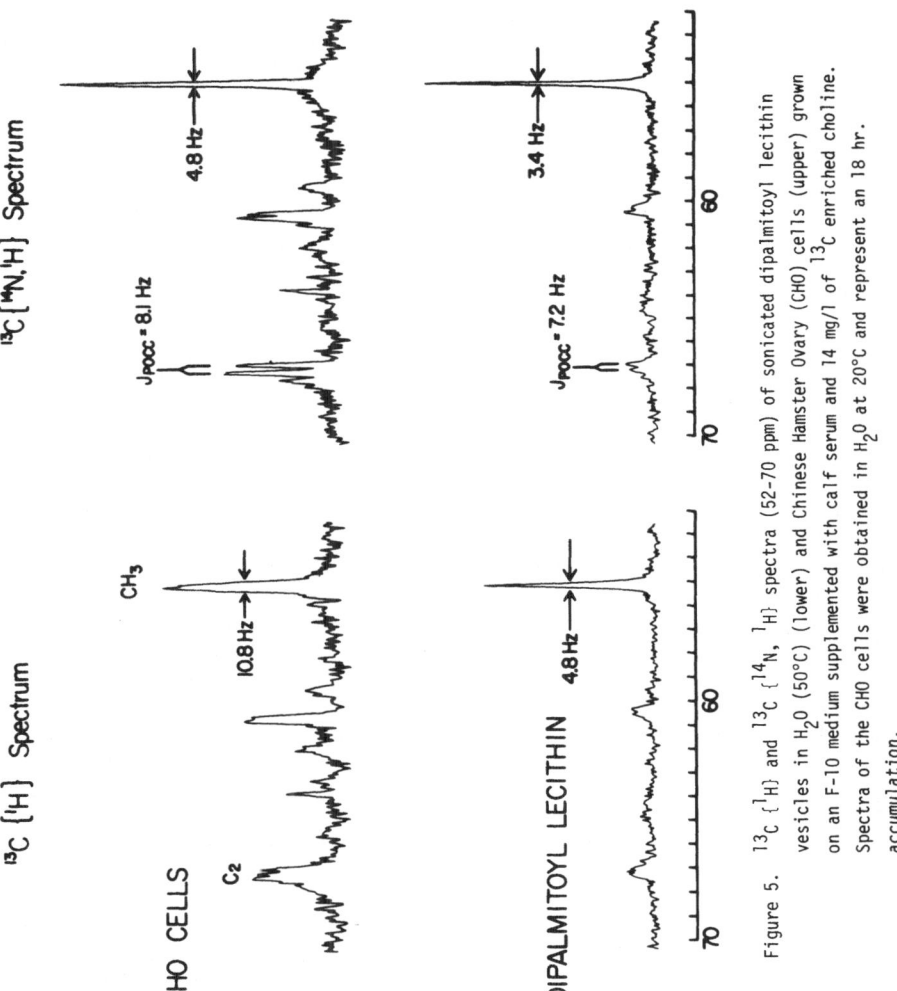

Figure 5. ^{13}C {^{1}H} and ^{13}C {^{14}N, ^{1}H} spectra (52-70 ppm) of sonicated dipalmitoyl lecithin vesicles in H$_2$O (50°C) (lower) and Chinese Hamster Ovary (CHO) cells (upper) grown on an F-10 medium supplemented with calf serum and 14 mg/l of ^{13}C enriched choline. Spectra of the CHO cells were obtained in H$_2$O at 20°C and represent an 18 hr. accumulation.

cellular systems but it is well to point out that the field is
still in its infancy and, because of the complexity of the
systems involved, we can only expect partial answers to the
simplest questions. We are usually talking about a heterogeneous
population of cells and we have a variety of membrane systems
which have an intrinsic asymmetry, a diverse population of structural and functional proteins distributed among membranes, ribo-
somes, chromatin, the cytosol, etc. The sheer complexity of the
system has led some investigators to approach the problem from
the other extreme -- label individual proteins, lipids, nucleic
acids, etc.; study them individually; and then as part of a
reconstituted macroscopic assemblage. This approach is likely
to pay big dividends, even if only for the study of the individual macromolecules as an adjunct to x-ray crystallography. In
the next section I would like to illustrate the approach with a
labeled enzyme we have been studying.

^{13}C LABELED ENZYMES: DIHYDROFOLATE REDUCTASE

Several recent studies of enzymes (5, 28-32) and proteins
(21,33,34) have shown that the incorporation of ^{13}C labeled
amino acids into macromolecules can provide an excellent, non-
perturbing NMR probe by which protein-ligand interactions can be
studied. In our own group at the national stable isotopes
resource at the Los Alamos Scientific Laboratory we have, in
cooperation with Professor R. L. Blakley of the University of
Iowa Medical School, initiated a series (35-38) of ^{13}C NMR studies
of the enzyme dihydrofolate reductase (DHFR) labeled with a
number of ^{13}C enriched amino acids. The enzyme catalyzes the
NADPH-linked reduction of dihydrofolate to tetrahydrofolate
which has an essential role in purine and pyrimidine biosynthesis.
Apart from its convenient size (∿20,000 daltons) for NMR studies,
interest in the structure of the DHFR and its complexes derives
from its clinical importance, the enzyme being the target of
antineoplastic and antibacterial chemotherapeutic agents such as
methotrexate and aminopterin. Although our studies are still in
the early stages, the results which are surveyed in the following
illustrate the potential and limitations of the method.

In Figure 6 are summarized the proton decoupled ^{13}C NMR in
the [γ-^{13}C]-Trp region for DHFR isolated from Streptococcus
faecium grown in medium containing [γ-^{13}C]tryptophan. The
enzyme S. faecium contains four tryptophan (Trp) residues at the
following positions in the amino acid sequence (39): Trp-6;
Trp-22; Trp-115; and Trp-160. In the crystal structure of the
Lactobacillus casei enzyme complexes (40,41), Trp-5 and Trp-21
correspond to positions 6 and 22 of the S. faecium enzyme in
highly conserved sequences and are buried in hydrophobic pockets
in the ternary DHFR-MTX-NADPH complex. The two other residues
of the Lactobacillus casei enzyme, Trp-133 and Trp-158, do not

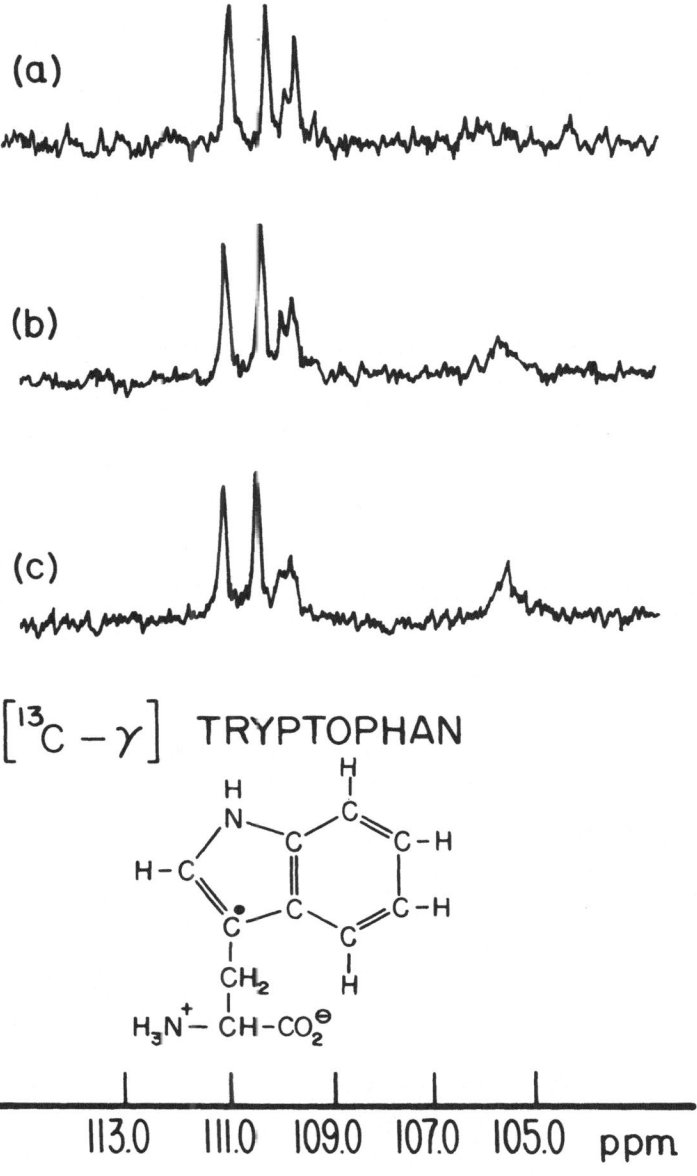

Figure 6. ^{13}C NMR spectra of [γ-^{13}C]tryptophan labeled
dihydrofolate reductase at various temperatures
a. 25°C.   b. 15°C.   c. 5°C.

occur in such highly conserved sequences; and the former is
partially and the latter is fully accessible to solvent. Urea
denaturation of the enzyme with consequent solvent exposure of
all Trp residues causes collapse of the ^{13}C spectrum to three
closely spaced narrow lines at 109.5, 109.7, and 109.9 ppm
(Figure 7). Thus, it is likely that the resonance at 110.4 ppm
in the native enzyme (Figure 6) is either Trp-115 or Trp-160.
Although a likely assignment for the other solvent exposed Trp
is the double resonance at ∿110 ppm, the peak at 111 ppm is
assigned to it for the reasons discussed below.

The resonance centered at ∿110 ppm and ∿106 ppm are more
informative in structural terms. First, the former is a double
peak with relative intensities of the components of 2:3. Since
the separation of these peaks is 0.2 ppm at both 25.2 MHz and
90 MHz, the most reasonable interpretation of this double
resonance is that it arises from the occupation of two distinct
environments by a single Trp residue. It is probably Trp-6 which
is deeply buried (41,42) in a hydrophobic pocket composed of
phenylalanine, leucine, valine, methionine, etc. side chains.
As described by London, et al. (37), the Trp probably samples
these two environments by a slow, restricted rotation about the
C_β-C_γ bond with some rotation about the C_α-C_β bond as well. A
slow rotation about these bonds would be expected from steric
interactions of the bulky indole side chain in the compact,
highly organized region of the DHFR structure occupied by Trp-6.

The remaining resonance at ∿106 ppm, which can be assigned
by difference to Trp-22, exhibits some unusual features. First,
the line is very broad for a quaternary resonance in an enzyme
of this molecular weight (half-width of ∿30 Hz at 25°C) and,
second, the linewidth increases with increasing temperature,
accompanied by an 0.8 ppm downfield shift of the resonance as
the temperature increases from 5 to 25°C. These data are
consistent with a slow exchange of the Trp-22 residue between
two distinct chemical environments with unequal populations, the
minor component (<10%) having an ∿5 ppm shift downfield of the
major resonance at ∿106 ppm. Because of the small population of
the former, its ^{13}C resonance will be undetectable. The temperature dependence of the line shapes of the 106 ppm resonance are
consistent with this interpretation, the lifetime of the major
species being ∿0.03 s at 15°C.

The ^{13}C spectra of DHFR as it gradually uncoils when
titrated with urea (Figure 7) provide some insight into the
nature of the chemical exchange process involving Trp-22. As the
urea concentration increases, the spectra are consistent with a
gradual depopulation of the site giving rise to the resonance
at ∿106 ppm. This is accompanied by a broadening of the ∿106 ppm
resonance and the apparent growth of a broader one in the region

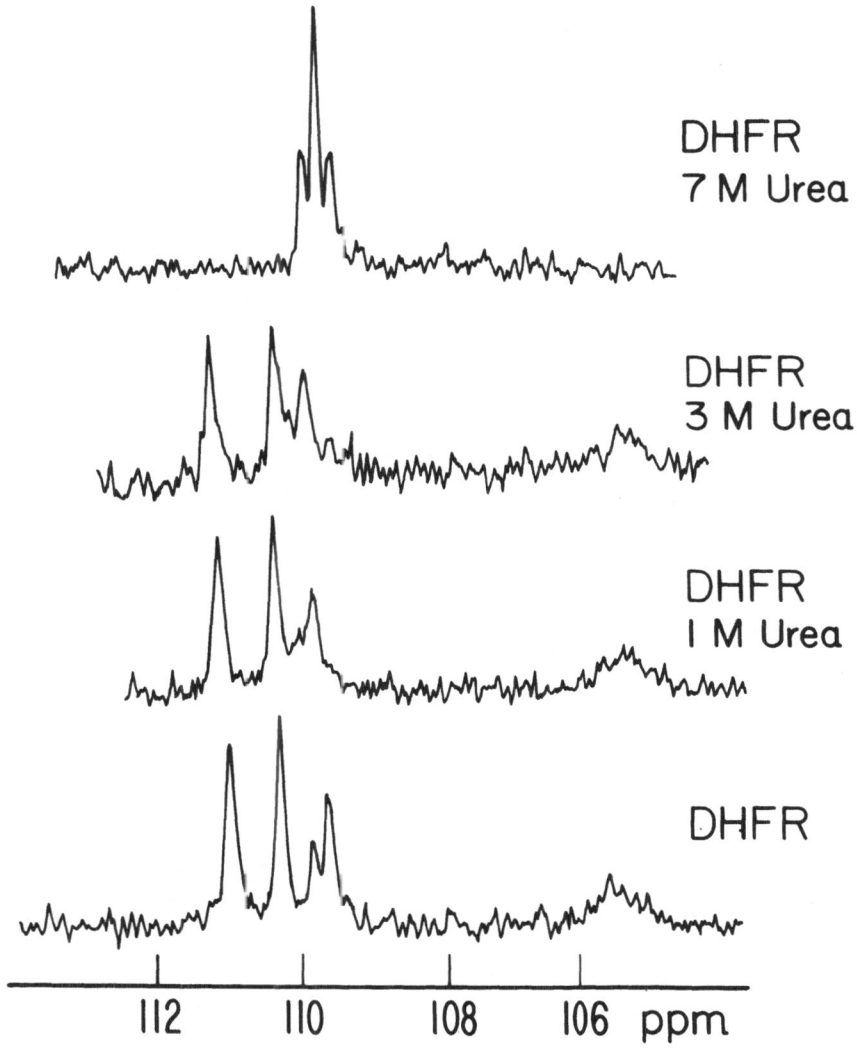

Figure 7. ^{13}C NMR spectra of [γ-^{13}C]tryptophan labeled
            dihydrofolate reductase (15°C) in various
            stages of denaturation with urea. Urea concentrations
            are as indicated on each spectrum.

of 110 ppm (note particularly the 3 M urea spectrum, Figure 7). It is generally accepted that urea denaturation of a protein results in a random coil configuration, exposing hydrophobically "buried" residues to the solvent. Although the structures of intermediate states on the route to denaturation are uncertain, it seems reasonable that these would correspond to progressively open structures associated with the gradual disruption of β-pleated sheets and α-helices. This suggests to us that: the Trp-22 site characterized by the ∿106 ppm resonance is a predominately hydrophobic one in which the unusual chemical shift is induced by electric field effects of nearby immobilized polar or charged groups; the site corresponding to the ∿110 ppm resonance is a solvent exposed one; and that Trp-22 samples these two environments by breathing motions of DHFR which may be localized to the region of the inner cavity surface where Trp-22 resides or may involve the conversion of the whole structure from one well-defined conformation to another.

Continuing in this vein, it is interesting to speculate on the structural basis of this chemical exchange process in terms of the discussion of the structure of DHFR complexes by D. A. Matthews (42). Trp-22 (Trp-21 in the Lactobacillus casei enzyme) resides on the inner surface of the large cavity in which MTX and the nicotinamide (NA) ring of NADPH are bound. In the DHFR-MTX-NADPH complex, the Trp-22 indole side chain is in Van der Waal's contact with the carboxamide portion of the NA ring and is in a hydrophobic pocket completely inaccessible to solvent. In the MTX or NADPH binary complexes, approximately three-quarters of the indole side chain is buried in the pocket with about one-quarter of the solid angle of the indole ring being exposed to solvent. Matthews (42) suggests that in the uncomplexed enzyme, the vacant, active site would permit solvent access to approximately half of the solid angle around the indole side chain of Trp-22. Thus, in the progression DHFR → Binary Complex → Ternary Complex, the Trp-22 residue becomes progressively more immobilized in a hydrophobic pocket. These structural considerations, broadly interpreted, are entirely consistent with the preceding discussion of the ^{13}C spectra of the native and urea denatured proteins which are at the extreme structural disorganization at the surface of the active site cavity. The other extreme of structural organization is provided by the binary and ternary complexes, DHFR-NADPH, DHFR-MTX, and DHFR-NADPH-MTX, whose ^{13}C spectra are summarized in Figure 8. As evident from the spectra, binding of NADPH in the binary complex narrows the Trp-22 ∿106 ppm resonance somewhat compared to the uncomplexed enzyme whereas in the binary DHFR-MTX complex the corresponding resonance narrows sharply. As expected from Matthews' evaluation of the structures, the resonance in question is also narrow in the ternary DHFR-NADPH-MTX complex. In these systems, as well as in the urea treated enzyme, spectral changes occur also in the resonances assigned

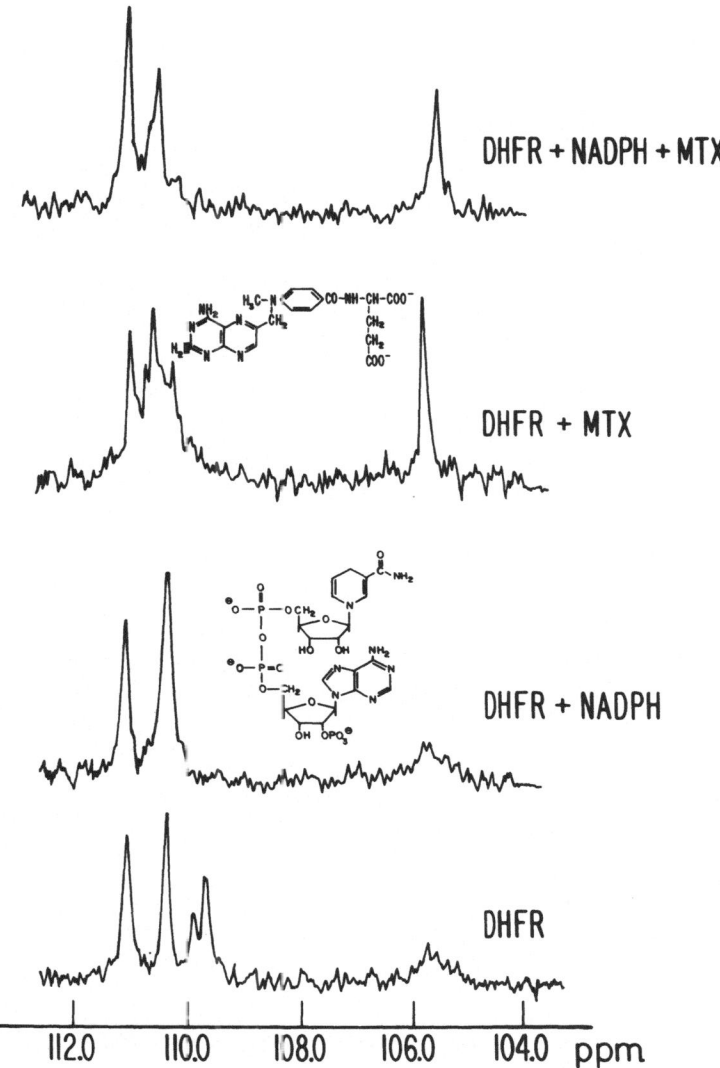

Figure 8. ^{13}C NMR spectra of a series of binary and ternary
            complexes of dihydrofolate reductase (15°C) labeled
            with [γ-^{13}C]tryptophan. From bottom to top the
            complexes are as follows: uncomplexed enzyme,
            enzyme-NADPH binary, enzyme-methotrexate binary,
            enzyme-NADPH-methotrexate ternary.

to Trp-6 but these will not be considered in detail here. We
note only that in the urea treated enzyme not only does the
Trp-6 resonance broaden but also the population of the conformation corresponding to the downfield component of the resonance
increases markedly.

Finally, with respect to the assignments we have made of the
Trp C_γ resonances, the results of some preliminary nitroxide spin
label studies with the analog of NADPH, are of interest (Figure
9). As evident from the spectra of DHFR containing the spin
label NAPH-SL (43), the resonances assigned to Trp-115 and
Trp-160 are unaffected by the paramagnetic center. This is not
unexpected since these residues are more than 15 Å away from the
expected location of the nitroxyl radical (41,42). On the other
hand, both Trp-6 and Trp-22 are near the active site. The indole
ring of the latter is in Van der Waal's contact with the carboxamide ring of NADPH in the crystalline binary DHFR-NADPH and the
ternary DHFR-MTX-NADPH complexes whereas the backbone atoms of
Trp-6 are near the pyrimidine ring of MTX but the side chain
points away from the active site. The relative paramagnetic
broadening evident from the spectra in Figure 9, $\leq$ 12 Hz for
Trp-6 and > 25 Hz for Trp-22, is consistent with this general
structure for solutions of the enzyme as well.

CONCLUDING REMARKS

This brief and highly selective survey presents an optimistic
outlook for the utility of stable isotope labeling in the study
of architecture and dynamics of biological systems by NMR spectroscopy. It is appropriate at this juncture, however, to
emphasize some of the limitations of the method. As Matthews
points out, and our discussion of DHFR illustrates, the most
useful insights at this point in time will be obtained from the
method when it is used to study systems whose crystal structures
are known. In part, this derives from our inadequate understanding of the origin of ^{13}C (and ^{15}N) shifts in macromolecules and
macromolecular assemblages. This point is illustrated well by
our other studies of [^{13}C-methyl]methionine labeled DHFR (36).

In Figure 10 are summarized the ^{13}C NMR spectra in the
S-methyl region for [^{13}C-Me]methionine enriched DHFR in the
presence of a variety of substrates and effector analogs. There
are seven methionine residues in DHFR from Streptococcus faecium
and at least six S-methyl ^{13}C resonances can be resolved in
several of the spectra (36). The urea denatured enzyme exhibits
a single resonance at 15.32 ppm. For the complete set of analogs
studied, the S-methyl resonances of DHFR span a chemical shift
range of 3.4 ppm. It has been suggested (44) that it is protein
folding which immobilizes amino acid residues and that in turn
gives rise to large ^{13}C chemical shift dispersions in proteins.

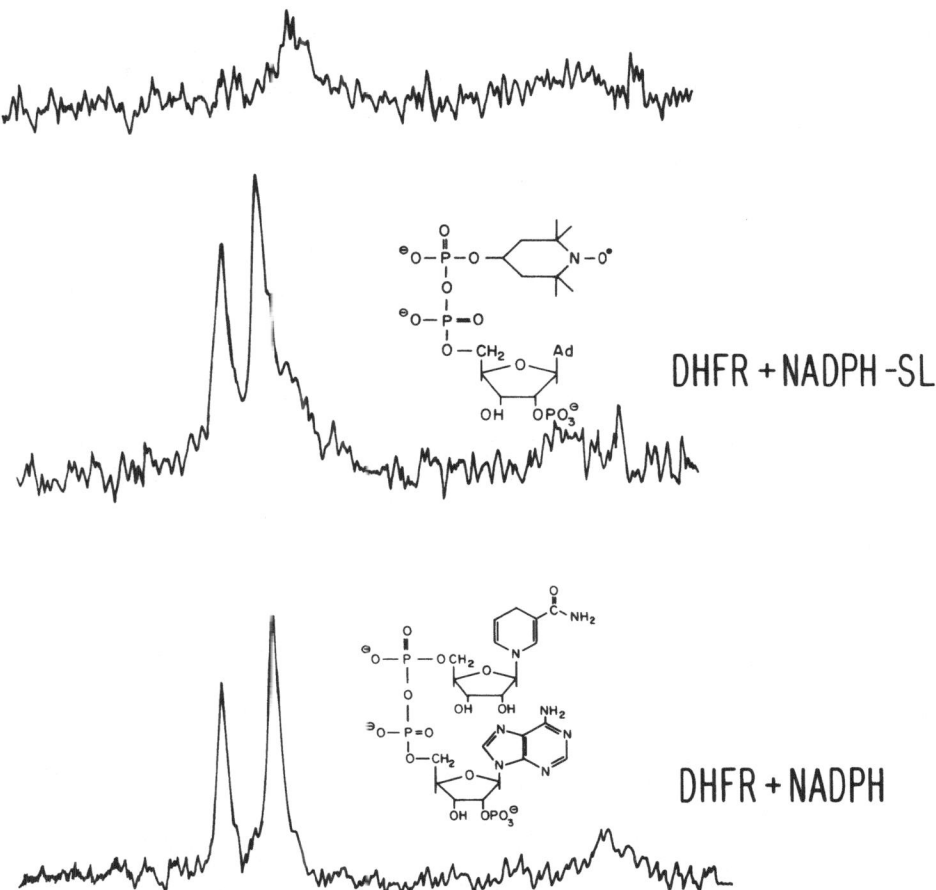

Figure 9. ^{13}C NMR spectra of dihydrofolate reductase (15°C) labeled with [γ-^{13}C]tryptophan. The middle spectrum is a binary complex with the NADPH-spin label compound shown. The top spectrum is a partially relaxed spectrum of the same spin labeled binary complex. In this spectrum only those rapidly relaxing resonances which are affected by the spin label are observed; all other peaks are nulled. The bottom spectrum is a binary complex with NADPH used as a control.

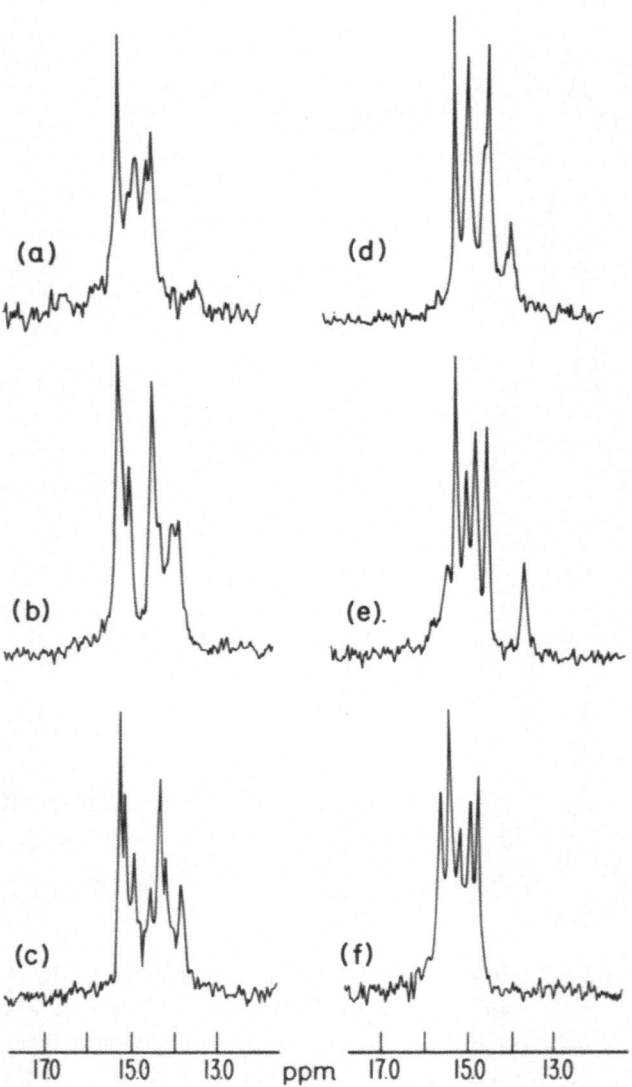

Figure 10. ^{13}C NMR spectra of ternary complexes of dihydrofolate reductase
labeled with [methyl-^{13}C]methionine. (a) Enzyme-dihydrofolate-NADP$^+$, 15°C;
(b) Enzyme-methotrexate-NADP$^+$, 25°C; (c) Enzyme-methotrexate-NADPH, 25°C;
(d) Enzyme-methotrexate-PADPR, 25°C; (e) Enzyme-diaminopyrimidine-NADPH, 20°C;
(f) Enzyme-dichloromethotrexate-NADPH, 25°C.

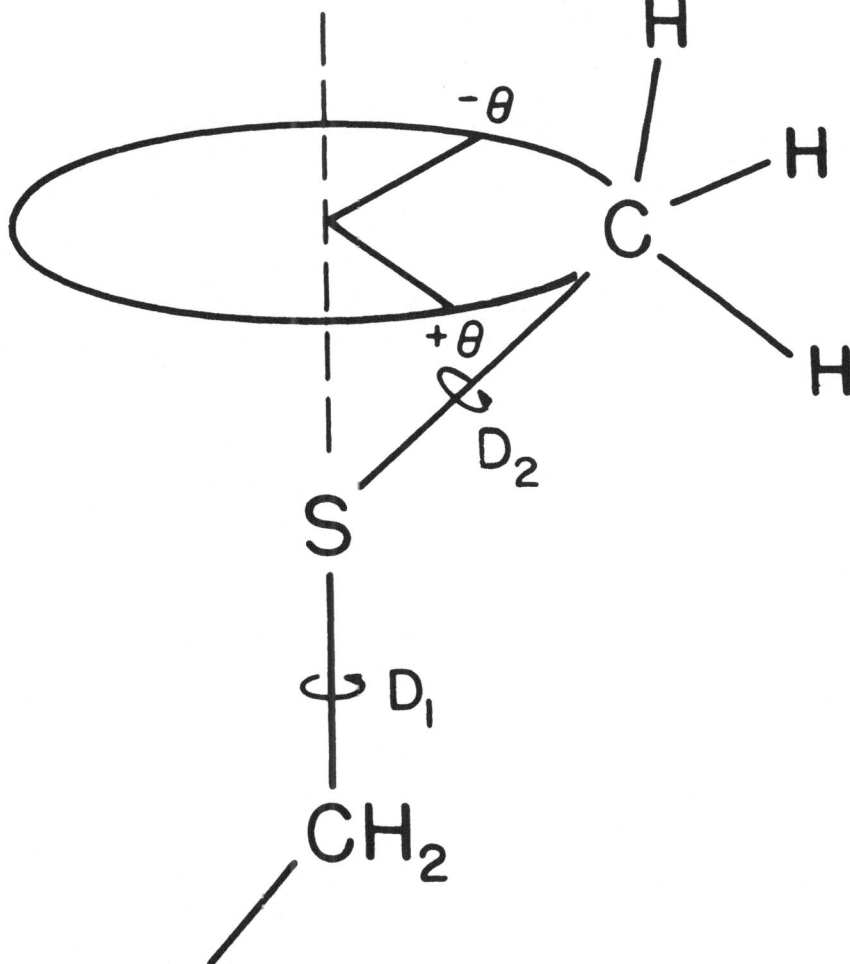

Figure 11. Illustration of the possible motions around the sulfur atom in methionine. The combination of motions about D_1 (restricted) and D_2 (unrestricted) leads to relatively narrow lines for the methyl ^{13}C resonance in protein methionyl side chains of dihydrofolate reductase.

Because a proton bearing ^{13}C in an immobilized residue should exhibit a resonance which is extensively broadened by dipolar interactions, it has also been suggested that structural pertur- bations in proteins are best monitored in ^{13}C NMR through a study of "non-protonated" quaternary carbons which should exhibit narrower lines and a higher degree of resolution. Although this is certainly true of the $^{13}C_\gamma$ in the tryptophan residues of DHFR just described, the [^{13}C-Me]methionine residues of the same enzyme provide a remarkable exception to the generalization: not only are the $^{13}CH_3$-S shifts large but the lines are narrow.

London (45) has provided a theoretical explanation for this "exception" which may be encountered more frequently in macromolecules than previous theoretical treatments would suggest. The model he has developed is illustrated in Figure 11. It allows unrestricted rapid diffusion (D_2) about the S-CH_3 bond coupled to restricted amplitude diffusion ($\pm \theta = 90 - 180°$) about the CH_2-S bond. The model is physically reasonable since the rotation of the methyl group does not require the creation of free volume and therefore should not be restricted by inter- residue interactions. Rotation about the CH_2-S bond, however, does require sweeping additional free volume and, therefore, should be restricted by inter-residue contacts. The combination of the two motions coupled to rotational diffusion of the protein as a whole imparts rapid pseudo-isotropic motion to the ^{13}C methyl group with attendant narrow lines. Yet this model is not incon- sistent with a partial immobilization of the methionyl side chains by the inter-residue contacts which cause the chemical shift dispersion. In this context, two things should be noted:
(1) The $^{13}CH_3$-S linewidths, Nuclear Overhauser Enhancements, and T_1 values could not be fit using a free internal dif- fusion model (45), that is to say the treatment is not a "grasping" force fit of the data; and
(2) There is a semi-quantitative correlation (Table 1) between the restriction of the amplitude of diffusion about the CH_2-S bond (short T_1 values) and the inter-residue induced chemical shifts from the (urea induced) 15.32 ppm "random coil" configuration of the protein.

As gratifying as this exercise is in resolving an apparent conflict between our intuitions about what induces chemical shifts and restricted motions in macromolecules, we are still left in a quandary about the origin of the ^{13}C chemical shifts not only of the methionyl residues but also of the Trp residues of DHFR. There is a missing data base that may not be provided even in the study of environmental and structural perturbations of amino acids and peptides -- it simply may not be possible to artificially construct those shift perturbations that enzymes and proteins supply so naturally. This underlines the need at this point in time to fold the results of x-ray crystallography

TABLE 1.   CHEMICAL SHIFTS AND RELAXATION TIMES OF THE $^{13}CH_3$
RESONANCES IN $[^{13}C-Me]$METHIONINE ENRICHED DHFR
COMPLEXES.

Complex	Resonance Shift (ppm)	NT_1 Value (msec)
Enzyme Alone	15.35	590
	14.99	470
	14.66	500
Enzyme - MTX	15.37	720
	15.24	400
	14.95	410
	14.74	330
	14.57	460
	14.18	370
Enzyme - MTX - NADP	15.36	710
	15.24	560
	15.03	470
	14.68	350
	14.44	480
	13.96	420

into the design of NMR experiments on isotopically labeled macromolecules and macromolecular assemblages. Consistent with the
main theme of this conference, I would also point out that isotope
labeling in conjunction with the use of properly designed shift
and relaxation reagents can carry us far in mapping enzyme-ligand
interactions. We have briefly discussed this latter point in the
outline of the $[^{13}C_\gamma]$Trp labeled DHFR experiments and have conducted much more extensive studies of the interaction of labeled
analogs with $[^{13}C-Me]$methionine enriched DHFR but time does not
permit a discussion of the results.

*   This work performed under the auspices of the U. S. Depart-
        ment of Energy.
(1)   Dickerson, R. E.: 1972, Ann. Rev. Biochem. 41, pp. 815-842.
(2)   Perutz, M.: 1978, Scientific American 239, pp. 92-125.
(3)   Blow, D. M.: 1976, Accounts Chemical Research 9, pp. 145-
        152.
(4)   Kraut, J.: 1977, Ann. Rev. Biochem. 46, pp. 331-358.
(5)   Bachovchin, W. W. and Roberts, J. D.: 1978, J. Am. Chem.
        Soc. 100, pp. 8041-8047.
(6)   However, it appears that a determination of the root mean
        square displacements of the atoms may provide a method for
        mapping the dynamic features of a macromolecule (7,8).
(7)   Phillips, D. C. and Sternberg, M.: 1979, J. Molec. Biol.
        in press.
(8)   Frauenfelder, H., Petsko, G. A., and Tsernoglau, D.: 1979,
        Nature, submitted.
(9)   Matwiyoff, N. A. and Needham, T. E.: 1972, Biochem. Biophys.
        Res. Commun. 49, pp. 1158-1163.
(10)   Rossi-Bernardi, L. and Roughton, F. J.: 1967, J. Physiol.
        189, p. 1.
(11)   Kilmartin, J. V. and Rossi-Bernardi, L.: "CO_2: Chemical,
        Biological, and Physiological Aspects",Foster, R. E., Edsall,
        J. T., Otis, H. B., and Roughton, F. J. W., eds., Symposium
        at Haverford College, Haverford, PA, NASA SP-188, Library
        of Congress Catalog No. 77600831.
(12)   Roughton, F. J. W.: 1970, Biochem. J., 117, p. 801.
(13)   Kilmartin, J. V. and Rossi-Bernardi, L.: 1969, Nature, 222,
        p. 1243.
(14)   Kilmartin, J. V. and Rossi-Bernardi, L.:1971 Biochem. J.
        124, p. 31.
(15)   Kilmartin, J. V., Fogg, J., Luzzana, M., and Rossi-Bernardi,
        L.: 1973, J. Biol. Chem. 248, p. 7039.
(16)   Arnone, A.: 1974, Nature, 247, pp. 143-145.
(17)   Matthew, J. B., Morrow, J. S., Wittebort, R. J., and Gurd,
        F. R. N.: 1977, J. Biol. Chem. 252, pp. 2234-2244, and
        references therein.
(18)   Wittebort, R. J., Hayes, D. F., Rothgeb, T. M., and Gurd,
        R. S.: 1978, Biophys. J. 24, pp. 765-778.
(19)   Silverman and co-workers (20), on the basis of the measure-
        ment of CO_2 kinetics in red cell suspensions by ^{18}O exchange,
        suggest that the intrinsic activity of carbonic anhydrase is
        the same in the cell and in homogeneous aqueous solution;
        but that the rate of access of CO_2 to carbonic anhydrase in
        intact red cells is slower than the catalytic hydration
        rate.
(20)   Tu, C., Wynns, G. C., McMurray, R. G., and Silverman, D. N.:
        1978, J. Biol. Chem., 253, p. 8178.
(21)   London, R. E., Gregg, C. T., and Matwiyoff, N. A.: 1975,
        Science 188, pp. 266-268.
(22)   London, R. E., Hildebrand, C. E., Olson, E. S., and

Matwiyoff, N. A.: 1976, Biochem. 15, pp. 5480-5486.
(23) See, for example: "The Molecular Basis of Membrane Function", Tosteson, D. C.: Editor, Prentice-Hall, Inc.,
New Jersey, 1969; Chapman, D.: 1975, Biomembranes, 7, p.1,
and references therein.
(24) Bocian, D. F. and Chan, S. I.: 1978, Ann. Rev. Phys. Chem.,
29, p. 307.
(25) London, R. E., Walker, T. E., Matwiyoff, N. A., and Wilson,
D. M.: 1979, Chem. Phys. Lipids, in press.
(26) Govil, G. and Smith, I. C. P.: 1973, Biopolymers 12, p.
2589.
(27) O'Connor, J., Nunez, H., and Barker, R.: 1979, Biochem.
in press.
(28) Browne, D. T., Kenyon, G. L., Packer, E. L., Sternlicht,
H., and Wilson, D. M.: 1973, J. Am. Chem. Soc. 95, p. 1316.
(29) Browne, D. T., Kenyon, G. L., Packer, E. L., and Wilson,
D. M.: 1973, Biochem. Biophys. Res. Commun. 50, p. 42.
(30) Hunkapiller, M. W., Smallcombe, S. H., Whitaker, D. R.,
and Richards, J. H.: 1973, J. Biol. Chem. 248, p. 8306;
ibid: 1973, Biochem. 12, p. 4732.
(31) Hunkapiller, M. W., Smallcombe, S. H., and Richards, J. H.:
1975, Org. Mag. Res. 7, p. 262.
(32) Browne, D. T., Earl, E. M., and Otvos, J. D.: 1976, Biochem.
Biophys. Res. Commun. 72, p. 398.
(33) Chaiken, I. M., Cohen, J. S., and Sokolski, E. A.: 1974,
J. Am. Chem. Soc. 96, p. 4703.
(34) Eakin, R. T., Morgan, L. O., and Matwiyoff, N. A.: 1975,
Biochem. J. 152, p. 529.
(35) Cocco, L., Blakley, R. L., Walker, T. E., London, R. E.,
and Matwiyoff, N. A.: 1977, Biochem. Biophys. Res. Commun.
76, p. 183.
(36) Blakley, R. L., Cocco, L., London, R. E., Walker, T. E.,
and Matwiyoff, N. A.: 1978, Biochem. 17, p. 2284.
(37) London, R. E., Groff, J. P., and Blakley, R. L.: 1979,
Biochem. Biophys. Res. Commun. 86, p. 779.
(38) London, R. E., Groff, J. P., and Blakley, R. L.: work in
progress.
(39) Gleisner, J. M., Peterson, D. L., and Blakley, R. L.: 1974,
Proc. Nat. Acad. Sci. 71, p. 3001.
(40) Matthews, D. A., Alden, R. A., Freer, S. T., Xuong, N.,
and Kraut, J.: 1979, J. Biol. Chem., in press.
(41) Matthews, D. A., Alden, R. A., Bolin, J. T., Freer, S. T.,
Hamlin, R., Xuong, N., Kraut, J., Poe, M., Williams, M.,
and Hoogsteen, K.: 1977, Science 197, p. 452.
(42) Matthews, D. A.: 1979, Biochem. 18, p. 1602.
(43) Cocco, L., Blakley, R. L., and London, R. E.: work in
progress.
(44) Oldfield, E., Norton, R. S., and Allerhand, A.: 1975,
J. Biol. Chem. 250, p. 6368.
(45) London, R. E. and Avitable, J.: 1978, J. Am. Chem. Soc.
100, p. 7159.

CONTACT AND DIPOLAR NMR SHIFTS: THEORY AND APPLICATIONS

William DeW. Horrocks, Jr.

Department of Chemistry, 152 Davey Laboratory
The Pennsylvania State University, University Park, PA
16802.

Abstract: The fundamental equations and basic qualitative
ideas regarding paramagnetic nmr shifts are reviewed. Particular
attention is paid to lanthanide systems. Including the problem
of nonaxiality. Modes of metal-ligand spin transferral are
discussed as are theoretical models of spin-delocalization in
ligands. The evaluation of dipolar nmr shifts from magnetic
susceptibility anisotropy data is treated along with other methods
for the separation of contact and dipolar contributions. The
theoretical evaluation of magnetic anisotropy from ligand field
models is considered with particular emphasis on low-spin d^5
iron(III) hemin systems and on lanthanide ion complexes.
Particular attention is paid to the theoretical and experimental
aspects of the temperature dependencies. Mention is made of the
use of lanthanide ions and cobalt(II) and substitutional or
extrinsic shift probes for the study of biological macromolecules.

Introduction

The purpose of this contribution is to review the basic
qualitative ideas and theoretical concepts regarding paramagnetic
shifts and then to discuss certain aspects and applications of
this phenomenon with emphasis being placed on biologically
pertinent areas. For an extensive account of the development of
this subject up to 1973 the reader is referred to a monograph
coedited by the present author (1). By definition paramagnetic
substances contain unpaired electrons. Since electronic magnetic
moments are much larger than nuclear magnetic moments, the
presence of unpaired electrons in a molecule produces large
internal magnetic fields which modify (augment or decrease) the
externally applied magnetic field of the nmr experiment. This

55

*I. Bertini and R.S. Drago (eds.), ESR and NMR of Paramagnetic Species in Biological and
Related Systems, 55–87.*
Copyright © 1979 by D. Reidel Publishing Company

has the effect of producing in such molecules what are known as
isotropic shifts $(\Delta H/H)^{iso}$ of tens or hundreds of ppm given by
eq. 1.

$$\left(\frac{\Delta H}{H}\right)^{iso} = \delta^{para} - \delta^{dia} \tag{1}$$

where δ^{para} and δ^{dia} are the chemical shifts (ppm) in the paramagnetic and analogous diamagnetic environments. Isotropic
shifts are produced either by a contact or dipolar interaction
or a combination of both such that eq. 2 holds

$$\left(\frac{\Delta H}{H}\right)^{iso} = \left(\frac{\Delta H}{H}\right)^{con} + \left(\frac{\Delta H}{H}\right)^{dip} \tag{2}$$

where the superscripts con and dip stand for contact and dipolar
(pseudocontact) respectively. These shift mechanisms will be
disucssed in turn, but with special emphasis on dipolar shifts.

Contact Shifts

For the paramagnetic systems of principal interest here,
namely transition metal and lanthanide ion complexes, the unpaired electrons are almost exclusively contained in metal d- or
f-orbitals. Contact shifts result from a direct electron-
nuclear hyperfine interaction and will vanish unless there is a
finite amount of unpaired electron spin-density at the nucleus
in question. Thus contact shifts result from a through-bond
transmittal of spin-density from the metal orbitals. Even a
minisucle amount of spin-density at a particular nucleus can
give rise to a large contact shift owing to the great magnitude
of the electronic magnetic moment. The contact shift will occur
only when there is a net internal magnetic field at the resonating
nucleus owing to polarization of the total spin angular momentum,
S, by the applied magnetic field of the nmr experiment. Contact
shifts, $(\Delta H/H)^{con}$, will thus be proportional to the expectation
value of the total spin: $\langle S_z \rangle$, in the direction (z) of the
applied magnetic field. Since $\langle S_z \rangle$ is itself directly proportional to the applied magnetic field, contact shifts will be the
same in ppm when measured on spectrometers operating at different
frequencies (magnetic field strengths). The electron-nuclear
hyperfine interaction constant A_i is proportional to the unpaired electron spin-density on the $i\underline{th}$ nucleus as given by
eq. 3.

$$A_i = \left(\frac{8\pi}{3}\right) \gamma_N hg\beta |\Psi(0)|_i^2, \tag{3}$$

where $|\Psi(0)|^2$ is the probability of finding the unpaired
electron at the ith nucleus, γ_N is the nuclear magnetogyric

ratio and β is the Bohr magneton. The most general form of the contact shift equation is given by eq. 4.

$$\left(\frac{\Delta H}{H}\right)^{con} = \frac{A_i}{\hbar\gamma_N H}\ \langle S_z\rangle \tag{4}$$

where A_i is given in ergs. A-values are often stated in other units which are related as follows: A(ergs); A/h(hertz); A/ℏ (radians s^{-1}) A/gβ (gauss), where g is the spectroscopic splitting factor. In the last case the appropriate g-value is not necessarily that of the free electron (2.0023) so the use of gauss as a unit for A is to be discouraged. Expressing (A/h) values in hertz is probably the most generally satisfactory approach.

A more useful form of eq. 4 results when $\langle S_z\rangle$ is evaluated. The results depend, of course, on the details of the electronic structure of the particular metal ion involved. For a spin-only metal ion, evaluation of $\langle S_z\rangle$ results in eq. 5.

$$\left(\frac{\Delta H}{H}\right)^{con}_i = -\left(\frac{A}{h}\right)_i\ \frac{2\pi g\beta S(S+1)}{\gamma_N\ 3kT} \tag{5}$$

It can be seen from eq. 5 that the temperature dependence of $(\Delta H/H)^{con}_i$ arises from the temperature dependence of $\langle S_z\rangle$. Simple spin-only cases have magnetic susceptibilities which follow the Curie law and their contact shifts will also. In the more general case where the total orbital angular momentum, L, does not vanish, and for systems with S>1/2 where zero-field splittings occur, the evaluation of S_z leads to more complex expressions. It is important to realize that while contact shifts will follow the behavior of $\langle S_z\rangle$, the magnetic suscepti-bility will behave as the appropriate average of $\langle L_\alpha + 2 S_\alpha\rangle$, $\alpha = x, y, z$. Thus contact shifts and susceptibilities need not have identical temperature dependencies. In many cases eq. 5 is obeyed fairly well, however since solution nmr data are generally available over only a limited temperature range, deviations from linear T^{-1} behavior are difficult to detect experimentally.

Analogy with Heteronuclear Spin-coupling. - Before proceeding with further discussions of $\langle S_z\rangle$ and spin delocaliza-tion it is instructive, perhaps, to compare the contact shift phenomenon with another type of nmr experiment which may be more familiar. The electron-nuclear hyperfine interaction can be considered as the coupling of a set of equivalent nuclei to an unpaired electron where the electron is a nonequivalent "nucleus" which posesses a huge magnetic moment. As an example consider the resonance of the set of six equivalent protons of the benzene radical anion, $C_6H_6^{-}$, being split into a doublet by coupling to the single unpaired electron. Since A/h is -10.5 MHz, one would observe two signals separated by 10.5 MHz centered at the corresponding diamagnetic resonance position. For a 100 MHz spectrometer this corresponds to shifts (upfield and downfield) of 52,500 ppm. If one were, again in a hypothetical experiment to observe the split proton nmr signals of a methyl radical the peaks would be separated by 64.6 MHz, while for the hydrogen atom itself the corresponding splitting would be 1420 MHz. These experiments are hypothetical, of course, since the relatively short electronic spin-lattice relaxation times cause the two hypothetical peaks to collapse to a single resonance in the same way that a heteronuclear spin-decoupling experiment causes a spin-spin multiplet to collapse. The two electronic spin-states (parallel ↑ and antiparallel ↓ to the magnetic field) have slightly different populations owing to the electronic Zeeman effect and the coalesced signal will be shifted slightly (from dead center) toward the signal corresponding to the more populous electronic spin state. The separation between the exact center and the actual position is, of course, the contact shift. It can still be quite significant, for the $C_6H_6^{\cdot-}$ example it amounts to a 277 ppm shift at 300°K. The great sensitivity of nmr to small quantities of spin-density can be appreciated when it is realized that in the $C_6H_6^{\cdot-}$ example only 1/6th of an unpaired electron occupies a p_π orbital on the carbon atom adjacent to the proton and that this orbital is orthogonal to a C-H σ-bond. Spin density at the hydrogen nucleus arises by an indirect spin-correlation mechanism.

Contact Shifts in Lanthanide Complexes. – Since trivalent lanthanide ions, Ln(III), are particularly important paramagnetic shift probes we will consider the form of the contact shift equation for these ions. Golding and Halton (2) have presented a general treatment. Owing to the large value of the spin-orbit coupling constant for Ln(III) ions, their electronic states are characterized by the values of the total angular momentum quantum number, J. The expectation value, $\langle S_z \rangle_J$ is given by eq. 6.

$$\langle S_z \rangle_J = \left(\frac{-\beta H}{3kT}\right)\left[\frac{(g-\gamma)gK(J+1)}{(2-\gamma)} + \frac{2kT(g-\gamma)(g-2)}{\lambda(2-\gamma)}\right] \quad (6)$$

where g is given by eq. 7.

$$g = 1 + \frac{J(J+1)-L(L+1)+S(S+1)}{2\,J(J+1)} \quad (7)$$

γ is an orbital reduction factor, $\lambda = \pm \zeta/2S$, ζ is the one-electron spin-orbit coupling constant and the remaining symbols have their usual significances. Since in many cases there are a number of thermally populated J-multiplet levels the appropriate sum over these must be taken to evaluate S_z properly. This sum is given by eq. 8.

$$\left\langle S_z \right\rangle = \frac{\sum_J \left\langle S_z \right\rangle_J (2J + 1) \exp (-E/kT)}{\sum_J (2J + 1) \exp (-E/kT)} \qquad (8)$$

where $E = \lambda J(J + 1)/2$. Golding and Halton (2) have evaluated $\left\langle S_z \right\rangle$ values for the series of Ln(III) ions at 300°K with the results shown in Table I. For comparison the ^{17}O nmr shift results for the Ln(III) aqua ions reported by Lewis et al (3)

Table I. Evaluation of $\left\langle S_z \right\rangle$ for Lanthanide Ions; $\gamma = 1.0$

Configuration	ion	ground term	(cm^{-1})	$\left\langle S_z \right\rangle$
f^1	Ce(III)	^2F	737	0.98
f^2	Pr(III)	^3H	879	2.97
f^3	Nd(III)	^4I	1030	4.49
f^4	Pm(III)	^5I	1190	4.01
f^5	Sm(III)	^6H	1361	-0.06
f^6	Eu(III)	^7F	1436	-10.68
f^7	Gd(III)	^8S	1755	-31.50
f^8	Tb(III)	^7F	1965	-31.82
f^9	Dy(III)	^6H	2189	-28.55
f^{10}	Ho(III)	^5I	2436	-22.63
f^{11}	Er(III)	^4I	2700	-15.37
f^{12}	Tm(III)	^3H	2817	-8.21
f^{13}	Yb(III)	^2F	3288	-2.59

are plotted vs $\left\langle S_z \right\rangle$ in Fig. 1.

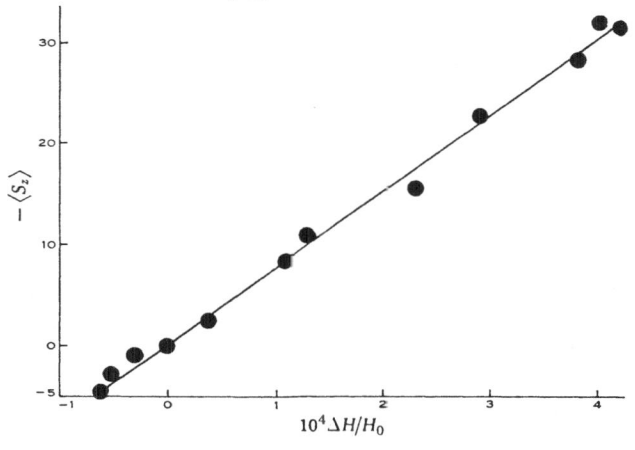

Fig. 1.— Calculated $\langle S_z \rangle$ values against experimental ^{17}O n.m.r. shifts from ref. 3.

The correlation is excellent, and suggests that for ligand atoms
directly coordinated to a paramagnetic Ln(III) ion that the
contact interaction is the dominant shift mechanism. As will be
shown later, dipolar shifts generally dominate for ligand nuclei
further removed from Ln(III) ions. The preceding treatment is
presented to emphasize the fact that the evaluation of $\langle S_z \rangle$ via
a simple analytical expression such as eq. 5 is the exception and
for rigorous work each system should be treated separately.

Mechanism of Spin Transferral to the Ligand in Transition
Metal Complexes. - The observation of nmr contact shifts for
ligand nuclei constitutes evidence for metal-ligand covalency.
The quantity and sign of transferred spin-density is not, however, always simply related to metal-ligand covalency. An in-
depth treatment of this subject is beyond the scope of this
article and the reader is referred to a chapter by La Mar (4) for
a more detailed discussion. Several qualitative features helpful
in gaining insight into the phenomenon of spin delocalization will
be mentioned, however. A subtle quantum mechanical phenomenon
known as spin-correlation is particularly important. In essence
this phenomenon results in electrons with parallel spins ($\uparrow \uparrow$)
being slightly less repulsive toward one another than electrons
with anti-parallel spins ($\uparrow \downarrow$). Thus, other things being equal,
parallel spins will, on average, be closer to one another
spacially than opposed spins. This simple qualitative picture is
quite useful in predicting the signs of coupling constants and
the sense of unpaired electron spin delocalization. Let us
consider the several situations which arise in octahedral
transition metal complexes where unpaired electron spin-density
resides initially in metal d-orbitals.

(a) A filled ligand (L) orbital interacts with a singly
occupied metal (M) d-orbital. In this case L→M charge transfer
takes place (donation of a share of paired (σ or π) electrons).
The metal orbital is considered to contain $\uparrow$ spin therefore only
$\downarrow$ spin can be donated because of the Pauli principle. This leaves
net $\uparrow$ spin behind on the ligand to be delocalized. Spin-transferral of this type is directly related to metal-ligand covalency
(4).

(b) A filled L orbital interacts with a vacant d-orbital,
but other metal orbitals contain $\uparrow$ spin. Here again L→M charge
transfer results, but transfer of $\uparrow$ spin is favored by spin-
correlation. This has the effect of leaving net $\downarrow$ spin behind
on the ligand. Since spin-correlation effects are dominant here,
the quantity of spin transferred is only indirectly related to
M-L covalency.

(c) <u>M→L π-type charge transfer</u>. This will result in either direct delocalization of ↑ spin to a vacant low-lying ligand π-MO or, if the metal dπ orbitals are filled, delocalization of ↓ spin owing to electron correlation effects on the metal with ↑ spin in metal d_σ antibonding orbitals.

The sign of the spin-densities delocalized onto the ligand for octahedral transition metal complexes with various d^n, n = 1-9 electronic configurations are summarized in Table II (4). It should be pointed out that in lower-symmetry situations, where rigorous σ-π separation is absent, it may be very difficult to determine the mechanism of spin-transferral from simple arguments.

Table II

ANTICIPATED SIGN OF NET SPIN DENSITY ON LIGAND IN OCTAHEDRAL ML_6 COMPLEXES[a]

Spin transfer mechanism	d^1	d^2	d^3	d^4	d^5	d^6	d^7	d^8	d^9
L → M σ	↓	↓	↓	↓	↑(↓)	↑(0)	↑(↓)	↑	↑
L → M π	↓	↓	↑	↑	↑	↑(0)	↑(b)	b	b
M → L π	↑	↑	↑	↑	↑	↑(0)	↑(↓)	↓	↓

[a] Given for high-spin configuration. If different for low-spin case, spin density sign is given in parentheses, with (0) designating diamagnetic configuration.
[b] L → M π change transfer not possible, since t_{2g} orbitals are filled.

<u>Delocalization of spin-density on ligands</u>. It is not generally feasible to carry out calculations on an entire complex which are of sufficient sophistication to yield meaningful results regarding spin-delocalization. Therefore most attempts to understand spin-delocalization have concentrated on calculations on a molecular fragment. The link between the observed contact shifts and the electronic structure of a complex comes <u>via</u> the electron-nuclear hyperfine interaction constant (eq. 3). It is the relative values of A_i for different ligand nuclei which are most useful in deducing the mechanism of M-L spin-transfer and the distribution of spin-density throughout the ligand molecule. Two examples of calculations of the sort will be discussed briefly.

Among the earliest and most extensively studied paramagnetic molecules were bis chelates of nickel(II) of the aminotroponemineate ligands (5) (see Table III). These unsaturated ligands exhibit alternating up- and downfield isotropic shifts characteristic of spin-delocalization in a π-MO, with spin-density getting to the protons by an indirect spin-polarization mechanism, quantitated by the famous relationship

$$A_i = Q_{CH}\rho_c, \quad Q_{CH} = -6.3 \times 10^7 \, Hz \tag{9}$$

where ρ_c is the spin-density in the p_π carbon orbital adjacent to the hydrogen in question. Table III shows the observed and calculated spin densities for the aminotroponeimineate neutral radical which corresponds to spin delocalization in the highest filled π-MO. The valence bond (VB) treatment is the more

Table III

Observed and Calculated Spin Density Distribution in Aminotroponeimine[a]

Position	ρ_{obs}	ρ_{calc}[b]	
		Valence bond	Hückel MO
α	+0.041	+0.38	+0.19
β	−0.021	−0.23	+0.002
γ	+0.057	+0.57	+0.20

[a] In the bis chelate of Ni(II).
[b] Calculated for a whole spin in the HFO.

sophisticated and correctly predicts negative spin-densities at the β-position. The Hückel method merely predicts very little spin-density at this position, but is incapable of incorporating spin-polarization effects. A modified Hückel procedure due to McLachlan (6), however is capable of including spin-polarization effects. The results of VB calculation on a γ-substitutional aminotroponeimineate ligand is shown in Table IV. These results

Table IV

Observed[a] and Calculated[b] Spin Densities for γ-Azophenyl-aminotroponeimineate Radical

Position	ρ_{calc} (VB)	ρ_{obs}	ρ_{calc}/ρ_{obs}
α	0.3677	0.0383	9.6
β	−0.2253	−0.0210	10.7
o	0.0752	0.0076	9.9
m	−0.0378	−0.0026	14.4
p	0.0627	0.0087	7.2

suggest that approximately 1/10th of an unpaired spin is delocalized into each ligand.

Another calculational method which has had some success,
particularly in the interpretation of results for delocalization
of spin in σ-systems is the INDO method of Pople (7). For
example, using the phenyl and substituted phenyl radicals as
models for σ-spin delocalization in pyridine and various picolines
coordinated to nickel (II), the ^{1}H and ^{13}C contact shifts were
accounted for (9) as shown in Table V. A large number of data
are interpreted with essentially no user-adjustable parameters.
In these radical models the unpaired electron resides in an MO
with no p_π components, nevertheless the often-noted reversal of
sign of proton isotropic shift for a methyl group replacing a
hydrogen is reproduced by this theory. It is also able to
account reasonably well for the alternating signs of the ^{13}C
resonance shifts which are brought about by spin-polarization
effects.

Dipolar Shifts

As a paramagnet tumbles in fluid solution a magnetic moment
is induced parallel to an applied magnetic field. If the
paramagnet is anisotropic the magnitude of the induced magnetic
moment will fluctuate as the complex tumbles. Individual ligand
nuclei will experience internal magnetic fields described by the
lines of force surrounding the induced moment. If the magnitude
of the induced moment does not vary as the complex tumbles
(magnetic isotropy) the internal fields experienced by the
surrounding nuclei will average to zero. If, however, as happens
in a magnetically anisotropic molecule, the magnitude of the
induced moment depends on the orientation of the molecule in the
magnetic field, then the internal fields will not, in general,
average to zero and there will be a net nmr shift known as a
dipolar or pseudocontact shift. The magnitude of the shift
depends on the anisotropy in the molecular susceptibility tensor
and upon the geometrical positioning of the resonating nucleus
with respect to the metal ion and the principal magnetic axes
of the complex. The familiar "ring current" effect which accounts,
in large measure, for the low-field resonance position of aromatic
protons arises from an analogous effect of diamagnetic anisotropy.
Since paramagnetic anisotropies are often several orders of
magnitude larger than their diamagnetic counterparts, dipolar
shifts, $(\Delta H/H)^{dip}$ can be quite large. They are given in general
by eq. 10,

Table V

NORMALIZED ^{1}H AND ^{13}C ISOTROPIC SHIFTS, $\Delta\nu_i$,[a] FOR PYRIDINE-TYPE BASES COORDINATED TO Ni(acac)$_2$, INDO/2 HYPERFINE CONSTANTS, A_i,[b] CALCULATED FOR THE PHENYL RADICAL ANALOGS,[c] AND RATIOS OF THESE QUANTITIES

Nuclei[d]	Phenyl			p-Tolyl			m-Tolyl			o-Tolyl		
	$\Delta\nu_i$	A_i	$\Delta\nu_i/A_i$	$\Delta\nu_i$	A_i	$\Delta\nu_i/A_i$	$\Delta\nu_i$	A_i	$\Delta\nu_i/A_i$	$\Delta\nu_i$	A_i	$\Delta\nu_i/A_i$
α-H	−943	+18.7	−48.4	−394	+19.2	−20.5	−930	+17.4	−53.4	−293	+17.4	−16.8
α'-H							−930	+19.3	−48.2			
α'-CH$_3$										+48	−1.9	−25.3
β-H	−286	+6.1	−46.9	−71	+6.5	−10.9	−360	+5.7	−63.1	−146	+6.6	−22.1
β'H										−73	+5.3	−13.8
β'-CH$_3$							−72	+1.6	−45.0			
γ-H	−86	+3.9	−17.9	+36	−1.2	−30.0	−214	+4.3	−49.8	−24	+4.2	−5.7
γ-CH$_3$				+1000	−5.1	−196						
α-C	+1000	−4.7	−213				+1000	−4.9	−204	+1000	−4.0	−250
α'-C							+1000	−5.0	−200	−488	−3.2	+153
α'-CH$_3$										−585	+6.2	−94
β-C	−3689	+10.8	−342	−1930	+10.4	−186	−3214	+10.7	−300	−1659	+10.7	−155
β'-C							−3000	+10.4	−288	−1122	+10.8	−104
β'-CH$_3$							−430	+0.82	−524			
γ-C	+371	−2.6	−143	+180	−2.3	−78.2	+430	−2.8	−154	+98	−2.6	−37.7
γ-CH$_3$				−144	+1.4	−103						

[a] Taken from ref **8** with the α-C shift normalized to +1000. [b] Gauss. [c] Standard geometries were assumed with 120° or tetrahedral bond angles and the following bond distances: C–C(aromatic), 1.40 Å; C–CH$_3$, 1.51 Å; C–H(aromatic), 1.08 Å; C–H(methyl), 1.10 Å. [d] The primes refer to nuclei on the same side of the molecule as the methyl group.

$$\left[\frac{\Delta H}{H}\right]^{dip} = -\frac{1}{3N}\left(\chi_z - \frac{1}{2}(\chi_x + \chi_y)\right)\left\langle\frac{3\cos^2\Theta-1}{r^3}\right\rangle$$

$$-\frac{1}{2N}\left(\chi_x - \chi_y\right)\left\langle\frac{\sin^2\Theta\cos2\phi}{r^3}\right\rangle \qquad (10)$$

where χ_x, χ_y, and χ_z are the principal molecular susceptibilities, N is Avogadro's number and r, Θ, and ϕ are the usual spherical polar coordinates.

<u>Evaluation of dipolar shifts from magnetic susceptibility anisotropy</u>. – Since we proposed in 1966 (10) that dipolar shifts should be evaluated from susceptibility anisotropy rather than g–tensor anisotropy, we have been involved in single crystal susceptibility anisotropy measurements aimed at the interpretation of dipolar nmr shifts. Our approach has been corroborated by independent theoretical work (11). As an example, our study (12) of [Co(acac)$_2$(pyridine)$_2$], acac = 2,4–pentanedianato anion; will be described briefly. The structure (13) and coordinate system for this complex is shown in Fig. 2. The susceptibility aniso-tropies were measured in single crystals by Krishwan's critical

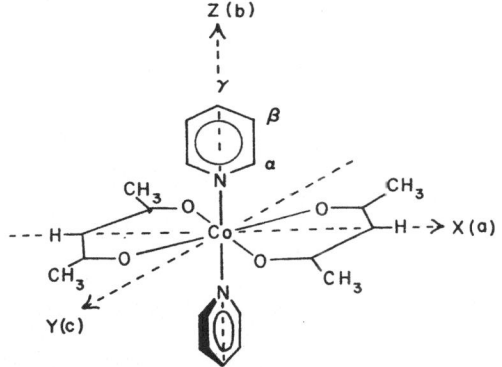

Fig. 2. Coordinate System for [Co(acac)$_2$(pyridine)$_2$]

torque method (14) which along with a $\bar{\chi}$ measurement yielded χ_x = 9836, χ_y = 11346, χ_z = 7526 VVk/mol. The calculated geometric factors, G(Θ,r) = $(3\cos^2\Theta-1)r^{-3}$, the observed ^{1}H and [13]C(8, 15) resonance shifts (ppm) and ratios thereof for the cobalt complex and the analogous nickel(II) complex are given in Table VI. Note that anisotropy in the x,y plane will be effectively averaged to zero by rotation of the pyridine ligand around the Co–N bond axis. It should be noted that the dipolar shifts calculated are non–neglegible when compared at the observed isotropic shifts. When the contact contributions for the cobalt complex are evaluated using eq. 2, and their ratios compared with the corresponding observed isotropic shift ratios for the nickel complex, the agreement is quite good.

Table VI

GEOMETRIC FACTORS AND ISOTROPIC, DIPOLAR, AND CONTACT
SHIFTS FOR $M(acac)_2(py)_2$ SYSTEMS AT 293°K

Atom or ratio	$G(\theta,r)^a$	$\Delta\nu_{Ni}^{iso\,b,c}$	$\Delta\nu_{Co}^{iso\,b,c}$	$\Delta\nu_{Co}^{dip\,b,d}$	$\Delta\nu_{Co}^{con\,b}$
α-H	2.33×10^{-2}	-87.7	-32.9^e	39.5	-72.4
β-H	1.06×10^{-2}	-26.6^e	-5.0	18.1	-23.1
γ-H	9.20×10^{-3}	-8.0	9.4	15.6	-6.2
α-C	5.48×10^{-2}	93.0	199	92.5	106
β-C	2.10×10^{-2}	-344	-229	35.7	-265
γ-C	1.67×10^{-2}	34.5	73.8	28.3	45.5
β-H/α-H		0.30	0.15		0.32
γ-H/α-H		0.09	-0.29		0.09
γ-H/β-H		0.30	-0.32		0.27
β-C/α-C		$-3.7(-1.7)^f$	-1.2		-2.5
γ-C/α-C		$0.37(0.45)^f$	0.37		0.43
γ-C/β-C		$-0.10(-0.27)^f$	-0.32		-0.17

^{a}In $\mathring{A}^{-3}$; X-ray data of ref. 13. ^{b}In ppm. ^{c}Data obtained from ratios of ref 8. ^{d}Calculated using eq. 10. ^{e}Datum used to place ratios of ref 8 on an absolute scale. ^{f}From ref. 8.

Since 6-coordinate nickel(II) is expected to have little magnetic anisotropy and hence neglegible dipolar shifts, this agreement is taken as evidence for similarity in the spin-delocalization mechanisms in the two complexes (σ-spin delocalization in the coordinated pyridine ligand), even though cobalt(II) has an unpaired electron in an orbital of π-symmetry.

Separation of contact and dipolar shifts – The ratio method. Since it is usually not feasible to resort to single crystal magnetic anisotropy measurements, it is of considerable importance to have methods available by which to separate dipolar and contact shift contributions to observed isotropic shifts without this information. Some time ago we proposed (16, 17) a procedure, known as the "ratio method," for achieving this. In order to apply it, several requirements must be met. Isotropic shift data must be available for two closely related systems, one with magnetic anisotropy and the other without, octahedral cobalt(II) and nickel(II) complexes respectively are the classic examples. The complexes must exhibit axial or effective axial symmetry, geometric factors must be calculable, and the mode of spin-delocalization in the ligands must be identical in both cases. This last criterion requires that the ratios of the contact shifts be the same for both complexes and allows the following equations to be set up, again we use the Co, Ni pair as an example.

$$\frac{\Delta H_i^{iso}(Ni)}{\Delta H_o^{iso}(Ni)} = \frac{\Delta H_i^{iso}(Co) - R_i X_o}{\Delta H_o^{iso}(Co) - X_o} \qquad (11)$$

where X_o is the dipolar shift of the standard nucleus (o) and R_i is the ratio $G(\theta,r)_i/G(\theta,r)_o$. Equations represent n-1 equations where n is the number of observed isotropic shifts. The single unknown is X_o so the problem is usually overdetermined. If good agreement for X_0 values obtain for various different isotropic shift pairs, then some confidence can be placed in the result. The results presented in the previous section for the [M(acac)$_2$-py$_2$] (12) and earlier (17) for the [M(bipyridine)$_3$]$^{2+}$ systems, M = Co, Ni, serve to substantiate the method.

Another method of use in the separation of isotropic shift components is that of spin isolated nuclei (18). If it can be established that a particular nucleus for which a shift is observed is isolated from unpaired electron spin density, then its shift can be simply ascribed to a dipolar interaction. Further, if the system exhibits effective axial symmetry and the geometric factors for the isolated nucleus and others may be calculated, then dipolar shifts for all the nuclei may be deduced and, by difference, the contact shifts. Further details on this and other aspects of the analysis of isotropic shifts can be found in a chapter by the present author (19).

Theoretical evaluation of susceptibility anisotropy Hemin systems. - Van Vleck's equation (20) provides the fundamental link between electronic structure (wave function) and paramagnetic susceptibility (eq. 12). It is through this equation that we have

$$
^M\chi_\alpha = \frac{N\beta^2 \sum_n \left\{ \frac{\langle n|\hat{L}_\alpha - 2\hat{S}_\alpha|n\rangle^2}{kT} - 2\sum_{m\neq n} \frac{\langle n|\hat{L}_\alpha + 2\hat{S}_\alpha|m\rangle^2}{E_n - E_m} \right\} \exp(-E_n^{(0)}/kT)}{\sum_n \exp(-E_n^{(0)}/kT)} \tag{12}
$$

carried out our evaluations of susceptibility anisotropy from a theoretical point of view. Our treatment (21) of the low-spin d^5 iron(III) hemin prosthetic group such as found in metmyoglobin cyanide, MbCN, and ferricytochrome c,, Cyt c, will serve as our first example. Heme proteins containing such iron prophyrin moieties generally show well-resolved nmr spectra with significant isotropic shifts for many of the resonances. It is clearly of interest to separate these shifts into their contact and dipolar components, however susceptibility anisotropy data is not available. Information concerning the g-tensor is, however, available (23, 24) in a number of cases and our approach is to fit the observed g-values by a theoretical treatment which is also capable of evaluating the susceptibility. The ground electronic configuration is t$_{2g}$^{5}, which may be treated as a single vacancy problem. The one-electron orbital energies: E(d$_{yz}$) = 0; E(d$_{xz}$); E(d$_{xy}$) are the ligand field parameters of the problem (22). Addition of spin-orbit coupling results in a 6 x 6 determinant, the roots of which (in pairs) represent the energies of the three lowest-lying Kramers doublets. The parameters were adjusted to fit the measured g-values and the principal magnetic susceptibilities were evaluated using the

eigenvectors of the problem and Van Vleck's equation.   The results
for MbCN and Cyt c are given in Table VII where the calculated
susceptibilities are separated into first order Zeeman (FOZ)
and second order Zeeman (SOZ) contributions.   The FOZ contribution

Table VII

CALCULATED MAGNETIC PROPERTIES OF FERRICYTOCHROME c AND METMYOGLOBIN CYANIDE AT 298 °K

FOZ, first-order Zeeman effect; SOZ, second-order Zeeman effect.

System, parameters	Quantity*	$g.\chi_x$	$g.\chi_y$	$g.\chi_z$	$[\chi_z-\frac{1}{2}(\chi_x+\chi_y)]$	$[\chi_x-\chi_y]$
Ferricytochrome c,						
$\lambda = -420$ cm^{-1}	g**	1.24	2.24	3.02		
Orbital energies: 0,600	χ(FOZ)	508	1565	2799	1762	-1057
1300 cm^{-1}; Kramers	χ(SOZ)	579	500	407	-133	79
doublets: 0,748,1480 cm^{-1}	χtotal	1087	2065	3206	1629	-978
Ferricytochrome c,						
$\lambda = -340$ cm^{-1}	g**	1.26	2.24	3.06		
Orbital energies: 0,500	χ(FOZ)	538	1524	2807	1776	-986
1150 cm^{-1}; Kramers	χ(SOZ)	658	586	492	-130	72
doublets: 0,623,1295 cm^{-1}	χtotal	1196	2110	3299	1646	-904
MbCN, $\lambda = -340$ cm^{-1}	g***	0.88	1.82	3.43		
Orbital energies: 0,300	χ(FOZ)	285	999	3378	2736	-714
1300 cm^{-1}; Kramers	χ(SOZ)	978	837	370	-538	141
doublets: 0,483,1456 cm^{-1}	χtotal	1263	1836	3748	2198	-573

   * χ values in VVk/mole, see ref. 4
   ** Observed g values for ferricytochrome c are: g_x 1.25, g_y 2.25, g_z 3.06, from ref. 23
   *** Observed g values for MbCN are: g_x 0.93, g_y 1.89, g_z 3.45, from ref. 24

arises from the first term in the numerator of eq. 12 and
represents the contribution of g-tensor anisotropy.   The SOZ
contribution comes from the second term in the numerator and
can be seen to be non-neglegible.   The axial and equatorial
anisotropy factors for use in evaluating dipolar shifts are
also given in Table VII.   Tables VIII and IX give some nmr data
and calculated shifts for Cyt c and MbCN, respectively.   The

Table VIII

NMR DATA AND CALCULATED SHIFTS FOR FERRICYTOCHROME c AT 298 °K

Proton	δ^{diam}*	$\left(\dfrac{\Delta H}{H}\right)^{dip}_{ax}$**	$\left(\dfrac{\Delta H}{H}\right)^{dip}_{eq}$**	$\left(\dfrac{\Delta H}{H}\right)^{con}$***	$\delta^{obs\dagger}$
CH$_3$ b$_1$	$-$ 2.1	3.9	-2.9	-32.9	-34.0
CH$_3$ b$_2$	$-$ 3.8	3.9	-2.3	-29.1	-31.3
CH$_3$ b$_3$	$-$ 3.5	3.9	2.3	-13.0	-10.3
CH$_3$ b$_4$	$-$ 3.4	3.9	2.3	-10.0	$-$ 7.2
Meso H d	-10.5	10.1	-1.4		
Meso H e	-10.5	10.1	1.4	3.4, 5.6	4.4, 6.6
Imidazole C2–H		-19.9	1.5		
Imidazole C5–H		-18.9	1.4		
Imidazole N3–H	-10.4	-12.8	0.1	8.1	-15.0
Methionine 80–CH$_3$	3.3	-17.8	1.2	36.8	23.4

Table IX

NMR DATA AND CALCULATED SHIFTS FOR METMYOGLOBIN CYANIDE AT 298 °K

For details see footnotes to Table II.

Proton	δ^{diam}	$\left(\dfrac{\Delta H}{H}\right)^{dip}_{ax}$	$\left(\dfrac{\Delta H}{H}\right)^{dip}_{eq}$	$\left(\dfrac{\Delta H}{H}\right)^{con}$	δ^{obs*}
$CH_3 b_1$	− 3.7	5.2	0.6	−20.3	−18.2
$CH_3 b_2$	− 3.7	5.1	−1.7	−26.6	−26.9
$CH_3 b_3$	− 3.7	5.2	1.7	−15.6	−12.4
$CH_3 b_4$	− 3.7	5.2	1.7	−16.6	−13.4
Meso H d	−10.5	13.3	4.8	0.3, ?	7.9, ?
Meso H e	−10.5	12.9	−4.8	3.9, 4.1	1.7, 1.5
Imidazole C_2–H	—	−23.9	5.2	—	—
Imidazole C_5–H	—	−26.5	2.1	—	—
Imidazole N_3–H	−10.4	−16.5	0.3	5.4	−21.2

atom labeling scheme is indicated in Fig. 3. Space does not permit a detailed dicussion, however our results reveal that

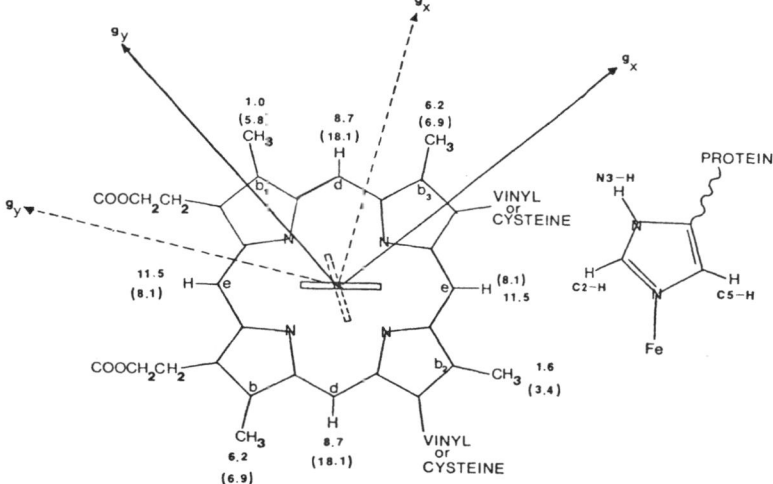

Fig. 3.   Total dipolar shifts, ppm. Cyt c(MbCN).   Axes: Cyt c: solid; MbCN: dashed.

dipolar shifts are significant and in-plane anisotropy and SOZ effects are not neglegible.

Temperature dependencies of dipolar shifts in hemin systems. These are of interest as a potential means of diagnosis of the dominant shift mechanism. We extended (25) our calculations to a variety of low-spin hemin types listed in Table X which indicates the g-values that were fit, the orbital energies, the fractional thermal population of the first excited Kramers doublet (therm. pop.), and the ratio R of the total axial anisotropy to that calculated for the FCZ contribution (g-values) of the lowest Kramers doublet alone. Plotted in Fig. 4 are the total and FOZ-anisotropies for each of the hemin types. Variation in the

Table X

Protein type	Example	g_x	g_y	g_z	$\dfrac{E_3(d_{xz})}{cm^{-1}}$	$\dfrac{E_2(d_{xy})}{cm^{-1}}$	Therm. Pop. (294 K)	R
type m	Metmyoglobin cyanide	0·93	1·89	3·45	300	1300	0·086	0·72
type c	Ferricytochrome c	1·25	2·25	3·15	480	1152	0·048	0·85
type b	Cytochrome b	1·47	2·26	2·95	615	1373	0·029	0·98
type h	Hemoglobin (histidine)	1·67	2·26	2·80	794	1710	0·014	1·12
type o	Hemoglobin, Myo-globin, hydroxy form	1·85	2·17	2·55	1197	2750	0·002	1·31
type p	Cytochrome P450	1·93	2·26	2·41	1729	2702	0·000	1·38

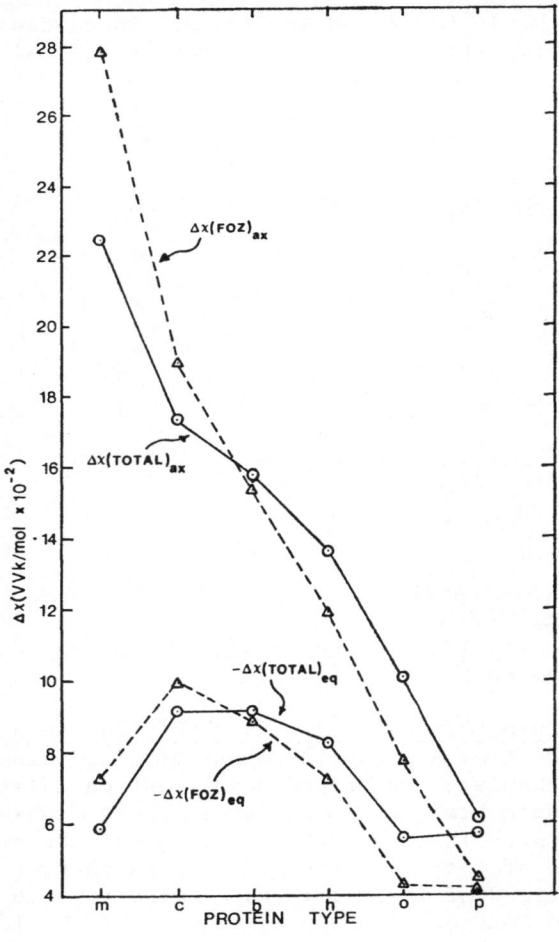

Figure 4

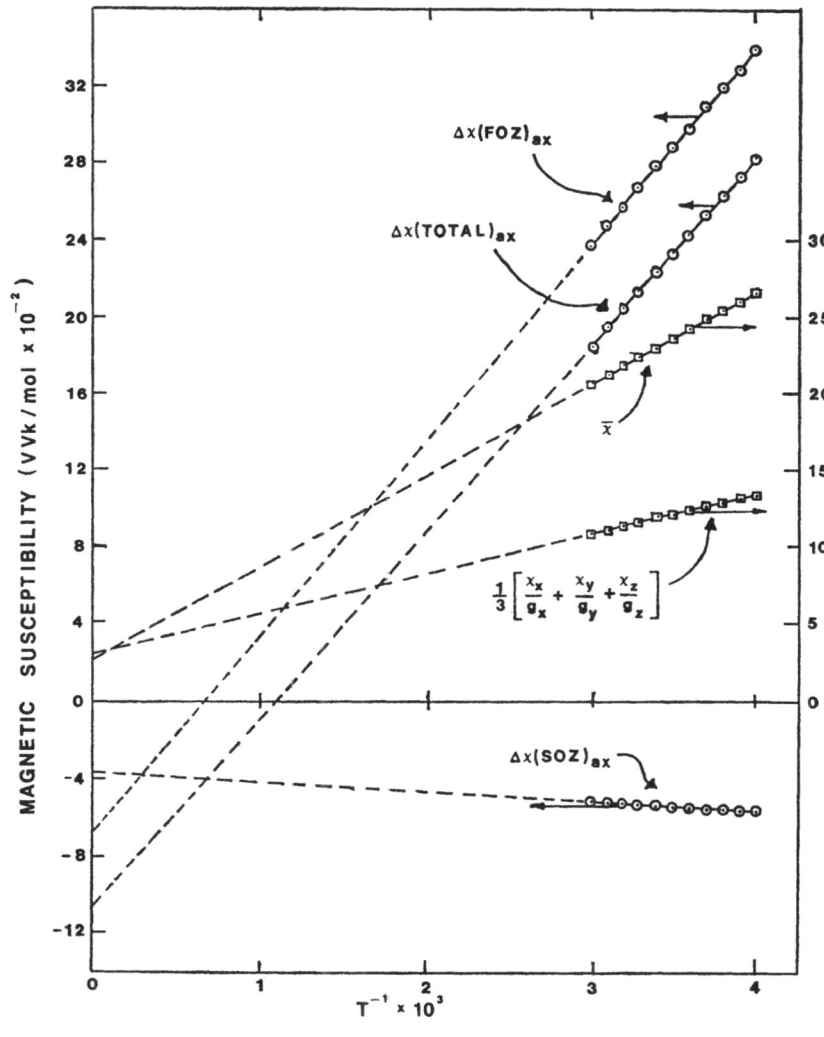

Figure 5

relative importance of axial and equatorial contributions across
the series is revealed.   In Fig. 5 are plotted vs. T^{-1} various
quantities of interest for MbCN.   General points to note are
the good leanearity with T^{-1} in the temperature range accessible
to nmr and the nonzero intercepts at $T^{-1} = 0$.   Space limitations
preclude further discussion.

   Lanthanide systems. - Lanthanide ions represent potentially
powerful replacement ion probes for use on calcium-binding
proteins and nucleic acids, moreover certain lanthanide complexes
have been extensivley exploited as nmr shift reagents.   I will
very briefly touch on several of our contributions to this area.
In 1971 Sipe and I (26) surveyed the proton resonance shifting
abilities of Ln(dpm)$_3$ complexes toward various organic substrates
for the entire series of lanthanide elements.   The results,
presented in Fig. 6, show the now-familiar trend of up- and
downfield shift reagents.   In order to establish the basic shift

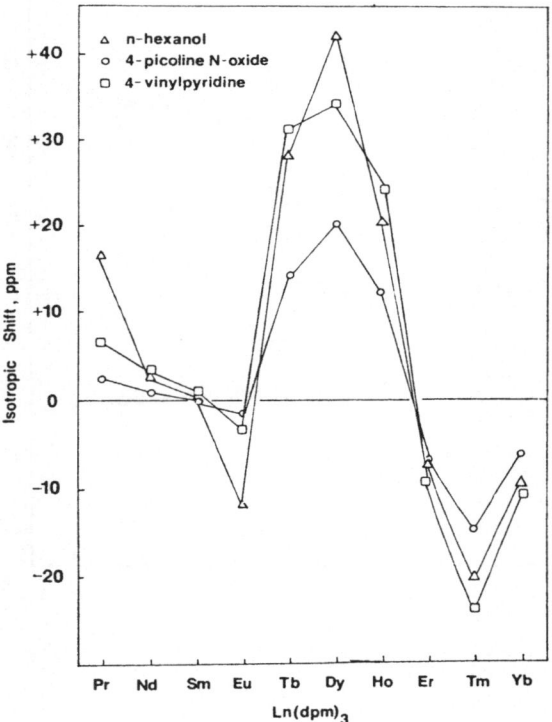

Fig. 6.   Observed proton isotropic shifts for the most shifted
resonances of deuteriochloroform solutions of n-hexanol (H-1),
4-picoline N-oxide (H-2), and 4-vinyl-pyridine (H-2) in the
presence of Ln(dpm)$_3$ with R_p = 0.125 at 30°C.   From ref. 16.

mechanism we solved the crystal structure of [Ho(dpm)$_3$(4-pico-
line)$_2$] (27) and carried out single crystal magnetic susceptibility anisotropy measurements on the entire isomorphous series
(28) with the results given in Table XI. The dipolar shifts
evaluated from this data are shown in Fig. 7. The error bars

Table XI

Principal Crystal Susceptibilities and Molecular Anisotropy

Factors for Ln(dpm)$_3$(4-picoline)$_2$ Systems at 298°K.[a]

Ln	χ_a	χ_b	χ_c	$\chi_z - \frac{1}{2}(\chi_x + \chi_y)$	$\chi_x - \chi_y$
Pr	2779	6315	4865	-3180	-1204
Nd	3849	5903	4337	-1357	-1508
Sm	931	1249	973	-188	-271
Eu	5606	3610	4861	1502	1163
Tb	28268	62209	26589	-15834	-35818
Dy	33033	65868	45114	-24597	-19328
Ho	38811	54796	41768	-9995	-12680
Er	35984	27572	36801	3653	9326
Tm	22688	9072	30051	1823	21848
Yb	9102	3563	3093	3453	4401

[a] In VVk/mole

indicate the range of dipolar shift possibilities corresponding
to our lack of knowledge of the direction of two of the principal
magnetic axes. The solid points are the solution nmr shifts of
the 4-methyl proton resonances. The results clearly support a
dominant dipolar interaction as the source of the observed shifts.

The problem of non-axiality. – The highly non-anxial
nature of the solid state structures of shift reagent adducts
and of the susceptibility anisotropy results mentioned above
stands in sharp contrast to the success many workers in the shift
reagent field have had in correlating the observed shifts to the
function $(3\cos^2\theta-1)r^{-3}$ which implies axial symmetry. Briggs
et al. (29) and I (30) have proposed distinctly different models
to account for this phenomenon. Briggs et al (29) showed that if
a substrate molecule is freely rotating about an axis which passes
through the metal (or is confronted with an n-fold rotational
barier, n >3) that the dipolar shifts will be proportional to
$(3\cos^2\theta-1)r^{-3}$ where θ is the angle that the radius vector makes
with the axis of rotation. Inspection of several solid-state
structures of shift reagent adducts reveal that such barrier-free

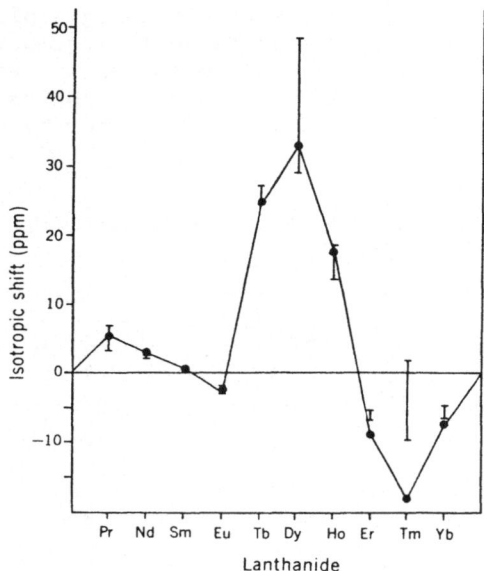

Figure   7

rotation cannot be expected to occur.

    I proposed (30) a model in which shift reagent adducts exist
in solution as an ensemble of rapidly interconverting fluctional
forms, each with its own distinct non-axial magnetic properties.
By means of a computer simulation I showed that the dipolar
shift ratios for a rigid substrate will approach those calculated
from an axial model if as few as 30 rapidly (on the nmr time
scale) interconverting fluctional forms are present. Stereo-
chemical nonrigidity is well-established for lanthanide
complexes making my model not unreasonable, however no direct
experimental test of this model has yet been devised. Further
details can be found in my article (30).

    Experimental and theoretical studies of the temperature
dependencies of dipolar shifts in lanthanide systems. In 1972
Bleaney (31) presented an elegant and concise theory which accounts
for the signs and approximate magnitudes of dipolar shifts across
the lanthanide series. According to Bleaney (31) the most
unequivocal prediction of his theory is that dipolar shifts
should vary at T^{-2} rather than T^{-1}. Since this prediction was
at variance with most of our experience with dipolar shifts and
single crystal anisotropy data, we undertook a theoretical
assessment of the problem by a different and somewhat more
general (and more cumbersome) approach (32). Independently
Golding and Pyykko reported (33) an analysis mathematically,

but not formally, equivalent to our own. Our theoretical
approach to the magnetic properties of lanthanide complexes is
identical to the one we used for transition metal species (34).
We employ a (L, S, M_L, M_S) basis including spin-orbit coupling.
We express the many-electron wave functions in Slater determinantal
form and, for comparison with Bleaney's work, express the matrix
elements of the crystal field between the one-electron f-orbitals
in terms of the phenomenological parameters, A_λ^K r^λ . Our
calculation encompasses the entire ground term including all of
the J-levels. Within the ligand field framework the only
additional approximations are those employed in the derivation of
VanVleck's equation. To illustrate our method the results of
our calculations (32) on the three ions of principal interest
as shift probes will be presented, namely Yb(III), Pr(III), and
Eu(III). We carried out calculations using the ligand field
parameters obtained from the much-studied lanthanide ethyl
sulfates of C_2^h symmetry. The calculated energy level diagrams
followed, for each ion, by the calculated susceptibility
anisotropy plotted vs T^{-1} are shown in Figs. 8-13 for Yb(III),
Pr(III), and Eu(III), respectively. The experimental susceptibility
data of Mookherji and Chachra (35) are included in Fig. 9. and
are in good agreement with theory. Several features emerge from
these calculations. Both FOZ and SOZ effects are important. At
higher temperatures $\Delta\chi_{total}$ appears to be quite linear in T^{-1} with
a large non-zero intercept at T^{-1} = 0. In Figs. 14 and 15 are
presented our experimental anisotropy measurements as a function
of T^{-1} for the 4-picoline shift reagent adducts of Pr(III) and
Yb(III). Again various temperature dependencies are found, none
of which are simply linear in T^{-2}. Finally, anisotropy data
from the literature (35, 36) for two axially symmetric systems
are plotted in Fig. 16 as $\Delta\chi T^2$ vs T for which Bleaney's theory
predicts a horizontal straight line. In our analysis (37) of
the reasons for the discrepency between the results of the two
theories it appears that it is Bleaney's neglect of ligand field
parameters of order higher than 2 and perhaps the breakdown of
his high-temperature approximation which are chiefly responsible.
In Fig. 17 we plot the results of our calculations on two different
Yb(III) systems as $\Delta\chi T^2$ vs T and again find significant deviations
from horizontal straight-line behavior. Stout and Gutowski (38)
have showed that our method agrees with an <u>exact</u> calculation for
a J = 3/2 system at 200°K to within 0.10% at which point Bleaney's
T^{-2} theory deviates by 1.49%. The principal conclusion to be
drawn from all of this is that temperature dependences are
likely to be complicated, often linear in T^{-1} with non-zero
intercepts and of little or no diagnostic value when it comes
to the determination of the shift mechanism.

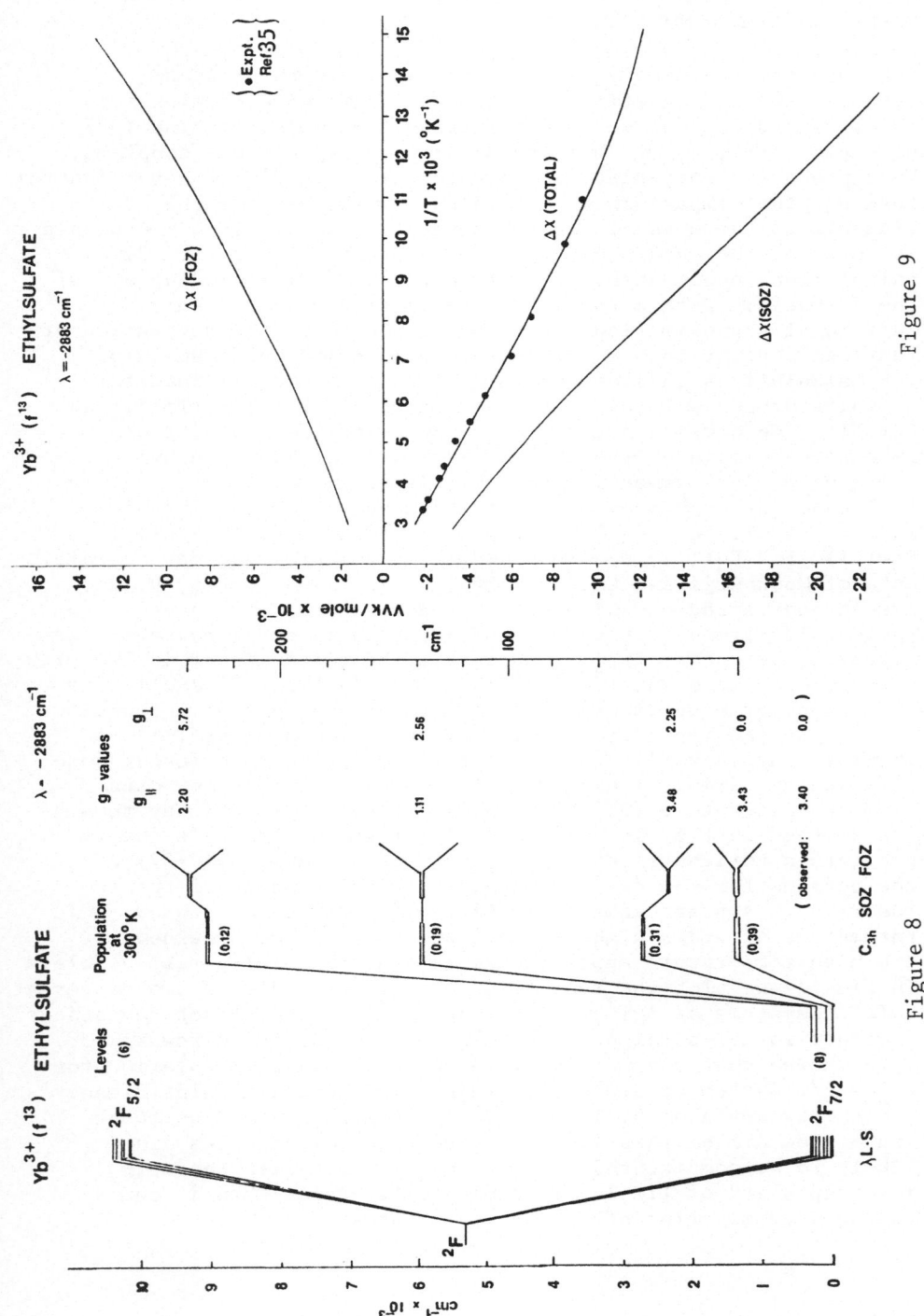

Figure 9

Figure 8

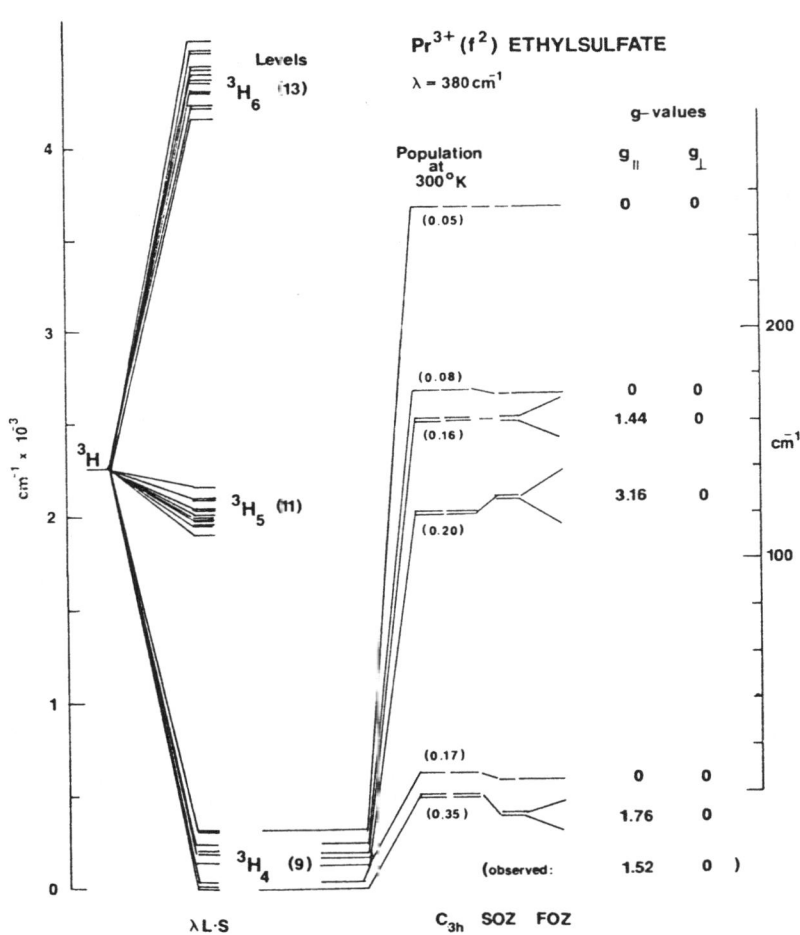

Figure 10

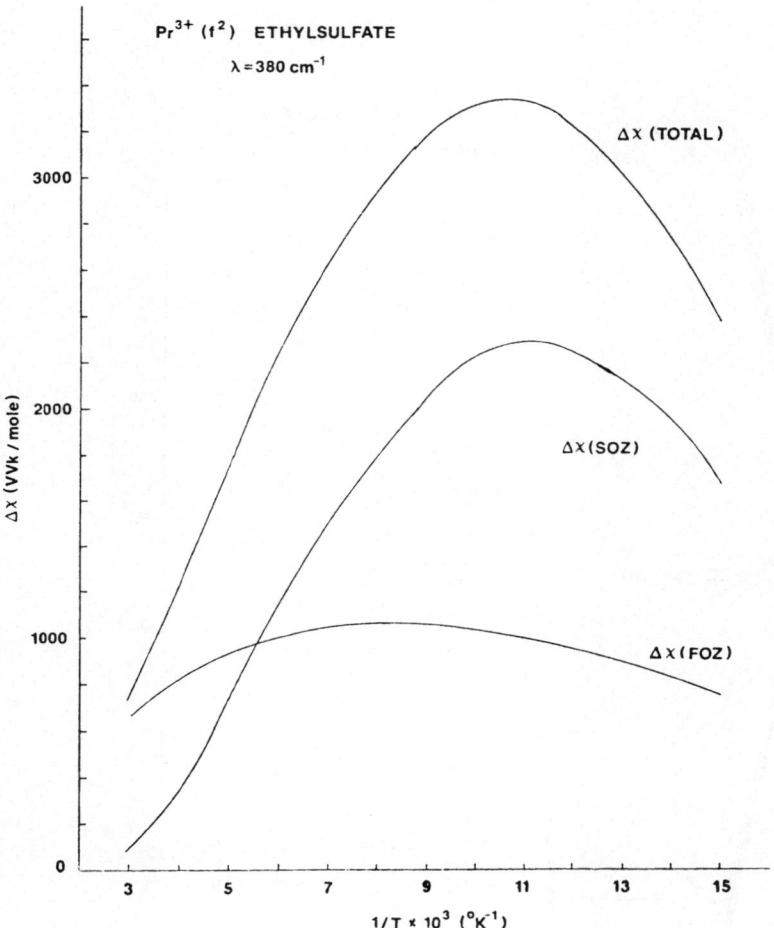

Figure 11

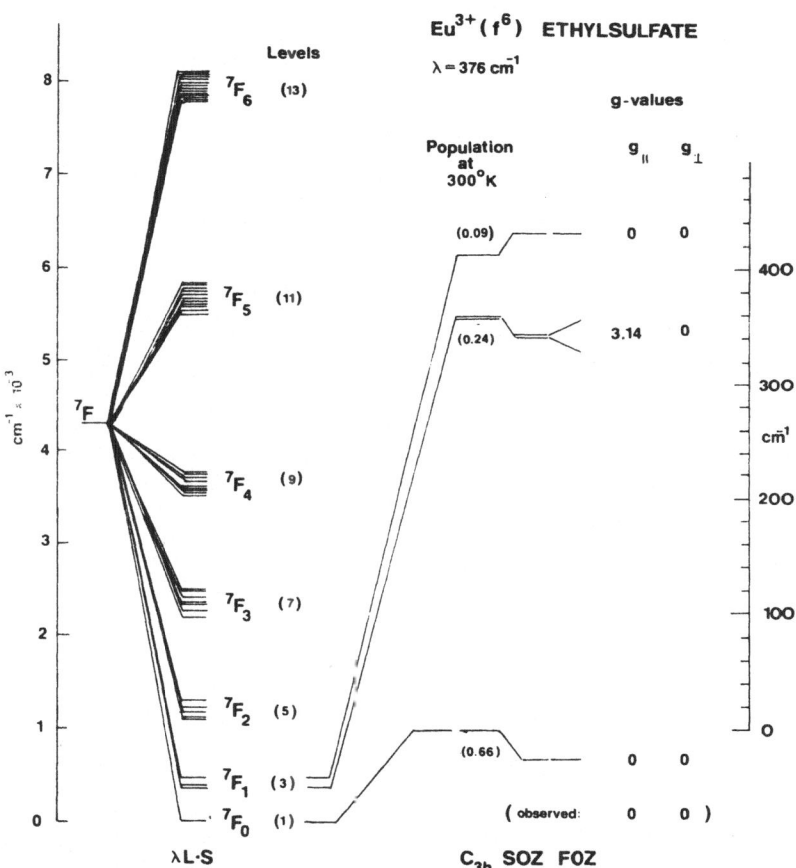

Figure 9. Calculated energy level diagram for the ^7F term of Eu^{3+}, f^6, using Yb^{3+} ethylsulfate parameters.

Figure 12

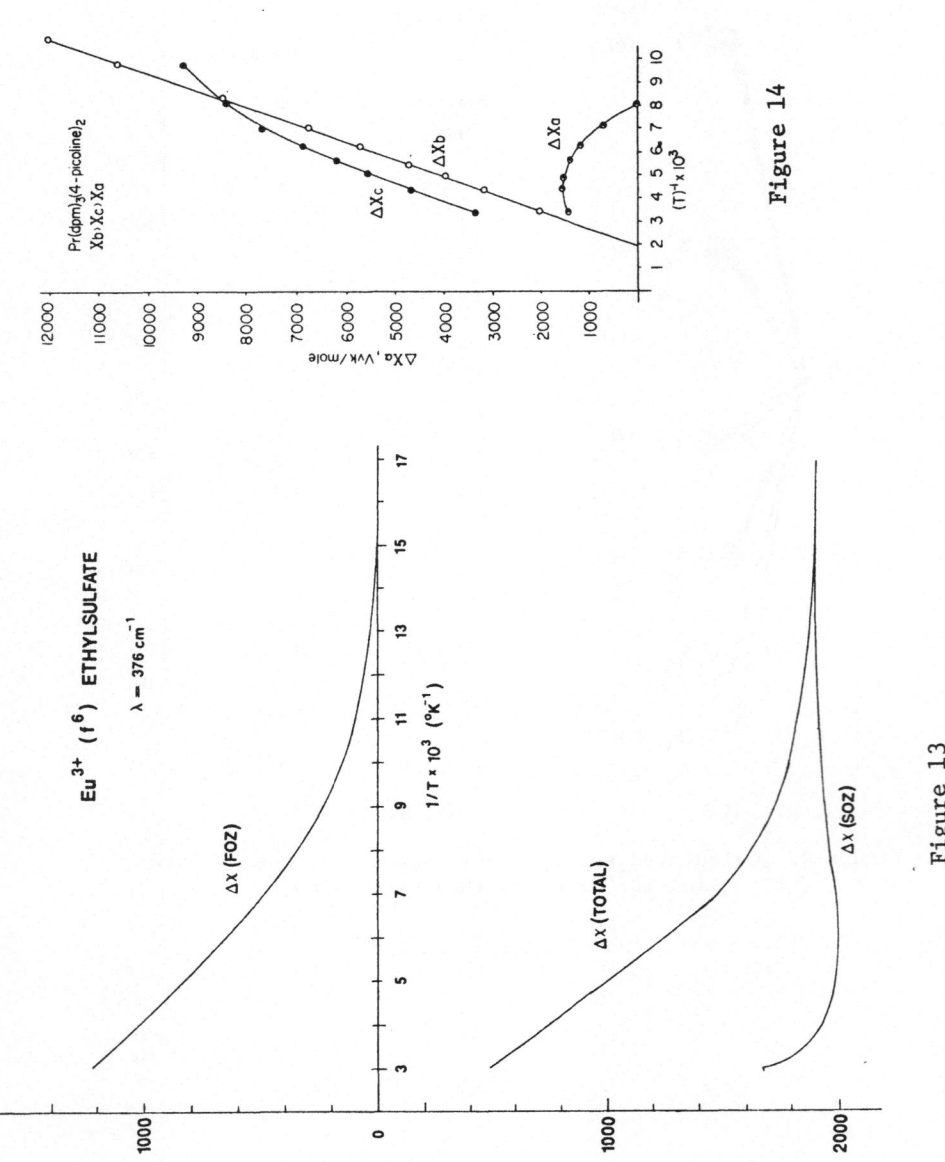

Figure 14

Figure 13

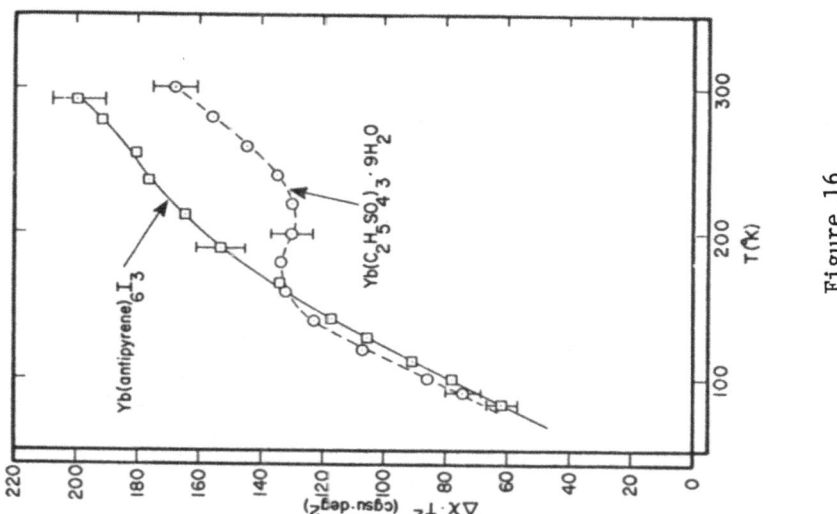

Figure 16

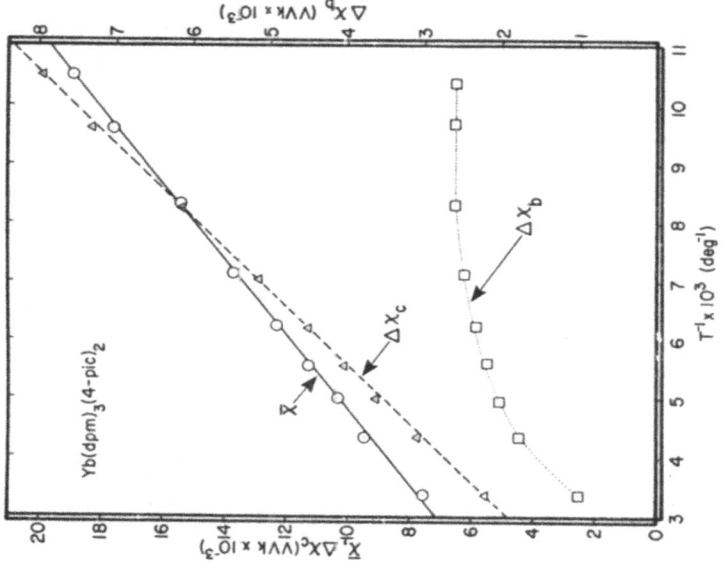

Figure 15

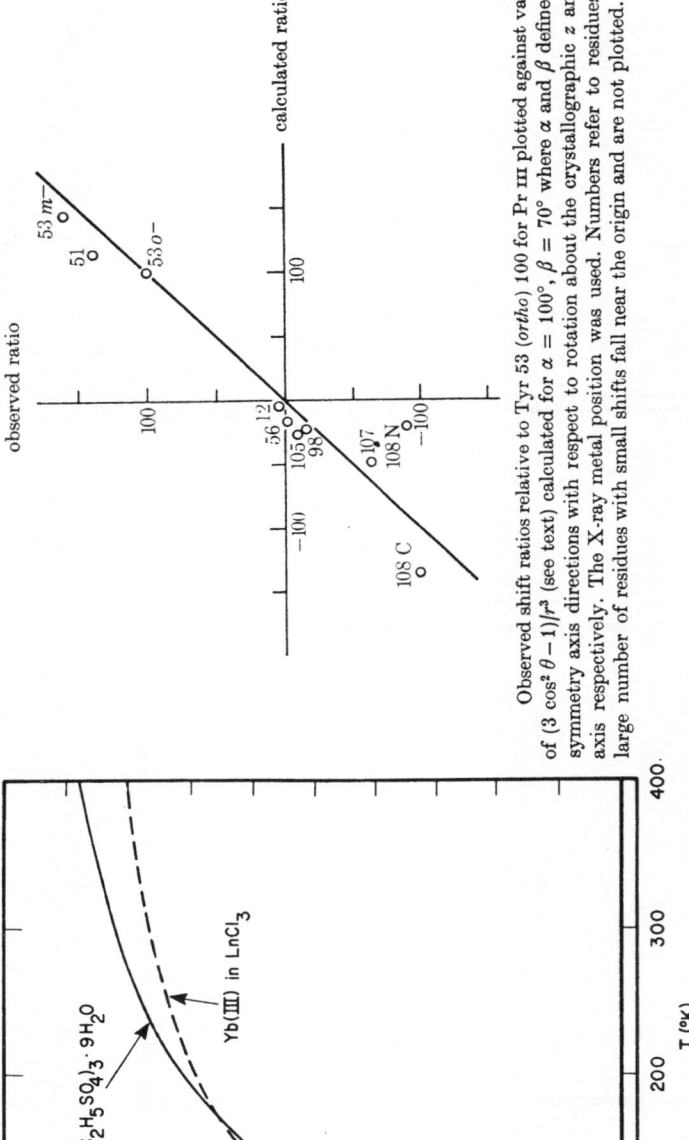

Observed shift ratios relative to Tyr 53 (*ortho*) 100 for Pr III plotted against values of $(3 \cos^2 \theta - 1)/r^3$ (see text) calculated for $\alpha = 100°$, $\beta = 70°$ where α and β define the symmetry axis directions with respect to rotation about the crystallographic z and x axis respectively. The X-ray metal position was used. Numbers refer to residues. A large number of residues with small shifts fall near the origin and are not plotted.

Figure 18

Figure 17

Metal ions as substitutional or extrinsic shift probes . - The most extensive use of paramagnetic shift probes has been in the heme protein area, aspects of which are discussed elsewhere in this volume; this topic will not be considered further here. Except for the early use of cobalt(II) by McDonald and Phillips (39) to perturb the proton nmr spectrum of lysozyme, most of the substitutional and extrinsic nmr probe work has been carried on at Oxford. I call your attention to the work of Williams and associates on the use of lanthanide ions to attempt to define the solution structure of lyso- zyme on the basis of Gd(III) induced broadening and shifts induced by the other lanthanides (40). Their observed shift ratios are set out in Table XII and the results for Pr(III) induced shifts are compared with a best computer fit to an axial model in Fig. 18 . Assignment of the shifted resonances was aided immeasurably by the fact that free and bound Ln(III) ions are in rapid exchange and shifted resonances can be extrapolated each to their diamagnetic positions.

Finally I show in Fig. 19 the nmr spectrum of cobalt(II) substituted azurin recorded by Hill et al (41). The paramagnetic cobalt(II) ion causes some extremely large shifts although assign- ment of the shifted resonances is difficult owing to the lack of rapid metal ion exchange. Unpublished single crystal magnetic anisotropy data of our own on bis(N-t-butylpyrrolate)cobalt(II), which contains a distorted tetrahedral CoN_4 coordination unit, predicts dipolar shifts of +23, -110, -30, +5, and +94 ppm for the following protons: t-butyl, 5-H, 7-H, 8-H, and 9-H respectively at 298K (see structure).

Clearly dipolar shifts cannot be ignored in tetrahedral cobalt(II) complexes.

Acknowledgement. The author's research in this area has been sponsored by the National Science Foundation.

Table XII   Gd III BROADENING RATIOS AND LANTHANIDE SHIFT RATIOS FOR SEVERAL
                              RESONANCES OF LYSOZYME

group	Gd III broadening ratio	shift ratio							
		Ce	Pr	Nd	Eu	Tb	Dy	Ho	Yb
Val 109 CH₃	2300	−55	73	−110	−190	—	—	−170	−200
Val 109 CH₃	2300	−180	−109	−380	−410	—	—	−330	−290
Ala 110 CH₃	1750	80	59	64	—	—	—	—	—
Trp 108 (C2H)	1200	—	−100	−55	−100	—	—	—	—
Trp 108 (N1H)	—	—	−90	—	−65	—	—	—	—
Ala 31 CH₃	163	−45	−64	−69	−83	−48	−54	−62	−52
Thr 51 CH₃	140	110	140	100	64†	61	150	110	120
Tyr 53 m −	—	—	160	—	160	—	—	—	120
Tyr 53 o −	100	100	100	100	100	100	100	100	100
Leu 56 CH₃‡	95	0	−2	12	0	−4	−4	−2	−18
Ile 98 γ-CH₃§	75	−8	−15	−11	−22	0	−18	−12	−15
Met 105 CH₃∥	35	−3	−8	5	−2	12	2	9	(18)
Met 12 CH₃	25	0	3	7	8	0	1	2	−2
Leu 17 CH₃	25	0	5	12	0	6	3	5	2
X CH	450	450	500	560	780	500	—	600	640
Y CH	40	50	45	56	56	56	—	40	45
Z ala CH₃	—	25	25	28	29	21	25	20	18

NOTE. these values are obtained at pH 5.3, 54 °C and are corrected for La III shift. No correction has been made for secondary binding sites, etc., but for all these resonances except Ile 98γ, this correction is generally less than ± 10 for the shift. X, Y, Z refer to unassigned resonances. The bold numbers represent values considered to be different from values with the other lanthanides by *well* outside the experimental error. The gaps in the table are values not yet obtained.

† Very temperature-dependent. The temperature dependence is small for Pr III and Yb III but not investigated for other lanthanides.

‡ The other CH₃ resonance of Leu 56 is similar.

§ After correction for binding at Asp 101 these values are *ca.* −40. The shift on the other CH₃ resonance is smaller.

∥ The ratio for Yb III is pH-dependent.

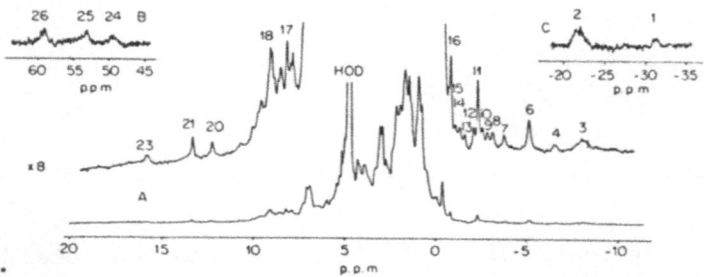

Figure 19.

The 270 MHz ^{1}H nmr spectrum of Co(II) azurin. pH* = 4.7; 1A 10 KHz spectral width, 21ºC; 1B lowest field lines, 23.8ºC; 1C highest field lines, 26.5ºC resonances are numbered 1 to 26 from high field to low field .
tables.

References

(1) La Mar, G. N., Horrocks, W. DeW., Jr., and Holm, R. H.: 1973, "NMR of Paramagnetic Molecules." Academic Press Inc. N.Y.

(2) Golding, R. M. and Halton, M. P.: 1972, Aust. J. Chem, 25, 2577-2581.

(3) Lewis, W. B., Jackson, J. A., Lemons, J. F., and Taube, H.: 1962, J. Chem. Phys., 36, 694.

(4) La Mar, G. N.: 1973, Chapter 3 in ref. 1.

(5) Eaton, D. R.: 1973, Chapter 5 in ref. 1.

(6) McLachlan, A. D.: 1960, Mol. Phys., 3, 233.

(7) Pople, J. A., Beveridge, D. L., and Dobosh, P. A.: 1967, J. Chem. Phys. 47, 2026.

(8) Doddrell, D., and Roberts, J. D.: 1970, J. Am. Chem. Soc., 92, 6839.

(9) Horrocks, W. DeW., Jr., and Johnston, D. L.: 1971, Inorg. Chem, 10, 1835-1838.

(10) Horrocks, W. DeW., Jr., Fischer, R. H., Jr., Hutchison, J. R., and La Mar, G. N.: 1966, J. Am. Chem. Soc., 88, 2436.

(11) Kurland, R. J. and McGarvey, B. R.: 1970, J. Mag. Reson, 2, 286-301.

(12) Horrocks, W. DeW., Jr., and Hall, D. DeW.: 1971, Inorg. Chem. 10, 2368-2370.

(13) Elder, R. C.: 1968, Inorg. Chem., 7, 1117.

(14) Horrocks, W. DeW., Jr. and Hall, D. DeW.: 1971, Coord. Chem. Rev. 6, 147.

(15) Morishima, I., Yonezawa, T., and Goto, K.: 1970, J. Am. Chem. Soc., 92, 6651.

(16) Horrocks, W. DeW., Jr., Taylor, R. C., and La Mar, G. N.: 1964, J. Am. Chem. Soc., 86, 3031.

(17)   Horrocks, W. DeW., Jr.:   1970, Inorg. Chem., 9, 690-692.

(18)   Happe, J. A., and Ward, R. L.:   1963, J. Chem. Phys., 39, 1211.

(19)   Horrocks, W. DeW., Jr., Chapter 4 in ref. 1.

(20)   VanVleck, J. H.: 1932, "The Theory of Electric and Magnetic Suscetibilities" The Clarendon Press, Oxford.

(21)   Horrocks, W. DeW., Jr., and Greenberg, E. S.:   1973, Biochim. Biophys. Acta, 322, 38-44.

(22)   Horrocks, W. DeW., Jr., and Burlone, D. A.:   1976, J. Am. Chem. Soc. 98, 6512.

(23)   Mailer, C. and Taylor, C. P. S.:   1972, Can. J. Biochem., 50, 1048-1055.

(24)   Hori, H.:   1971, Biochim. Biophys. Acta, 251, 227-235.

(25)   Horrocks, W. DeW., Jr., and Greenberg, E. S.:   1974, Mol. Phys., 27, 993-999.

(26)   Horrocks, W. DeW., Jr., and Sipe, J. P., III: 1971, J. Am. Chem. Soc., 93, 6800.

(27)   Horrocks, W. DeW., Jr., Sipe, J. P., III, and Luber, J. R.: 1971, J. Am. Chem. Soc., 93, 5258.

(28)   Horrocks, W. DeW., Jr., and Sipe, J. P., III: 1972, Science, 177, 994-996.

(29)   Briggs, J. M., Moss, G. P., Randall, E. W., and Sales, K. D.: 1972, J. Chem. Soc. Chan. Commun., 1180-1182.

(30)   Horrocks, W. DeW., Jr.:   1974, J. Am. Chem. Soc., 96, 3022-3024.

(31)   Bleaney, B.:   1972, J. Mag. Reson., 8, 91-100.

(32)   Horrocks, W. DeW., Jr., Sipe, J. P., III, Sudnick, D. R. in Sievers, R. E. ed.:   1973, "Nuclear Magnetic Resonance Shift Reagents" Academic Press, N.Y.   pp. 53-86.

(33)   Golding, R. M., and Pyykko, P.:   1973, Mol. Phys., 26, 1389-1396.

(34) Burlone, D. A.: 1975, Ph.D. Thesis, The Pennsylvania State University.

(35) Chachra, S. P. and Mookheji, A.: 1962, Indian J. Pure Appl. Phys., $\underline{7}$, 559.

(36) Gerlock, M., and Mackey, D. J.: 1970, J. Chem. Soc. A., 3030.

(37) Horrocks, W. DeW., Jr. and Greenberg, E. S.: 1974, Mol. Phys., $\underline{27}$, 993-999.

(38) Stout, E. W., Jr., and Gutowsky, H. S.: 1976, J. Mag. Reson. $\underline{24}$, 389-398.

(39) McDonald, C. C., and Phillips, W. D.: 1969, Biochim. Biophys. Res. Commun., $\underline{35}$, 43.

(40) Campbell, I. D., Dobson, C. M., and Williams, R. J. P.: 1975, Proc. Roy.Soc. London A345, 41-59.

(41) Hill, H. A. O., Smith, B. E., Storm, C. B., and Ambler, R. P.: 1976, Biochim. Biophys. Res. Commun., $\underline{70}$, 783-790.

SOLVENT NUCLEAR MAGNETIC RELAXATION DISPERSION
(NMRD) IN SOLUTIONS OF PARAMAGNETIC PROTEINS.
A Critical Analysis, by Example.

Seymour H. Koenig and Rodney D. Brown, III

IBM Thomas J. Watson Research Center, Yorktown
Heights, New York 10598

ABSTRACT

        Measurements of the magnetic relaxation rate of solvent pro-
tons in protein solutions, first at a single magnetic field and later at a
small number of fields over a limited range, have been used by many
investigators in attempts to determine the accessibility of solvent to
the paramagnetic sites of solute protein.   The data were generally
interpreted as changes in the hydration of paramagnetic ions as they
formed progressively larger complexes with enzymes, substrates, and
cofactors.   Because of the potential power of these techniques, and
because theory showed clearly that the relaxation should be strongly
magnetic field-dependent, we started, about a decade ago, to develop
and refine the instrumentation, the theory, and the applications of
NMRD, i.e., the magnetic field-dependence of solvent relaxation.   As
we studied NMRD in greater depth, we become progressively con-
vinced of the futility of obtaining hydration numbers from the data.
For demonstrable, but as yet poorly understood, reasons, the observed
magnitude of paramagnetic NMRD rates cannot be accounted for
theoretically.   On the other hand, we find that the use of NMRD
measurements as an indicator, much as with any other spectroscopic
tool, is a powerful technique for investigating many macromolecular
processes.

*I. Bertini and R.S. Drago (eds.), ESR and NMR of Paramagnetic Species in Biological and
Related Systems, 89–115.*

In these lectures, we first present (an extensive) set of NMRD data for solutions of two forms of Concanavalin A, an Mn^{2+}-containing protein that appears particularly amenable to theoretical analysis. We then review traditional theory of relaxation, as it is generally applied to solutions of macromolecules, and use it to analyze the data for Concanavalin A. We confirm our earlier claims that it is difficult to account for the magnitude of the observed relaxation in solutions of paramagnetic proteins. We then present a novel, simple derivation of the equations normally used to describe paramagnetic relaxation. This new derivation has the advantages that is easy to generalize to more complex conditions than that of two interacting point dipoles, and that the magnitude of the errors that are introduced by various simplifying assumptions become easy to estimate. We finally reexamine the data in terms of the new theoretical insights, and conclude with a discussion of the utility of NMRD measurements.

INTRODUCTION

For many years we have been concerned with the phenomena of the magnetic field-dependence of the magnetic relaxation rate $(1/T_1)$ of solvent nuclei in protein solutions, or, as we term it, Nuclear Magnetic Relaxation Dispersion, NMRD. We have taken extensive NMRD measurements on solutions of diamagnetic proteins (1-4), a rich subject in its own right that will not be considered here in any detail. The major phenomena in diamagnetic systems occur at magnetic fields corresponding to nuclear Larmor frequencies below about 10 MHz; these contributions must be understood, or at least measured, since they are often significant corrections to the major subject considered here: the contribution of protein-bound paramagnetic ions to the NMRD of protein solutions.

Our goals have been three-fold: to understand the underlying physical basis of NMRD experimental observations and to modify or clarify the theory, if possible, to accord better with the data; to look for significant problems in the biophysical chemistry of proteins which NMRD measurements might elucidate; and to develop and improve the "field cycling" instrumentation as dictated by experience and the requirements imposed by the systems investigated. At this point, when we speak of the paramagnetic contribution to the NMRD spectra of a protein solution, we typically mean data obtained at perhaps

20 values of magnetic field corresponding to proton Larmor frequencies from 10 kHz to 50 MHz, corrected for the NMRD of the diamagnetic part of the protein, and perhaps for the presence of paramagnetic aquoions as well.

The uncertainty in the value of $1/T_1$ in an NMRD spectra is generally less than 1%, the major limitation being due to problems relating to the status of the sample (temperature, solvent-loss, geometric problems, etc.). It has become routine to take NMRD spectra at several temperatures, for several values of sample parameter such as metal-content, pH, etc., before beginning to consider the implications of the extensive data so obtained.

We have measured NMRD spectra for complexes of protein with many paramagnetic ions: Fe^{3+}, Fe^{2+}, Ni^{2+}, Cu^{2+}, Co^{2+}, Gd^{3+}, and Mn^{2+} (5-11 and unpublished). The latter, Mn^{2+}, though quite complex, nonetheless is the most tractable and the most rewarding to work with, and the emphasis throughout will be on Mn^{2+}-protein complexes. We probably have the greatest accumulation of highly accurate NMRD data, taken over the greatest range of external conditions, of any laboratory, and have pondered the theory more than most. Our experience to date convinces us that the goals of the early workers in the field (12), to monitor changes in hydration number of, for example, Mn^{2+} as progressively higher order complexes of metal, enzyme, substrate, and cofactors are formed, are goals which cannot be achieved using NMRD data, even when data for ^{1}H and ^{2}H dispersion are included (13). However, putting things more positively, if NMRD spectra are regarded as are other spectroscopic data, namely as indicators of the state or states of a system, then NMRD can be a powerful technique in its own right for solving certain classes of problems.

In the following sections we first present NMRD data for solutions of two forms of Concanavalin A (Con-A) at two temperatures. We choose Con-A, one of the few proteins that contains Mn^{2+} ions in the native state, because it is the most ideal system for study that we have encountered (as will be become clear). Second, we write down the traditional Solomon-Bloembergen-Morgan (SBM) equations in the time-honored way, mainly for ease of reference, but also accompanied by some observations that may not be generally known. The problem is then posed: given our extensive and accurate data, and an accepted theory, how does one derive the values for the several parameters of the theory? This we proceed to do. As will be seen, the result obtained for the effective strength of the interaction of solvent protons with Con A-bound Mn^{2+} ions is inordinately large. Expressed either as the number of protons at 2.8 Å from the Mn^{2+}, or as r^{-6} for a single water of hydration, the result is too large by a factor of about five.

Following the presentation of data and its analysis in the traditional manner, we present a recent (re)derivation of the SBM equations, using a novel view of the physics of relaxation that gives one a more certain footing in the quagmire. We show how to account quite confidently for the various departures from ideality of the Mn^{2+} spin Hamiltonian and, in particular, how to account, simply and rigorously, for ligand-field effects in the limit of both low and high values of magnetic-field. Application of this added sophistication confirms and reinforces the original dilemma. The remainder of the paper is taken up with considerations of the implications of these results for the general case in which one wants to use the quantitative, experimentally derived, interaction strengths to obtain distance values.

CONCANAVALIN A DATA

Con A is a saccharide-binding protein that, because of its high selectivity and high affinity for certain saccharides, has become widely used in biochemical and clinical research applications (14). Concommitant with its utility has been a significant amount of physical biochemical research on the protein itself (15-21), so that Con A is one of the "better known" proteins. Its source is the jack bean and, though its physiological function is still not established, proteins with similar highly specific saccharide affinities are found widely distributed among the legumes (22).

The monomer of Con A has a molecular weight of 27,000, with a binding site S1 that can bind a variety of transition metal ions but generally contains some Mn^{2+} ions *in vivo*, and a site S2 that generally contains Ca^{2+}. Binding is known to be sequential, that is, metal ions can bind at S2 only after S1 is occupied and, as we have shown, adding two metals to the usual demetallized preparations of protein induces a conformation change, the kinetics of which we have investigated in some detail (11,19) using NMRD measurements. A major difference between the two conformations is the strength with which metals are bound; the final conformation is called "locked" because of the tenacity with which it holds Mn^{2+} and Ca^{2+}, in contrast to the "unlocked" conformation, for which the metal-binding is orders of magnitude weaker. The locked conformation binds saccharide whereas the unlocked does not (19,23).

Below about pH 7 in solution, Con A is mostly dimeric at millimolar concentrations, whereas above about pH 7 as well as in the crystalline state, it is tetrameric. However, though there are known to be small interactions among the monomeric units of the dimer, these are not relevant to the results considered below.

The X-ray structures of both conformations are known in some detail (15-17), as is the amino acid sequence; in particular it is known that the bound Mn^{2+} and Ca^{2+} ions share common carboxylic groups of two aspartic acid residues. As will be seen, the presence of Ca^{2+} influences the paramagnetic NMRD spectrum significantly. However, extensive EPR investigations, on solutions and single crystals, with and without Ca^{2+}, have shown that the spin Hamiltonian of the Mn^{2+} ion is altered but little by the presence of Ca^{2+} (20,21). In fact, the spin Hamiltonian for the different complexes is as close to that of a free $S = 5/2$ spin as for any protein system reported, with a zero-field splitting of ~ 200 Oe or $\sim 0.02 cm^{-1}$. Thus, Con A is a protein for which the chemical structure is well known in the crystalline state; for which the electronic structure of the Mn^{2+} ion is known to be similar in both solution and crystalline state; and for which correlations exist between the conformation in the crystalline state and in solution. All this combines to give one confidence regarding the chemical and electronic structure of Con A in solution.

The data in Figure 1 are from Brown et al. (11). They show the NMRD spectra, expressed as proton relaxivity R (in units of relaxation rate of solvent protons per mM Mn^{2+}-protein complex), of the binary Mn^{2+}-Con A complex (MP) and the ternary Ca^{2+}-Mn^{2+}-Con A complex (CMPL), as a function of magnetic field at two temperatures. The MP complex is in the unlocked conformation, and the CMPL complex is the locked conformation; however it is known that the conformation state has little influence on the NMRD spectra of the ternary complex (11).

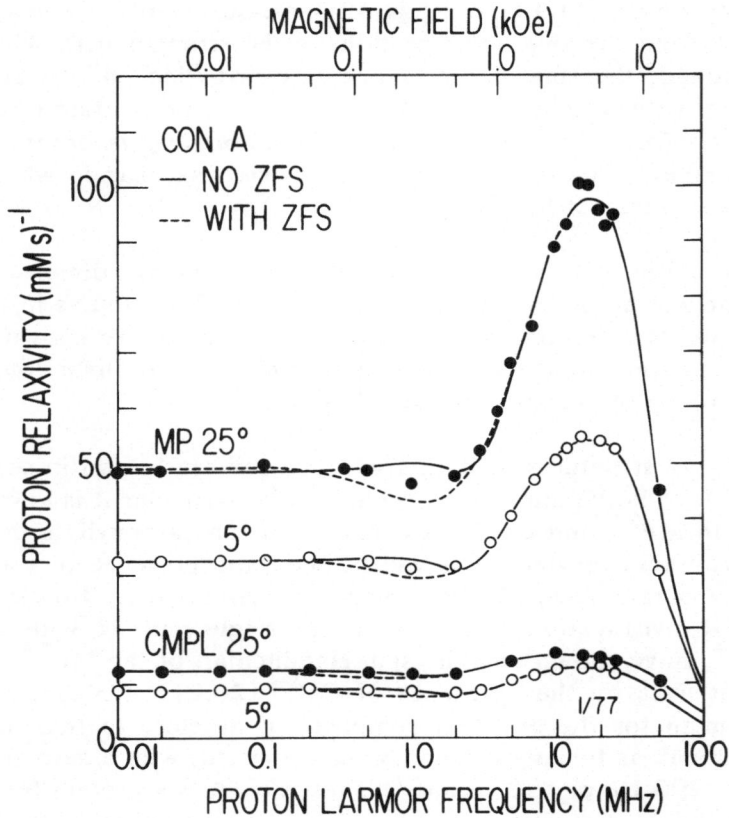

Figure 1. The solid circles show data for the magnetic field dependence of the spin-lattice relaxivity of solvent water protons at 25° in
solutions of the paramagnetic Mn^{2+}-Con A binary complex (MP) and
the Ca^{2+}-Mn^{2+}-Con A ternary complex (CMPL). The open circles
are analogous data at 5°. The solid lines through the data points are
the results of a least-squares fit of the usual Solomon-Bloembergen-
Morgan theory of relaxation to the data, ignoring ligand field splittings. The dashed lines result from a fit using the theory modified to
include ligand field effects, with the data between 0.1 and 5 MHz
ignored. The data are from Brown *et al.* (11).

Though the experimental results, Figure 1, were obtained from measurements of a sample at pH 6, the derived relaxivities are independent of pH over the range 5.2-6.4, so long as care is taken with the MP sample to ensure that the fractional occupancy of the primary Mn^{2+}-binding site (S1) is so low that there is negligible binding of Mn^{2+} at S2. Otherwise a ternary Mn^{2+}-Con A complex will slowly form, with the protein changing to the locked conformation. Additional details, including methods of sample preparation and data collection, and the definition of the two conformations and their relation to metal-binding, are given in Brown *et al.* (11).

TRADITIONAL SBM THEORY

Although the dipolar interaction Hamiltonian that produces relaxation of, say, a proton by a nearby paramagnetic ion in an external magnetic field H_o is symmetric in the two spins, the physical picture generally presented by SBM theory is not. Rather, one analyzes the spectral density of stochastic fluctuations of the local magnetic field at the proton produced by the paramagnetic ion. These fluctuations arise from the variations in spatial geometry of the two spins due to Brownian rotation of the complex, to exchange of proton-containing water ligands, and to changes in orientation of the paramagnetic moment (due to its own independent relaxation processes). Components of the Fourier spectrum of the local field at appropriate frequencies (energies) cause "flips" of the orientation of the proton moment with respect to H_o, thereby contributing to proton relaxation. The picture, then, is one in which the "big" spin relaxes the "little" spin.

There are three distinct contributions to the spin-lattice relaxation of the proton due to the dipolar interaction with a dissimilar spin: a term containing $\omega_I = \gamma_I H_o$, (the Larmor precession frequency of the proton) that arises from a proton flip along H_o together with a coupled change in orientation of the paramagnetic moment in the plane transverse to H_o; and two terms containing, respectively, $\omega_S \pm \omega_I$, where $\omega_S = \gamma_S H_o$, (the Larmor precession frequency of the paramagnetic moment) that arise from mutual flips of the two spins with simultaneous changes in their components along H_o. The noise power spectrum of the local field is calculated from the Fourier transform of its time-autocorrelation function, generally assumed exponential and characterized by a correlation time τ_C. The result for $1/T_{1M}$, the spin-lattice relaxation rate for the proton in the two-spin complex derivable from the results of Solomon (24), is

$$\frac{1}{T_{1M}} = \frac{B}{r^6}\left[\frac{6\tau_{C2}}{1 + (\omega_S - \omega_I)^2 \tau_{C2}^2} + \frac{\tau_{C2}}{1 + (\omega_S + \omega_I)^2 \tau_{C1}^2} + \frac{3\tau_{C1}}{1 + \omega_I^2 \tau_{C1}^2}\right],$$

$$(1)$$

where $B = 2\gamma_S^2\gamma_I^2\hbar^2 S(S + 1)/15 = 2.88 \times 10^{-31}$ cm^6 sec^{-2} for $S = 5/2$; r is the separation of the two spins; and the τ_{Ci} ($i = 1,2$) are given by

$$\frac{1}{\tau_{Ci}} = \frac{1}{\tau_{Si}} + \frac{1}{\tau_M} + \frac{1}{\tau_R}.$$

$$(2)$$

Here τ_R is one-third the orientational relaxation time of the complex due to Brownian rotational motion, and can be estimated, for example, using Stokes's law, or from measurements of the NMRD spectra of solutions of the apoprotein (4); τ_M is the lifetime of the proton in the complex. The τ_{Si} are the relaxation times of the paramagnetic moment, assumed to arise from interactions of the ion with the protein or lattice sufficiently strong to maintain its magnetization at thermal equilibrium despite its interaction with the proton.

Eq. 1 treats the paramagnetic moment as a point dipole located at the center of the ion, with energy levels at all values of H_o given by $m_S\gamma_S\hbar H_o$ where γ_S is a constant. It is here that the inclusion of ligand-field interactions would alter things. In place of each of the two terms in Eq. 1 containing τ_{C2} there would be a summation of many terms due to the multiplicity of transitions allowed by the more complex spin Hamiltonian of the ion; ω_S would now represent the energies of these transitions and become a complicated function of H_o.

It is generally the τ_{Si}, which are in the range $\sim 10^{-8} - 10^{-10}$s, that dominate Eq. 2, though τ_M and τ_R are not always negligible. It can readily be estimated, then, that the first two terms in Eq. 1 do not contribute to $1/T_{1M}$ unless one takes measurements at magnetic fields below a few MHz for the proton Larmor frequency. This region has been our domain for many years; the potential information content of these terms has long been accessible, but pursued mainly in our laboratory because of our unique instrumentation.

A word is in order regarding the possibility that the Mn^{2+}-proton contact interaction (often called scalar or hyperfine) may make contributions comparable to the dipolar term, Eq. 1. Many authors include an expression for this term and then argue that it must be small. We disregard it here for the simple reason that the magnitude of the square of the contact interaction is about 1% of that of the square of the dipolar interaction for the Mn^{2+}-aquoion. However, since the contact interaction is rotationally invariant (by definition) and the dipolar interaction is not, the correlation times for these two interactions for Mn^{2+} aquoions differ by a factor of about 100 in such a direction as to make the contributions of the two interactions to $1/T_{1M}$ about equal. For Mn^{2+}-protein complexes, however, the τ_{Si} are essentially the correlation times for both interactions, and therefore the contact contribution to $1/T_{1M}$ is about 1% that of the dipolar interaction.

It was discovered almost a decade ago that, for Mn^{2+} ions bound in macromolecular complexes (10,25-28), the τ_{Si} are themselves function of H_o, as was found earlier for the Mn^{2+} aquoion by Bloembergen and Morgan (29). They developed a theory of relaxation for paramagnetic ions in solution that is now generally used to describe the field dependence of the τ_{Si} in macromolecules as well, though without justification, that gives

$$\frac{1}{\tau_{S1}} = \frac{1}{5\tau_{S0}} \left[\frac{1}{1 + \omega_S^2 \tau_V^2} + \frac{4}{1 + 4\omega_S^2 \tau_V^2} \right] , \qquad (3a)$$

$$\frac{1}{\tau_{S2}} = \frac{1}{10\tau_{S0}} \left[3 + \frac{5}{1 + \omega_S^2 \tau_V^2} + \frac{2}{1 + 4\omega_S^2 \tau_V^2} \right] . \qquad (3b)$$

Here τ_V is the correlation time for the interaction that couples the paramagnetic moment to the lattice, presumed due to fluctuations in the ligand field of the ion bound in the protein complex. The theory was developed for paramagnetic aquoions; for Mn^{2+} aquoions the coupling is considered to result from fluctuations that produce deviations from octahedral symmetry of the ligand field of the aquoion; i.e., fluctuations about zero interaction (29,30). The lowest-order perturbation that contributes to relaxation would therefore have to be quadratic in strain. For Mn^{2+}-protein complexes there is generally a nonzero static ligand-field splitting, and the fluctuations would be about this value; the lowest-order perturbations would be linear in the fluctuating strains, rather than quadratic as in the aquoion. Nonetheless, Eqs. 3a and 3b are generally used to describe protein-bound Mn^{2+} ions; the major justification, we would argue, is that the form of Eqs. 3a and 3b is extremely general, in the sense that almost any realistic theory of motional narrowing gives results that can be described by the functional forms of Eqs. 3a and 3b (31). It is important to realize, however, that implicit in these equations are assumptions about the isotropy of the fluctuations; these assumptions lead to $\tau_{S1} = \tau_{S2}$ in the limit $\omega_S \rightarrow 0$. This becomes, for the protein case, a requirement that the fluctuations of the ligand field be reasonably isotropic, since the complex itself generally rotates too slowly to give isotropic averaging of the interaction of the paramagnetic ion with fluctuations of its ligand field.

For Mn^{2+} aquoions, $\tau_V \simeq 5 \times 10^{-12}$s (cf.,10), so that the dispersion of the τ_{Si} becomes important only above about 50 MHz (for protons). Bloembergen and Morgan (29) pointed this out for Mn^{2+} aquoions, where the effect was invoked to explain the ratio of T_1/T_2; it is observable because the contact interaction contribution to T_2 persists at high fields. In Mn^{2+}-macromolecular complexes, an increase in τ_{Si} with increasing field results in an increase in the last term in Eq. 1 for $\omega_I \tau_C < 1$, and a subsequent more rapid decrease when $\omega_I \tau_C > 1$, giving rise to the peak in Figure 1. This behavior was first seen and identified in Mn^{2+}-RNA complexes by Peacocke et al. (25) and Danchin and Gueron (26), and subsequently reported simultaneously for Mn^{2+}-protein complexes by Studebaker et al. (27) and Reuben and Cohen (28). We note in passing that τ_V for Mn^{2+}-protein complexes is about an order of magnitude longer than for the Mn^{2+} aquoion, and that the source of τ_V is not yet understood.

Finally, experiments on solvent nuclei do not measure $1/T_{1M}$ directly, but rather $1/T_{1P}$, the paramagnetic contribution to the spin-lattice relaxation rate $(1/T_1)$ of the solvent nuclei that are in rapid exchange with the paramagnetic site on the protons. These rates are related by (32)

$$R \equiv (T_{1P} \times [M])^{-1} = \frac{q}{55.6(T_{1M} + \tau_M)} = \frac{(q/\tau_M)}{55.6(1 + (T_{1M}/\tau_M))} ,$$

$$(4)$$

where [M] is the molarity of bound paramagnetic ions, and q is the number of water ligands (all traditionally regarded as equivalent) per bound ion. The assumption of equivalence is, of course, a potentially devastating assumption. A contribution to R due to outer-sphere effects, i.e., long range dipolar interactions between the Mn^{2+} ion and all water molecules beyond the first coordination sphere, is neglected for the present. Though its magnitude depends on how deep the Mn^{2+} ions are below the protein surface, it is generally small in Mn^{2+}-protein complexes; using the simplified equations given by Koenig and Brown (7), the outer-sphere contribution can be estimated at about 5% at low fields and much less at high fields.

COMPARISON OF DATA WITH TRADITIONAL SBM THEORY

We consider, mainly, the data for MP, and note that the NMRD spectra for the two temperatures (and indeed the data for CMPL as well) differ mainly in the extent of the relaxivity variation as a function of field. That is, the fields at which the spectra inflect and therefore the correlation times that determine the intrinsic shape of the spectra are essentially the same in all four cases; it is an increasing τ_M, first with decreasing temperature and second, with addition of Ca^{2+} ions, that reduces the extent of the field dependence of the relaxivity.

There are six unknown parameters in the theory outlined above: q, r, τ_M, τ_{S0}, τ_V, and τ_R. However, it is readily seen from the structures of Eqs. 1-4 that q, r, τ_M, and τ_R represent only three independent variables. Thus, from Eq. 4, it is clear that (q/τ_M) can be regarded as one variable if T_{1M} is measured in units of τ_M. This means, Eq. 1, that (r^6/τ_M) may be considered as a second independent variable. Finally, $(1/\tau_R) + (1/\tau_M)$ in Eq. 2 may be lumped together as a third. These, together with τ_V and τ_{S0}, give five independent variables. In order to obtain values from experiment for the five independent parameters, data must be obtained over a very wide range of magnetic fields, including the low field region where $\omega_S\tau_V << 1$, Eq. 3. Otherwise, τ_{S0} cannot be separated from τ_V, and a unique value for τ_M cannot be obtained. Though this last point is less obvious, it has been demonstrated explicitly (33) that τ_M determines the ratio of the relaxivity at the peak (near 15 MHz) to that in the low field limit. Thus, unless data are available over the ranges $\omega_S\tau_V >> 1$ and $\omega_S\tau_V$

$<< 1$, values for all the five independent parameters cannot be obtained from a theoretical analysis of data for the relaxation disperion $1/T_{1P}$. An independent value for τ_M must then be obtained from an ancillary experiment, or else some assumption must be made regarding its value.

The results of a least squares comparison of the data for MP with SBM theory, for $5°$ and $25°$, are shown in Figure 1 by the solid curves. Values obtained for the five independent parameters from the least-squares analysis are listed in Table I for the cases q = 1 and q = 2. These correspond to extremes of the two reasonable possibilities for Con A since X-ray data for CMPL (15,16) show four amino acid ligands for the bound Mn^{2+} ion, and two water molecule ligands. Probably only one water is in rapid exchange, since the other is in a position to hydrogen bond to a nearby valine residue. The situation presumably is similar for MP. As seen from Table I, with the exception of τ_M, the values derived for the parameters vary little for the two extreme assumptions. Therefore the more complex assumption of rapid but different exchange rates for the two water molecules should not alter the results very much either.

There are two major points to note regarding the results for MP in Table I: (1) the change in R with temperature is essentially all due to the temperature dependence of τ_M; and (2) the value $r \simeq 2.1$ Å is much less than the value 2.8 Å anticipated from the known Mn-O bond length and geometrical considerations (34). Herein lies the difficulty: the apparent interaction strength, as derived from this fit, is too large by a factor $(2.8 \div 2.1)^6 \simeq 5$.

Before reconsidering SBM theory in some depth, we consider the effect of the ligand field splitting (ZFS) on the values of the parameters of the theory. The first two terms in the brackets of Eq. 1 combine since $\omega_S >> \omega_I$, and it has been suggested that, in the presence of ZFS that is mainly uniaxial, replacement of the 7 in the numerator of the first term in brackets of Eq. 1 by 1.8 when the Zeeman energy is much less than the ZFS (10) is a reasonable approximation to a correct theory. We show below that this procedure is exact for $H_o = 0$. Since it is often said that when $H_o >> ZFS$, the splittings need not be considered, and since we have a modification of the theory that is exact for $H_o << ZFS$, we simply leave out the data in the intermediate region 20-1000 Oe, and fit the remaining data to a presumably valid theory. The result of this procedure is the dashed lines in Figure 1. The numerical values derived for the parameters are in Table I. It is clear that this procedure in no way resolves the quandry.

TABLE I

Values for the five parameters of the Solomon-Bloembergen-Morgan theory of relaxation obtained from a least-squares fit to $1/T_1$ dispersion data for solvent water protons in solutions of the Mn^{2+}-Con A binary complex (MP). Calculations were done either assuming no influence of ligand field splittings (7), or with the theory modified to include this interaction (1.8), as discussed in the text. Results are tabulated for each case assuming either one or two (identical) exchangeable water molecule ligands on the bound Mn^{2+} ions (q = 1,2 respectively).

	T(°C)	r(Å)		$\tau_M(\mu s)$		$\tau_R(ns)$		$\tau_{S0}(ps)$	$\tau_V(ps)$
q		1	2	1	2	1	2	1,2	1,2
FITS									
7	5	2.09	2.35	0.31	0.62	75	67	120	68
1.8	5	2.21	2.48	0.30	0.60	41	38	330	39
7	25	1.99	2.23	0.17	0.33	173	114	110	69
1.8	25	2.15	2.41	0.16	0.31	45	39	340	39

Before proceeding with a further comparison of data and theory, we present another approach to the SBM equations, an approach that makes the physical source of the various dispersive terms, Eq. 1, more apparent.

ANOTHER APPROACH TO SBM

We present below a derivation of Eq. 1 that results from a view diametrically opposite to the traditional one. We regard the precessing protons, either driven at resonance in a steady state or else precessing in a transient mode after, say, a $90°$ pulse, as producing a local field H_1' at their respective paramagnetic ions. This field arises from the rotating transverse magnetic moment of each proton, and is at frequency ω_I ; i.e., it is not noise. Since the value of $1/T_{1M}$ relates to the rate of energy loss of the precessing protons, one need only investigate the absorption of energy at frequency ω_I by the paramagnetic ions to derive the three dispersive terms in Eq. 1. As seen below, each becomes identified with absorption of energy from each of three components of H_1', one along H_0 and two rotating in opposite senses in the plane transverse to H_0.

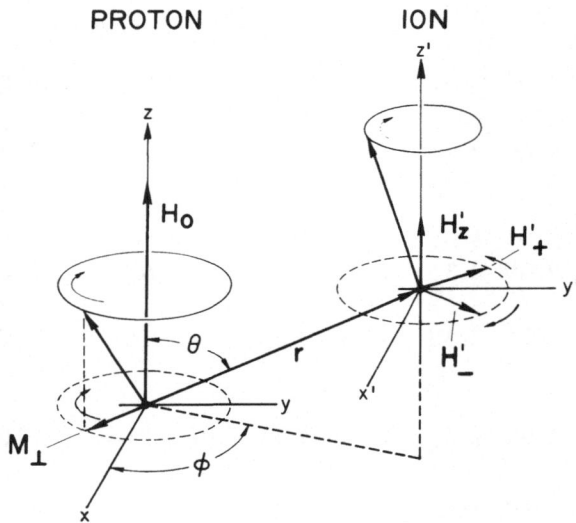

Figure 2. Illustration of the geometrical relationships of the two spins, and the various magnetic field vectors, distances, and angles referred to in the text.

In terms of the geometry of Figure 2, there is a rotating component of magnetization of the proton of magnitude $M_\perp$ that produces $\underset{\sim}{H}_1{}'$. For a coordinate frame fixed in the laboratory, and $\underset{\sim}{H}_o$ along the positive z axis,

$$M_x = M_\perp e^{i\omega_I t} \ , \tag{5a}$$

$$M_y = iM_\perp e^{i\omega_I t} \ . \tag{5b}$$

For γ_I positive, ω_I is also, and the transverse magnetization of the proton rotates in the sense $x \rightarrow -y$, or clockwise looking toward the origin from the positive z direction. The term $\underset{\sim}{H}_1{}'$ is a time-varying dipolar field given by

$$\underset{\sim}{H}_1{}' = [3(\underset{\sim}{M} \cdot \underset{\sim}{r})\underset{\sim}{r} - \underset{\sim}{M}]/r^3 \ , \tag{6}$$

where

$$\underset{\sim}{M} = M_x \underset{\sim}{x} + M_y \underset{\sim}{y} \tag{7}$$

and r, x, and y are unit vectors. The three oscillatory components of $\underset{\sim}{H}_1{}'$ derived from Eq. 6 are

$$H_z{}' = 3r^{-3} M_\perp \ e^{i\omega_I t} e^{i\phi} \sin\theta \cos\theta \ , \tag{8a}$$

$$H_+{}' = 3r^{-3} M_\perp \ e^{i\omega_I t} e^{2i\phi} \sin^2\theta \ , \tag{8b}$$

$$H_-{}' = r^{-3} M_\perp \ e^{i\omega_I t} (1 - 3\cos^2\theta) \ , \tag{8c}$$

where, in the usual way, θ and ϕ are the polar angles and

$$H_\pm{}' = H_x{}' \pm iH_y{}' \ . \tag{8d}$$

Note that even though the proton is rotating in a single sense, the field it produces at the paramagnetic ion contains both senses of rotation. Indeed (as seen below) $H_-{}'$, which rotates in the same sense as $M_\perp$ is on the average less than $H_+{}'$, and vanishes at the "magic angle."

The power absorbed by the paramagnetic ion from each of the three components, Eqs. 8a to 8c, can be readily computed. (We note that the power absorbed by each component can be considered separately since, as seen in the rotating frame of the paramagnetic ion, each drives the ion at a different frequency. The two rotating components appear at $\omega_S \pm \omega_I$, and the component parallel to H_o is at ω_I.) Assuming that in a time T_1 each proton, by exchange, is able to sample all orientations of r so that one can first average the power absorbed over all vaues of θ and ϕ, one then integrates over angles to find the average of the squares of each field component, since one-half this value times the imaginary part of the susceptibility is what enters into the expressions for absorption of power:

$$\langle |H_z'|^2 \rangle = 6M_\perp^2/5r^6 \ , \tag{9a}$$

$$\langle |H_+'|^2 \rangle = 24M_\perp^2/5r^6 \ , \tag{9b}$$

$$\langle |H_-'|^2 \rangle = 4M_\perp^2/5r^6 \ . \tag{9c}$$

The power absorbed from H_z' is a nonresonant absorption proportional to the imaginary part of the differential susceptibility, the "paramagnetic relaxation" of preresonance days. The presence of H_o does not alter things under typical conditions. The power absorbed from the two rotating terms is due to resonance absorption far from the resonance frequency ω_S. The expressions for the power absorbed become, after incorporating Eqs. 9a to 9c,

$$P_z = 3C\tau_{S1}/[1 + \omega_I^2 \tau_{S1}^2] \ , \tag{10a}$$

$$P_+ = 6C\tau_{S2}/[1 + (\omega_S + \omega_I)^2 \tau_{S2}^2] \ , \tag{10b}$$

$$P_- = C\tau_{S2}/[1 + (\omega_S - \omega_I)^2 \tau_{S2}^2] \ , \tag{10c}$$

$$C = \omega_I^2 M_\perp^2 \chi_{S0}/5r^6 \ . \tag{10d}$$

Here χ_{S0} is the static susceptibility of the paramagnetic ion. Eq. 10a comes directly from the imaginary part of the complex susceptibility along the z axis, $\chi_{S0}/[1 + i\omega_I\tau_{S1}]$. The expressions for $P_\pm$, however, involved two fine points. They follow from the steady-state solutions of the Bloch equations, as in Abragam, (35, pp. 44, ff.) when these equations are modified to allow for relaxation of the precessing magnetization toward the instantaneous direction of the magnetic field. Abragam carries this through only for the case $H_0 \rightarrow 0$ (10, p. 54) in order to demonstrate the transition from resonant to nonresonant absorption. However, this modification is necessary here because, off resonance and in the rotating frame, there is always a component of transverse magnetization along the driving field. This component of transverse magnetization does not relax to zero, with the time constant τ_{S2}, but rather toward an equilibrium value determined by χ_{S0} and the magnitude of the rotating field. The inclusion of this relaxation (which also leads to the expressions for $T_{1\rho}$ for H_1' not small) results in one of the factors ω_I (Eqs. 10b to 10d) which would otherwise be ω_S. (At resonance, this correction becomes irrelevant.) The second fine point is that the quantity H_1 of Abragam corresponds to our $H_{\pm}/2^{1/2}$.

Eqs. 10a to 10c are the essential results to be compared with SBM theory, Eq. 1, for our purposes. They result, in essence, from regarding the proton as precessing in an isotropic environment of complex magnetic susceptibility. Three loss terms contribute to $1/T_{1M}$, each of which involves absorption of energy from the local field at the paramagnetic ion produced by the transverse component of the magnetization of the nearby proton rotating at frequency ω_I. The $1/r^6$ dependence falls out quite naturally, as does the specific dependence of the different terms on the different τ_{Ci}. The implicit temperature dependence which is contained in χ_{S0} ultimately is canceled by a similar dependence in the susceptibility of the proton spin system, which affects its response to the driving rf field. Finally, if H_0 is sufficiently large that the susceptibility of the paramagnetic ion becomes field dependent, then this susceptibility substituted for χ_{S0} in Eq. 10d gives the temperature-dependent correction terms discussed by Gueron (36).

To obtain a complete theory for relaxation, the above considerations must be related to the usual derivation, a comparison perhaps best accomplished by considering Bloch's equation for the time dependence of the longitudinal magnetization M_0 of the proton spin system

$$\frac{d(M-M_o)}{dt} = -\frac{(M-M_o)}{T_{1M}} . \tag{11}$$

This is the defining equation for T_{1M}. Multiplying through by H_o, one obtains an expression for the rate of change of the excess energy of the system over its value at thermal equilibrium

$$\frac{d}{dt}(H_oM - H_oM_o) = -\frac{H_oM}{T_{1M}} + \frac{H_oM_o}{T_{1M}} \ . \tag{12}$$

The term $-(H_oM/T_{1M})$ on the right hand side is the rate at which energy is lost from the proton spin system, i.e., dissipated to the lattice, and is precisely what has been computed above. The term (H_oM_o/T_{1M}) is the rate at which energy is transferred to the spin system from the lattice at thermal equilibrium. This is what is calculated in the usual approach by considering fluctuations at equilibrium. Indeed, at equilibrium, $M = M_o$, the time rate of change of energy becomes zero, and the dissipation term becomes equal to the fluctuation term. (This is but one example of the dissipation-fluctuation theorem, developed by Einstein to relate diffusion and electrical mobility, that states, in essence, that the steady state dissipation of a system in response to a perturbation that produces a small deviation from equilibrium can be computed from fluctuations at equilibrium.)

A complete theory for T_{1M} may now be obtained by equating the total power absorbed by the paramagnetic ions with $M_oH_o/T_{1M} = \chi_{I0}H_o^2/T_{1M}$, and realizing that since each proton interacts only with its neighboring ion, $M_\perp^2$ is just two-thirds the square of the nuclear magnetic moment. Thus

$$M_\perp^2 = 2\gamma_1^2\hbar^2 I(I + 1)/3 \tag{13}$$

with $I = 1/2$ for protons. Substituting Eq. 13 into Eqs. 10a to 10c and equating the result to M_oH_o/T_{1M} gives the SBM results, Eq. 1, directly. Note that it is independent of I.

A further remark is in order regarding the relation of the two approaches to the SBM equations which, mathematically, relate $1/T_{1M}$ to the Fourier transform of the time-autocorrelation function of the spin components of the paramagnetic ion. In the traditional approach, the transform gives the noise power at the frequencies that produce transitions of the protons; the rates of these transitions are then related to $1/T_{1M}$ by rate equations. One is really computing M_oH_o/T_{1M} directly, from fluctuations. In the present approach, $1/T_{1M}$ is regarded as a dissipation term resulting from a linear interaction of the proton with the complex magnetic susceptibility of the ion. The energy loss to the ion, proportional to the imaginary part of its (frequency-dependent) susceptibility, is related to the fluctuations of the responding spins; the fluctuation-dissipation theorem shows the loss to be given by the Fourier transform of the time-autocorrelation function of the spin components. Thus the difference between the two approaches is whether the proton is regarded as responding to noise from the ion, or whether the ion is regarded as responding to a driving force from the proton.

IMPLICATIONS OF THE NEW VIEW

Mn^{2+} is an $S = 5/2$ ion. In the limit $H_0 \ll ZFS$, the normally 6-fold degenerate levels will split into three doublets in a cylindrically symmetric field that are further split by H_0. The only transition that the low frequency ω_I will excite with any probability is that of the doublet, with a splitting linear in H_0, often referred to as the $m_S = -1/2 \longleftrightarrow +1/2$ transition. Other transitions involve energies comparable to the ZFS energy. Therefore the total absorption of energy by the paramagnetic ion from the field produced by the precessing proton will be restricted to the fractional oscillator strength in the lowest level, which has the value $9/35$. This is the basis for the procedure used above when considering the influence of the ZFS on R.

In the intermediate range of fields, such that the Zeeman and ZFS terms in the spin Hamiltonian are comparable in energy, the absorption becomes quite complex. But computing the absorption spectra, averaged over all orientations of the different protein molecules, etc., is a perfectly tractable computational problem, and the proper replacement for the first two dispersive terms in Eq. 1 can be obtained for every H_0. Other problems, such as whether the lines may be regarded as Lorenzian far off resonance, that is, since $\omega_I \ll \omega_S$, are common to both approaches. This is a question as to whether the autocorrelation functions are exponential, something beyond the question of the validity of SBM theory. But with a clearer view of the source of these low-field terms, we feel confident that we have an expression valid for computing the resonance contribution to $1/T_{1M}$ in the limit $H_0 \ll ZFS$.

The fact that $S = 5/2$ brings in surprises in the other limit, $H_0 \gg ZFS$, as well, and this may well be the first presentation of these ideas. Traditional wisdom has it that in this limit the ZFS should not influence the dipolar interaction between the proton and paramagnetic ion; i.e., the term $3\tau_{S1}/(1 + \omega_I^2\tau_{S1}^2)$ that arises from the imaginary part of the nonresonance susceptibility of the paramagnetic ion. The real part of the susceptibility, as long as $kT \gg$ ZFS, is unaffected by the splitting, since the net susceptibility of a set of levels split by a nonmagnetic perturbation and equally populated thermally can readily be shown to remain unaltered by the perturbation. But the relaxation among the difficult levels can be altered substantially by the ZFS interaction and thus affect the dissipation.

The term in Eq. 1 that we are considering is in reality the sum of three terms, each with a different τ_{S1} ; the relative contribution of the three terms depends on the symmetry of the perturbation that is used to compute the paramagnetic relaxation. Thus, Rubinstein et al. (30) computed these three terms for the Mn^{2+} aquoion, for all values of H_o, and found that only one component of the three eigenvalues of the relaxation matrix contributed to the loss. Therefore the term we are considering reduced to the $S = 1/2$ case, for which it was initially derived. Moreover, not only did the three terms reduce to a single one, but it was the one that corresponded to the $m_S = -1/2 \longleftrightarrow +1/2$ transition again, and it could conveniently be compared with relaxation parameters obtained from EPR data. However, the symmetry of the interaction that relaxes the paramagnetic ions in a protein environment, a solid state environment, the interaction that has been characterized by τ_V and τ_{S0} so far, can be much lower than the near-spherical symmetry of that of the aquoion. In general, there can be three terms to consider, all with different dispersive frequencies, and only one of which, the $-1/2 \longleftrightarrow +1/2$ transition, will contribute much to relaxation. Thus, as with the resonant terms, the nonresonant term may also be reduced by the factor 9/35. This will, of course, only make matters worse. Finally, if the ligand field is significantly anisotropic, resonant contributions can arise from H_z', Figure 1, and complicate matters further. Preliminary computations by M. Margulies (unpublished) show this quite clearly. However, more realistic computations remain to be performed to clarify the magnitude of these effects.

CRITIQUE OF THE PARAMETERS

The Residence Lifetime, τ_M.

The data for MP at 5° and 25° are similar, differing mainly in the relative variations in the NMRD spectra; i.e., the two dispersion curves are similar in that they inflect and peak at the same field values. This is apparent in Table I in that τ_V and τ_{S0}, the two parameters that are most important in determining the form of the field dependence of the dispersion, are the same at both temperatures. It must be τ_M, (which, as seen from Eq. 4, damps the range of variation of T_{1M}), that is responsible for the decrease at 5° of the peak relaxivity relative to its low field value and relative to its value at 25°. Indeed, the calculations show that T_{1M} at the peak is so short that τ_M can be obtained (Eq. 4) by assuming $(1/T_1) \simeq q [M]/55.6 \, \tau_M$ at the peak. (We note, once again, that τ_M for CMPL is so long that it damps out almost the entire field dependence of the dispersion.) Meirovitch and Kalb (37) have obtained approximate values for τ_M (with ± 30% uncertainty) for an Mn^{2+}-Con A complex (not obviously the binary complex MP) and for CMPL, under conditions close to ours. Their results, from a comparison of solvent proton $1/T_1$ and $1/T_2$ (spin-spin relaxation rate) data at two high values of field, are consistent with ours within their large uncertainty. There can be little doubt, particularly from the damping of the range of variation of the relaxivity by τ_M effects, but also by the agreement with the (less accurate) results of Meirovitch and Kalb (37), that were obtained without low field data, that the values for τ_M, Table I, are correct.

The Parameters of the Mn^{2+} Ion, τ_{S0} and τ_V.

As discussed previously (33), approximate values for τ_{S0} and τ_V can be obtained from the qualitative features of the dispersion curves. The field value at which the relaxivity begins to rise substantially toward the peak ($\simeq 4$ MHz) corresponds roughly to the condition $\omega_S \tau_V = 1$. From this, one expects $\tau_V \simeq 6 \times 10^{-11}$ s, in agreement with the results in Table I. Similarly, the peak relaxivity condition corresponds to the condition $\omega_I \tau_C = 1/2^{1/2}$. Using this result and the known values of τ_M and τ_R (see below), one can estimate $\tau_{S0} \simeq 2 \times 10^{-10}$ s using Eqs. 3 and 4. Thus, the qualitative features of the dispersion curves are sufficient to give a reliable estimate of the values of τ_V and τ_{S0}; the least squares results, Table I, are essentially a refinement of this procedure.

There are no values for τ_{S0} and τ_V given in the literature for MP. However, Meirovitch et al., (21) have measured the electron spin resonance spectra of Mn^{2+} ions in solutions of Mn^{2+}-Con A and Ca^{2+}-Mn^{2+}-Con A, at both X (3 kOe) and Q (12 kOe) bands, and reported the linewidths of the resonances at Q-band. It is possible to show that the values for τ_{S0} and τ_V, Table I, are consistent with the data of Meirovitch et al., though the arguments are complicated and detailed, and are out of place in the present analysis. (The complexities relate to the fact that only the $- 1/2 \longleftrightarrow + 1/2$ transition is seen in electron spin resonance, and the field dependence of the relaxation time of this transition is quite different from that of other transitions; cf. Rubinstein, et al., 30.) It should also be noted that the values for τ_V are an order of magnitude longer than in the aquoion.

The Rotational Relaxation Time, τ_R.

The uncertainty in τ_R is intrinsically large for several reasons. An estimate of τ_R at $25°$, obtained either from Stokes' Law and corrected for anisotropy and concentration effects (33), or from dispersion measurements of (diamagnetic) Ca^{2+}-Zn^{2+}-Con A (3), is about 45 ns. This is for the dimer of 55,000 Daltons. CMPL is known to form tetramers above about pH 7 as mentioned earlier, but there is evidence, from NMRD measurements, that apo-Con A at pH 6 is tetrameric (Koenig et al., unpublished). Thus, the value anticipated for τ_R for MP at 25 $°$ is either about 45 ns, or twice that value.

The derivation of τ_R from the data is also somewhat uncertain. It is only at fields greater than about 20 MHz that τ_{S1} becomes sufficiently long for τ_R to contribute significantly to τ_{C1}, Eq. 3. Thus, the value of τ_R derives from data at only one field, 50 MHz. The final result is very sensitive to the contribution of τ_{S1} to τ_{C1} which, in turn, depends upon the precision of the functional form of Eq. 4. But this expression is an approximate representation of a complex process, involving three relaxation times for the Mn^{2+} ion (30), and though it should be a good approximation for describing the field dependence of τ_{C1}, the derivation of an accurate value for τ_R may require a more precise functional form than the present one for τ_{S1} at high fields.

These caveats notwithstanding, the values derived for τ_R are reasonable, though the expected variation with temperature (almost a factor of two decrease between 5 and 25°, assuming no change in association) is not found.

The Mn^{2+}-Proton Separation, r.

Freeman (34, Table VI-4) has tabulated the metal-oxygen and metal-nitrogen bond lengths for several transition metal-peptide complexes. With the exception of the apical water of hydration of Cu^{2+}, all bond lengths are within ± 5% of 2.1 Å, a value consistent with that obtained by summing the respective ionic radii. Though no analogous data were available for Mn^{2+} at the time, its ionic radius is a bit larger than that of the other ions. Using 2.1 Å, then, as a lower limit for the Mn-O distance, one computes 2.8 Å for the Mn^{2+}-proton distance, with the water ligand in the aquoion geometry. If this latter separation were 2.1 Å, as suggested by the results in Table I, the Mn-O distance would become 1.4 Å, or approximately the ionic radius of O^{2-} itself, a totally unrealistic value, as seen by comparison with the tabulations of Freeman.

DISCUSSION

We have not found a better Mn^{2+}-protein system than Con-A for which to attempt to compare NMRD data quantitatively with the theory of dipolar relaxation. The first aim, of course, is to derive the strength of the interaction between the protons and Mn^{2+}-ions, believing that since we know its value in advance, agreement of the two values (essentially finding that r = 2.8 Å for a reasonable value of q) will allow one to derive other useful information. This first goal, in our hands, and with extensive data, is a success, but not as anticipated. We find too large an apparent interaction, and as yet, we do not understand why. Nonetheless, given all the data, and all the pondering over the nature of the theory, we have the confidence to say that any and all deductions regarding the value of q made by other workers based on limited solvent relaxation data for solutions of Mn^{2+}-protein complexes, plus the assumption that r = 2.8 Å, are questionable.

One can go a bit beyond this by recalling some data rarely referred to. In particular, even without invoking r = 2.8 Å to interpret limited R data, it had been tacitly assumed by many authors that one could certainly obtain qualitative results by observing whether relaxivity went up or down when, say, substrate was added to a Mn^{2+}-protein complex. If R went up, the bridging might be protein-substrate-metal, for example, and if R went down, the metal might be in the middle. The view was dominated by the idea that if q, the number of water ligands of the metal, changed as a result of some interaction, then R would change proportionately; if q went up, so would R. However there are proton relaxation data for methemoglobin solutions that show that when fluoride displaces the water (or hydroxyl) ligand of the Fe^{3+} heme ion, R increases by an order of magnitude or more (cf. 38), quite contrary to traditional expectations. Moreover, for Mn^{2+}-carboxypeptidase, addition of the inhibitor L-phenylalanine causes R to increase below 10 HMz and decrease above 10 MHz (39); does addition of inhibitor increase or decrease the access of solvent to the Mn^{2+} ion?

The point we would make is that NMRD measurements on solutions of paramagnetic proteins are rich in the variety of spectra they yield. The results are complex, and our current understanding regarding how to relate theory and data is inadequate. However, neither the complexity nor the inadequacy can even be well demonstrated without taking data at over a wide range of magnetic fields. Moreover, the suggestion of Burton et al. (13), that NMRD measurements of solvent deuterons may help clarify the situation, does not work for Concanavalin A. We have taken fairly complete deuteron NMRD data for MP and CMPL (unpublished) and find that, rather than leading to a simplification of the situation, the data indicate that another complexity must be considered: partial occupancy of a water-ligand site, and a water molecule shared between the Mn^{2+} and Ca^{2+} ions in CMPL. Once again, the more data, the richer the phenomena.

Finally we should make reference to the ability of using NMRD spectra as an indicator of change. Such measurements have been particularly useful in studies of Concanavalin A, for which we have discovered the existence of two conformation states, measured the kinetics of the transition between them (11), and shown the relationship among metal-binding, conformation, and saccharide-binding activity: conformation determines activity (19). In another protein, Co^{2+}-substituted carbonic anhydrase, NMRD spectra were basic to the resolution of a long-standing problem: the explanation of the apparent pH-dependent activity of this enzyme (40). In both cases the spectra, more than only the quantitative values of R, were important; the spectra over a wide range of field were necessary, their magnitudes were of secondary importance. We would argue, then, that R data taken at only a few values and over a limited range of field are difficult to interpret, whereas NMRD spectra taken over a wide range of field is a potent technique for investigating many biochemical problems. And, of course, the technology exists for rapid and accurate measurements of NMRD of protein solutions.

1. Koenig, S.H., and Schillinger, W.S.: 1969, J. Biol. Chem. 244, pp. 3283-3289.
2. Lindstrom, T.R., and Koenig, S.H.: 1974, J. Magn. Reson. 15, pp. 344-353.
3. Hallenga, K., and Koenig, S.H.: 1976, Biochemistry 15, pp. 4255-4263.
4. Koenig, S.H., Bryant, R.G., Hallenga, K., and Jacob, G.S.: 1978, Biochemistry 17, pp. 4348-4358.
5. Koenig, S.H., and Schillinger, W.E.: 1969, J. Biol. Chem. 244, pp. 6520-6526.
6. Gaber, B.P., Brown, R.D., Koenig, S.H., and Fee, J.A.: 1972, Biochim. Biophys. Acta 271, pp. 1-5.
7. Koenig, S.H., and Brown, R.D.: 1973, Ann. N.Y. Acad. Sci. 222, pp. 752-763.
8. Fabry, M.E., Koenig, S.H., and Schillinger, W.S.: 1970, J. Biol. Chem. 245, pp. 4256-4262.
9. Wells, J.W., Kandel, S.I., and Koenig, S.H.: 1979, Biochemistry 18, pp. 1989-1995.
10. Koenig, S.H., Brown, R.D., and Studebaker, J.: 1971, Cold Spring Harbor Symp. Quant. Biol. 36, pp. 551-559.
11. Brown, R.D., III, Brewer, C.F., and Koenig, S.H.: 1977, Biochemistry 16, pp. 3883-3896.
12. Cf. the following reviews and references therein.
    Mildvan, A.S., and Cohn, M.: 1970, Advan. Enzymol. 33, pp. 1-70.
    Mildvan, A.S., and Engle, J.L.: 1972, Methods Enzymol. C 26, pp. 654-682.
    Mildvan, A.S., and Gupta, R.K.: 1978, Methods Enzymol. F 49, pp. 322-359.
13. Burton, D.R., Dwek, R.A., Forsen, S., and Gunnar, K.: 1977, Biochemistry 16, pp. 250-254.

14. Bittiger, H., and Schnebli, H.P.: 1976, *Concanavalin A as a Tool*, Wiley, New York, N.Y.

15. Hardman, K.D., and Ainsworth, C.F.: 1972, Biochemistry 11, pp. 4910-4919; Hardman, K.D., and Ainsworth, C.F.: 1976, Biochemistry 15, pp. 1120-1128.

16. Hardman, K.D.: 1973, *Metal Ions in Biological Systems*, Dhar, S.K., Ed., New York, N.Y., Plenum Press, pp. 102-123.

17. Edelman, G.M., Cunningham, B.A., Reeke, G.N., Jr., Becker, J.W., Waxdal, M.J., and Wang, J.L.: 1972, Proc. Natl. Acad. Sci. U.S.A. 69, pp. 2580-2584.

18. Goldstein, I.J., Hollerman, C.E., and Smith, E.E.: 1965, Biochemistry 4, pp. 876-883.

19. Koenig, S.H., Brewer, C.F., and Brown, R.D.: 1978, Biochemistry 17, pp. 4251-4260.

20. Meirovitch, E., Luz, Z., and Kalb (Gilboa), A.J.: 1974a, J. Am. Chem. Soc. 96, pp. 7538-7542.

21. Meirovitch, E., Luz, Z., and Kalb (Gilboa), A.J.: 1974b, J. Am. Chem. Soc. 96, pp. 7542-7546.

22. Sharon, N., and Lis, H.: 1972, Science 177, pp. 949-959; Lis, H., and Sharon, N.: 1973, Annu. Rev. Biochem. 42, pp. 541-574.

23. Harrington, P.C., and Wilkins, R.G.: 1978, Biochemistry 17, pp. 4245-4250.

24. Solomon, I.: 1955, Phys. Rev. 99, 559-565.

25. Peacocke, A.R., Richards, R.E., and Sheard, B.: 1969, Mol. Phys. 16, pp. 177-189.

26. Danchin, A., and Gueron, M.: 1970, Fourth International Conference on Magnetic Resonance in Biological Systems, Oxford (Abstract); and J. Chem. Phys. 53, pp. 3599-3609.

27. Studebaker, J., Brown, R.D., and Koenig, S.H.: 1970 Fourth International Conference on Magnetic Resonance in Biological Systems, Oxford (Abstract).

28. Reuben, J., and Cohn, M.: 1970, Fourth International Conference on Magnetic Resonance in Biological Systems, Oxford (Abstract); and 1970, J. Biol. Chem. 245, pp. 6539-6546.

29. Bloembergen, N., and Morgan, L.O.: 1961, J. Chem. Phys. 34, pp. 842-850.

30. Rubinstein, M., Baram, A., and Luz, Z.: 1971, Mol. Phys. 20, pp. 67-80.

31. Pyper, N.C.: 1971, Mol. Phys. 21, pp. 1-33.

32. Luz, Z., and Meiboom, S.: 1964, J. Chem. Phys. 40, pp. 2686-2692.
33. Koenig, S.H., Brown, R.D., and Brewer, C.F.: 1973, Proc. Natl. Acad. Sci. U.S.A. 70, pp. 475-479.
34. Freeman, H.C.: 1967, *Advances in Protein Chemistry*, Vol. 2, Anfinsen, Jr., C.B., Anson, M.L., Edsall, J.T., and Richards, F.M., ed., Academic Press, New York, pp. 257-425.
35. Abragam, A.: 1961, *The Principles of Nuclear Magnetism*, Oxford Univ. Press (Clarendon), London.
36. Gueron, M.: 1975, J. Magn. Resonance 19, pp. 58-66.
37. Meirovitch, E., and Kalb (Gilboa), A.J.: 1972, Biochim. Biophys. Acta. 303, pp. 258-263.
38. Cf. Gupta, R., and Mildvan, A.S.: 1975, J. Biol. Chem. 250, pp. 240-243.
39. Quiocho, F.A., Bethge, P.H., Lipscomb, W.N., Studebaker, J.F., Brown, R.D., and Koenig, S.H.: 1971, Cold Spring Harbor Symp. Quant. Biol. 36, pp. 561-567 (see Fig. 6).
40. Koenig, S.H., Brown, R.D. III, and Jacob, G.S.: 1979, Symposium on Biophysics and Physiology of Carbon Dioxide, Regensburg, Germany (Proceedings in press).

MULTIPLE PULSE ^{1}H NMR EXPERIMENTS FOR STRUCTURAL STUDIES OF
HEMOPROTEINS

Kurt Wüthrich and Regula M. Keller

Institut für Molekularbiologie und Biophysik
Eidgenössische Technische Hochschule
CH-8093 Zürich-Hönggerberg, Switzerland

ABSTRACT

Applications of various different multiple pulse ^{1}H NMR experiments for studies of hemoproteins are described. This includes spin decoupling difference spectroscopy, saturation transfer studies, measurements of selective nuclear Overhauser effects in the presence of spin diffusion, and resolution enhancement by the use of two-dimensional spectroscopy. Structural information on hemoproteins obtained with the use of these multiple pulse techniques is briefly surveyed.

1. INTRODUCTION

1.1. Structure and Biological Functions of Hemoproteins

Hemoproteins are conjugated proteins which consist of one or several polypeptide chains and one or several heme groups. The heme groups are iron porphyrin complexes (Fig. 1) which can be combined with the polypeptide chains through one or several covalent or coordinative bonds and through a multitude of weaker interactions. High resolution NMR is an attractive technique for studies of structure-function relations in hemoproteins. The NMR spectral properties are markedly influenced by the heme groups, which act as natural shift and relaxation probes [1,2]. As a consequence the ^{1}H NMR spectra of hemoproteins usually contain numerous well separated resonances, which has long made hemoproteins attractive for tests of new NMR approaches for studies

117

*I. Bertini and R.S. Drago (eds.), ESR and NMR of Paramagnetic Species in Biological and
Related Systems, 117–134.*
Copyright © 1979 by D. Reidel Publishing Company

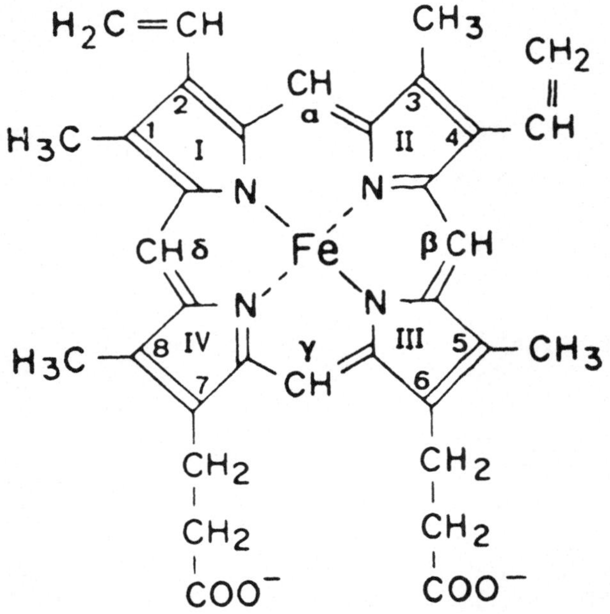

Figure 1. Structure of protoheme IX, the iron complex
with protoporphyrin IX. The four pyrrole rings are
numbered I to IV, the peripheral β-positions 1 to 8,
and the four meso-positions α to δ. Protoheme IX is
found e.g. in hemoglobin, myoglobin and cytochrome b_5.
In the heme groups of c-type cytochromes one or both
of the vinyl substituents in positions 2 and 4 are
substituted by thioether linkages with cysteinyl residues of the polypeptide chain, $-CH(CH_3)-S-$.

of protein structure.

Hemoproteins are involved in many vital processes in living
organisms, and the heme groups appear to be an integral part of
the active centers in all the hemoproteins known to-date. Prominent among their biological functions are those of hemoglobin,
which is the oxygen transporting protein in the blood, myoglobin,
which binds and stores oxygen in the muscles, various cytochromes, which act as electron transferring oxidation-reduction
carriers, and enzymatically active hemoproteins which control
diverse biochemical reactions. Many biochemical reactions of
hemoproteins involve changes of the oxidation and spin states of

the heme iron, e.g. the oxygenation of myoglobin and hemoglobin

$$Mb^{II}(Fe^{2+}, S = 2) + O_2 \rightleftharpoons Mb^{II}O_2(Fe^{2+}, S = 0) \qquad [1]$$

and the electron transfer in cytochromes, e.g.

$$Cyt\ c^{III}(Fe^{3+}, S = 1/2) + e^- \rightleftharpoons Cyt\ c^{II}(Fe^{2+}, S = 0)\ [2]$$

Four different electronic configurations of the heme iron are commonly encountered (1). Three of these are paramagnetic, i.e. the high spin ferric (Fe(III), S = 5/2), low spin ferric (Fe(III), S = 1/2) and high spin ferrous (Fe(II), S = 2) forms, whereas the low spin ferrous state (Fe(II), S = 0) is diamagnetic.

In investigations of structure-function relationships in hemoproteins, prominent questions bear on the molecular conformations in the different states of a reaction cycle such as [1] and [2] and on the pathways through which electrons are transferred to and from the heme iron during biological oxidation-reduction reactions. This latter question may be closely related to the electronic structure of the heme groups, which is accessible to NMR investigation (2). The hemoglobins and the cytochromes of the c-type present two of the largest presently known families of homologous proteins available for studies of evolution. It is an additional interesting use of NMR to complement comparative studies of evolutionarily related proteins with data on structural features which are not readily amenable to observation with other techniques.

1.2. ^{1}H NMR Spectra of Hemoproteins

The heme groups in hemoproteins affect the ^{1}H NMR spectra in two ways. Firstly, the spectrum contains the resonances of the heme protons, many of which have chemical shifts which are not usually encountered for protons in polypeptide chains (1). Secondly, through their local magnetic fields the heme groups may act as naturally built-in shift or relaxation probes on nearby protons of the polypeptide chains (2). Both these effects may lead to the appearence of well separated resonance lines with unusual chemical shifts in the spectra of hemoproteins. As an illustration Fig. 2 shows the ^{1}H NMR spectrum of ferricytochrome c-552 from Euglena gracilis (3).

In diamagnetic hemoproteins the ring current field of the heme group (1,2,4-6) gives rise to outstanding conformation dependent chemical shifts. The ring current field opposes the external magnetic field H_0 in the areas above and below the ring plane

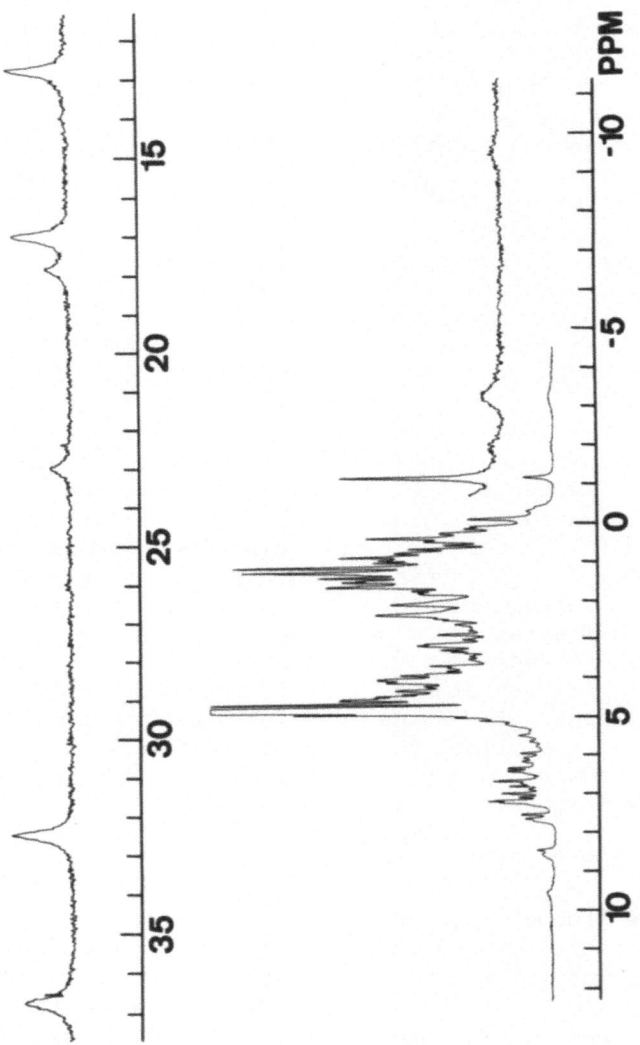

Figure 2. Fourier transform 360 MHz ^{1}H NMR spectrum of
a solution in ^2H_2O of ferricytochrome c-552 from
Euglena gracilis, protein concentration 0.003 M, p^{2}H
7.0, T = 29° C. The high field region from -1 to -11 ppm
and the low field region from 10 to 38 ppm are represented with a 5-fold expanded vertical scale. The well
separated lines in these regions correspond in intensity
to 1 to 3 protons each. Nearly all the protons of the
polypeptide chain are seen between 0 and 8 ppm. The
strong line at 4.8 ppm comes from the residual solvent
protons (from ref. 3).

and leads therefore to upfield shifts of the resonances of protons located near the heme plane in the three-dimensional protein
structure. Protons in the plane of the ring are subject to downfield shifts. Upfield shifts as large as -6 ppm may be caused for
protons located near the heme plane. Since the ring current shifts
are related with the relative coordinates of the heme and the observed protons, they are suitable probes for studies of the polypeptide conformation in the heme crevice (1,2,4).

In paramagnetic hemoproteirs, interactions of nuclear spins with
the unpaired electrons in paramagnetic heme groups give rise to
the so-called hyperfine shifts of the NMR lines and may largely
affect the spin relaxation times (1,2). Hyperfine interactions in
paramagnetic compounds are extensively treated in other reports
contained in this volume. Suffice it here to mention that among
different contributions to the hyperfine shifts for protons, δ_{hf},
two fundamentally different terms can be distinguished (7). These
are the contact shifts, δ_c, and the pseudocontact shifts, δ_{pc}.

$$\delta_{hf} = \delta_c + \delta_{pc} \tag{3}$$

The contact shift δ_c arises from Fermi contact coupling and the
pseudocontact shift δ_{pc} from dipolar coupling. Because of the
delocalization of the unpaired electrons contact shifts may be
observed for the protons of the heme and the axially bound
ligands, and the extent of the shifts is related by eq. [4] to
the electron spin density ρ_C^{π} localized on the aromatic carbon
atom to which the observed proton or aliphatic substituent is
bound (1,2,7,8).

$$\delta_c = Q\rho_C^{\pi} \tag{4}$$

In contrast, pseudocontact shifts depend on through-space interactions and can be related to the magnetic susceptibility tensor
of the heme iron and the coordinates of the observed protons
relative to the heme iron (1,2,7). Hence, if the two terms in
eq. [3] can be determined separately, δ_c provides information on
the unpaired electron spin density distribution in the active
center of paramagnetic hemoproteins and δ_{pc} on the electronic
g-tensor and the spatial structure of the heme crevice. Studies
of these different aspects of the molecular and electronic structures are of utmost interest in view of structure-function relations in hemoproteins.

The nuclear spin relaxation times in diamagnetic hemoproteins
correspond to those observed in non-heme proteins of comparable
size (2). In the low spin ferric state the paramagnetic hemes

have, because of the short electronic relaxation time T_{1e} of
approximately $2 \cdot 10^{-12}$ s (1), only little influence on the proton
relaxation. As a consequence even the largely hyperfine shifted
resonances of the heme group protons are usually observed as
sharp lines, as illustrated by the spectral region from 10 to
38 ppm of ferricytochrome c-552 (Fig. 2). In high spin ferrous
and high spin ferric hemoproteins, the resonances of the heme
groups and numerous nearby protons of the polypeptide chain are
usually quite drastically broadened by electron-nucleus interactions via dipolar coupling and scalar coupling (9,10), and by
Curie-spin relaxations (11,12). Measurements of relaxation enhancements by the paramagnetic hemes can provide substantial
support for the analysis of pseudocontact shifts in studies of
the protein conformation (2,13).

2. MULTIPLE PULSE EXPERIMENTS

Most of the early ^{1}H NMR experiments with hemoproteins were recorded in the continuous wave mode at resonance frequencies from
60 MHz to 220 MHz (1,2). In these spectra the spin-spin coupling
fine structure of the resonances was not usually resolved and
hence even the use of spin decoupling for the identification of
the component resonances in the spin systems of individual heme
substituents or amino acid side chains (2) was rather limited.
Improved spectral resolution is now obtained with Fourier transform spectrometers operating at higher fields, the use of digital
data manipulation (14-16) and J-resolved two-dimensional NMR
(17,18). With the increased resolution and the use of suitable
multiple pulse experiments a more detailed analysis of the spectra
can be obtained based on the identification of numerous complete
spin systems and individual resonance assignments for the heme
groups and parts of the polypeptide chains. In this section applications of different multiple pulse experiments with hemoproteins
are described.

2.1. Difference Spectra

The effects of selective double resonance irradiation in the [1]H
NMR spectra of proteins are usually small differences between two
complex spectra obtained with and without double irradiation.
Double irradiation effects are therefore often best seen in
difference spectra (19-21) and most of the results in this paper
are presented in this form.

The difference spectra obtained in our experiments are differences between spectra recorded on the one hand with selective irra-

diation on the desired line and on the other hand "off-resonance",
e.g. in an empty spectral region. To minimize spurious signals
due to Bloch-Siegert shifts the "off-resonance" frequency is
selected as closely as possible to resonance and low power levels
are used in the decoupling channel. To minimize effects of spectrometer instabilities, the following sequence of operations is
typically used (20,21): First, a free induction decay (FID) with
double resonance effects is accumulated in part A of the computer
memory. Second, a FID with off-resonance irradiation is accumulated in part B of the computer memory. This sequence is repeated
to improve the signal-to-noise ratio. After the data accumulation
the difference between the memory contents A and B is formed,

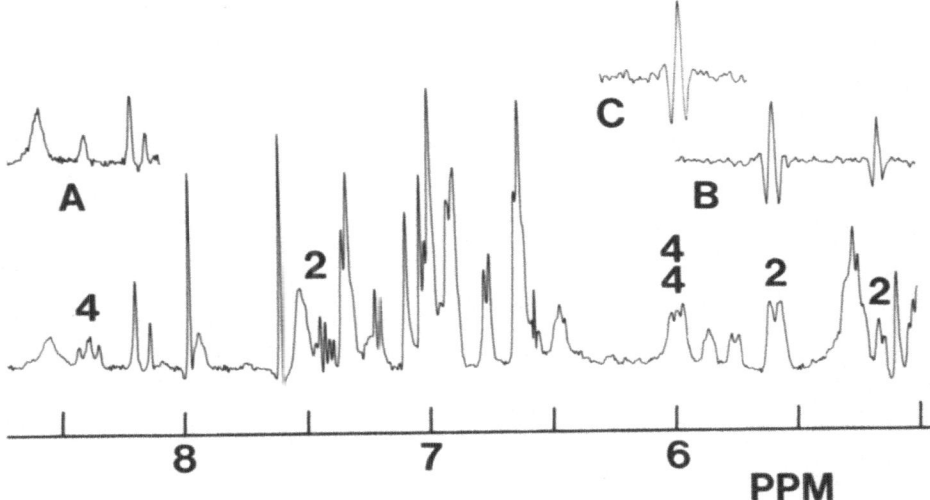

Figure 3. Region from 5 to 9 ppm in the Fourier transform ^{1}H NMR spectrum at 360 MHz of trypsin-solubilized
pig ferrocytochrome b$_5$, protein concentration 0.006 M,
p^{2}H 7.0, T = 25^0 C. The resolution was improved by
multiplication of the FID with a phase shifted sine
bell (16). The numbers 2 and 4 indicate the component
resonances of the AMX spin systems of the vinyl groups
in positions 2 and 4 of the heme (Fig. 1). The inserts
A-C show the spin decoupling experiments used to identify the two AMX spin systems. A. Decoupling of the
multiplet at 8.38 ppm obtained with irradiation at
5.97 ppm. B. Decoupling difference spectrum obtained
with irradiation at 7.49 ppm. C. Decoupling difference
spectrum obtained with irradiation at 8.38 ppm. No resolution enhancement was used for the difference spectra.

possibly treated with a resolution enhancement routine and
Fourier transformed.

2.2. Spin Decoupling

Spin decoupling is used primarily for the identification of the
individual components in the spin systems of the heme substituents (Fig. 1) and the amino acid residues of the polypeptide
chains (2). As an illustration Fig. 3 shows the experiments used
to assign the AMX spin systems of the vinyl substituents of protoheme IX in cytochrome b_5. The chemical shifts and spin-spin coupling constants in the two spin systems of Fig. 3 are typical for
vinyl substituents of heme groups (22), which differ markedly
from those observed for the common amino acid residues.

The spin decoupling difference spectra in Fig. 3 were recorded
with the following pulse sequence:

$$\left[\text{Obs.} - T_{ac}^{A} - T_d - \text{Obs.} - T_{ac}^{\text{off-res.}} - T_d \atop \omega^A \qquad\qquad\qquad \omega^{\text{off-res.}} \right]_n \qquad [5]$$

The experiment started with an observation pulse, Obs. The de-
coupling field ω^A, or $\omega^{\text{off-res.}}$, respectively, was applied during
the acquisition time T_{ac}. The FID's acquired during T_{ac}^A and
$T_{ac}^{\text{off-res.}}$ were stored in separate parts of the computer memory.
1000 transients were accumulated before the difference between
the two FID's was computed and Fourier transformed. The delay
time, T_d, was adjusted so that the spin system could recover to
equilibrium before the following observation pulse was applied.

2.3. Saturation Transfer

Saturation transfer experiments can generally be used to relate
resonances which are connected by rate processes, provided that
these processes occur on a suitable time scale (2). Additional
information is obtained on the kinetics of the reactions involved.
Attractive applications with hemoproteins include the identification of corresponding lines in different oxidation or spin states
of a hemoprotein and kinetic studies of the reactions connecting
the different forms of the protein (equations [1] and [2]). In an
early experiment of this type Redfield and Gupta (23) established
correlations between reduced and oxidized horse cytochrome c. As
an illustration Fig. 4 shows a saturation transfer in a half oxidized solution of Crithidia oncopelti cytochrome c-557, which resulted in the identification of corresponding lines in the oxi-

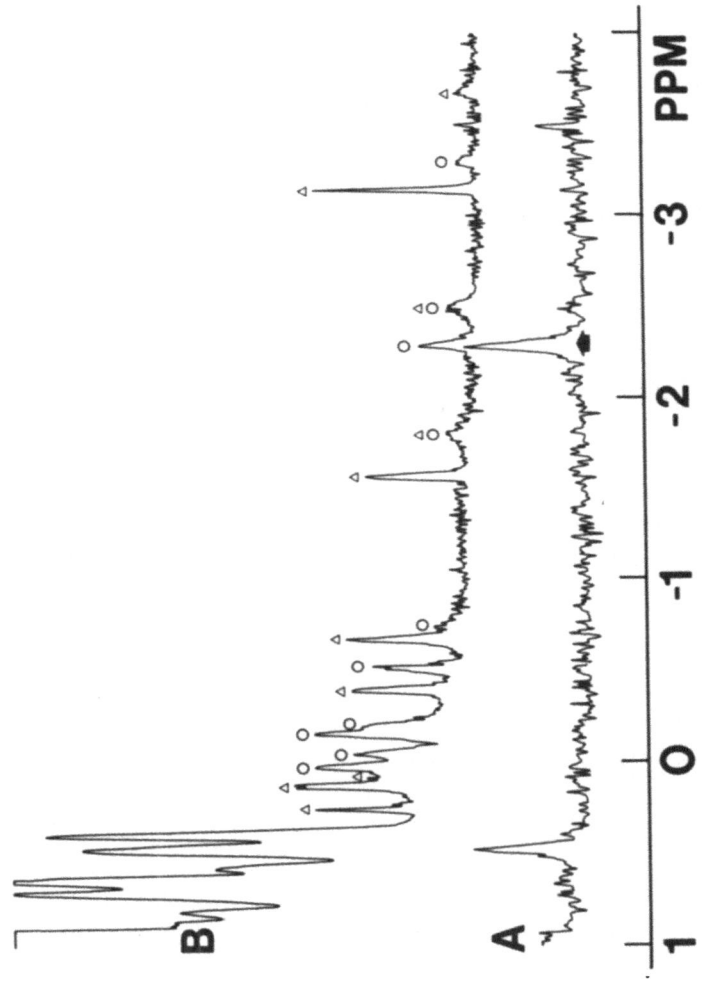

Figure 4. Saturation transfer experiments in a partially oxidized solution of cytochrome c-557 from Crithidia oncopelti containing 50 % each of the reduced and the oxidized protein; total protein concentration 0.001 M, p^{2}H 6.9, T = 26^0 C. A. Difference spectrum recorded with the pulse sequence [6], where the selective irradiation ω^A was on the ferricytochrome c-557 line at -2.3 ppm. The corresponding line in the reduced protein is seen to be at 0.43 ppm. B. spectral region from -4 to 1 ppm of the 360 MHz ^{1}H NMR spectrum obtained after resolution enhancement with a phase-shifted sine bell (16). Resonances of the oxidized protein are indicated by circles, those from the reduced protein by triangles.

dized and reduced protein.

The saturation transfer difference spectrum of Fig. 4 was obtained with the following pulse sequence:

$$\left[\omega^A - \text{Obs.} - T_{ac}^A - T_d - \omega^{off-res.} - \text{Obs.} - T_{ac}^{off-res.} - T_d \right]_n \quad [6]$$

The observation pulse was preceded by a selective saturation irradiation on resonance A, or off-resonance, respectively. The delay
time T_d was sufficiently long for the system to recover to equilibrium spin populations before the following experiment was
started.

2.4. Nuclear Overhauser Enhancement (NOE) and Spin Diffusion

The nuclear Overhauser effect (NOE) is the fractional change in
intensity of one NMR resonance when another resonance is irradiated, and has long been a valuable tool for structural studies of
small molecules (9,24). NOE experiments are of interest for studies of spatial molecular structures since under suitable conditions NOE's are related to the inverse sixth power of the nonbonding distance between irradiated and observed nuclei. Theoretical considerations (9,25) indicate a more favorable situation
for NOE studies of macromolecules at high fields than for the
more conventional experiments with small molecules under the conditions of extreme motional narrowing. This is because the limiting magnitude of the NOE in macromolecules is larger by a factor
2 and the NOE's are almost completely determined by intramolecular
dipole-dipole interactions. On the other hand spin diffusion is
of considerable importance in proteins, causing the NOE's to be
less specific and hence less useful (26-28). The adverse effect
of spin diffusion is readily seen in Fig. 5. Following selective
preirradiation of a well separated one-proton line at -1.9 ppm
(Fig. 5B) in the spectrum of reduced horse cytochrome c, the conventional steady state NOE difference spectrum contains a large
number of lines. Preirradiation of a different line at -3.7 ppm
yielded a nearly identical spectrum (Fig. 5C). Hence in these
spectra the distance information contained in the NOE's is partially masked by spin diffusion.

In contrast to the conventional steady state NOE's which have
successfully been used mainly by organic chemists for work with
small molecules (24), the initial build-up rates of NOE's are,
even in the presence of spin diffusion, simply related to the
inverse sixth power of the distance between observed and preirradiated nuclei (9,27-29). In the following we describe two
techniques for measurements of initial NOE build-up rates which

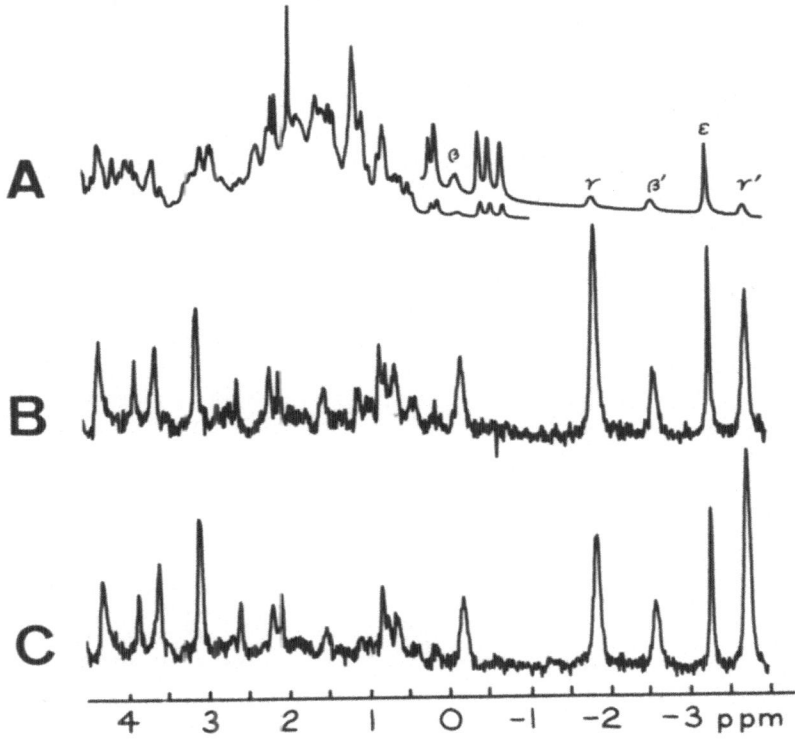

Figure 5. 360 MHz ^{1}H NMR spectra of horse ferrocytochrome c, protein concentration 0.008 M, p^{2}H 6.8, T =
49° C. A. Normal Fourier transform spectrum between -4
and 4.5 ppm. The upfield region from -4 to 0 ppm is represented with a 4-fold expanded vertical scale. The
locations of the resonances of the β-CH$_2$, γ-CH$_2$ and
ε-CH$_3$ protons of Met-80 are indicated. B. Steady-state
NOE difference spectrum obtained with preirradiation of
the line at -1.9 ppm. C. Steady-state NOE difference
spectrum obtained with preirradiation of the line at
-3.7 ppm. The traces B and C are represented with a
128 fold expanded vertical scale (from ref. 28).

are suitable for the determination of proton-proton distances in
macromolecules.

2.4.1. Transient NCE's. Transient NOE difference spectra obtained
with the pulse sequence [7] are a particularly straightforward
technique for measurements of NOE build-up rates (28).

$$\left[180^{\circ}(\omega^A)-t_1-\text{Obs.}-T_{ac}^A-T_d-180^{\circ}(\omega^{\text{off-res.}})-t_1-\text{Obs.}-T_{ac}^{\text{off-res.}}-T_d\right]_n \quad [7]$$

This experiment is initiated by a selective short pulse on resonance A. For good sensitivity a 180° pulse obtained with a pulse
length of typically 10 nsec would be optimal. The observation
pulse follows after a waiting time t_1 during which the NOE's are
built up in the absence of a radio-frequency field. The reference
spectrum without NOE, which is obtained by preirradiation in an
empty region of the spectrum, is recorded after a delay time T_d
during which the spin system is allowed to recover.

Fig. 6 shows a series of transient NOE difference spectra for reduced cytochrome c which were recorded with different waiting
times t_1. The selective preirradiation was on the same line as in
the steady state NOE difference spectrum of Fig. 5B, i.e. at
-1.9 ppm. It is seen that the transient NOE's are highly selective. The preirradiated line corresponds to a γ-methylene proton
of Met-80 which is bound to one of the axial coordination sites

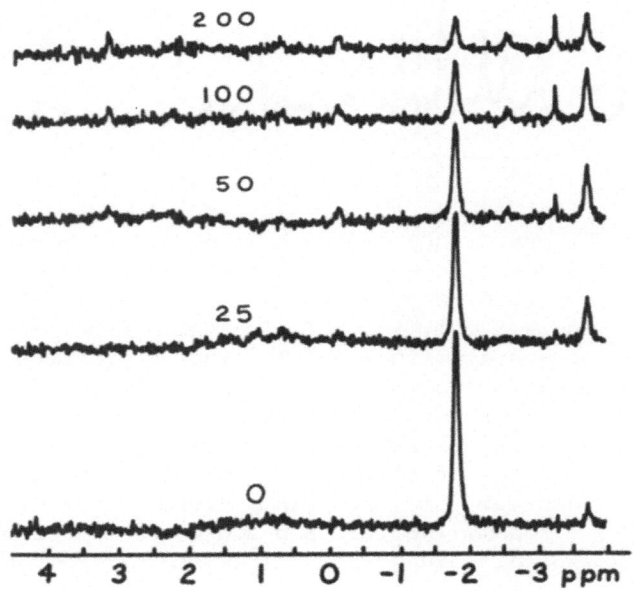

Figure 6. Transient NOE difference spectra between -4
and 4 ppm of the ferrocytochrome c solution of Fig. 5
obtained with the pulse sequence [7], where the pulse
ω^A was on the γ-methylene proton line of Met-80 at
-1.9 ppm. The numbers indicate the waiting time t_1 in
ms (from ref. 28).

of the heme iron (30) (Fig. 7). Transient NOE's are observed for
all the protons of Met-80. In the difference spectra recorded at
different times t_1 after the 180° pulse, the intensity of the
pulsed line decreases due to relaxation, while the intensities
of other lines build up by spin diffusion. At $t_1=0$, the line at
-3.7 ppm, which corresponds to the geminal γ-proton (Fig. 7), has
already emerged because of spin diffusion during the 15 ms of the
pulse duration. The intensity of this line reaches a maximum between 50 and 100 ms after the pulse and then decreases again as
relaxation becomes dominant. With a time lag of between 25 and
50 ms, four lines appear at -3.3, -2.7, -0.2 and 3.1 ppm, which
correspond respectively to the ε-methyl group, the two β-methylene
protons and the α-proton of Met-80 (Fig. 7). The time evolution
of the transient NOE's thus delineates the pathway along which
spin diffusion occurs in the side chain of Met-80. In contrast
to the steady-state NOE experiment of Fig. 5, the two transient
NOE experiments with preirradiation of the γ-methylene protons
at -1.9 ppm or -3.7 ppm, respectively, give different results,
thus manifesting the specific spatial arrangement of the Met-80
side chain (Fig. 7) in the protein structure (28).

Transient NOE's are particularly useful when the spectra contain
well separated lines, which is rather typical for hemoproteins.
On the other hand, in crowded spectral regions a reasonable compromise between high selectivity of the presaturation pulse and
a workable signal-to-noise ratio in the transient NOE spectra is
generally difficult to obtain. For such situations the following
experiment was found to be more useful in practice.

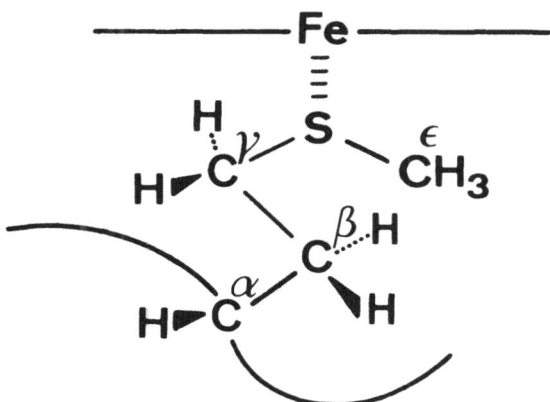

Figure 7. Structure of the axial heme iron ligand
Met-80.

2.4.2. Truncated driven NOE's (TOE). TOE difference spectra are
obtained with the pulse sequence [8].

$$\left[t_1(\omega^A) - \text{Obs.} - T_{ac}^A - T_d - t_1(\omega^{\text{off-res.}}) - \text{Obs.} - T_{ac}^{\text{off-res.}} - T_d \right]_n \qquad [8]$$

A selective low power radio frequency field is applied to resonance A for a period of time t_1, which is followed immediately by
the observation pulse. After a waiting time T_d sufficient for the

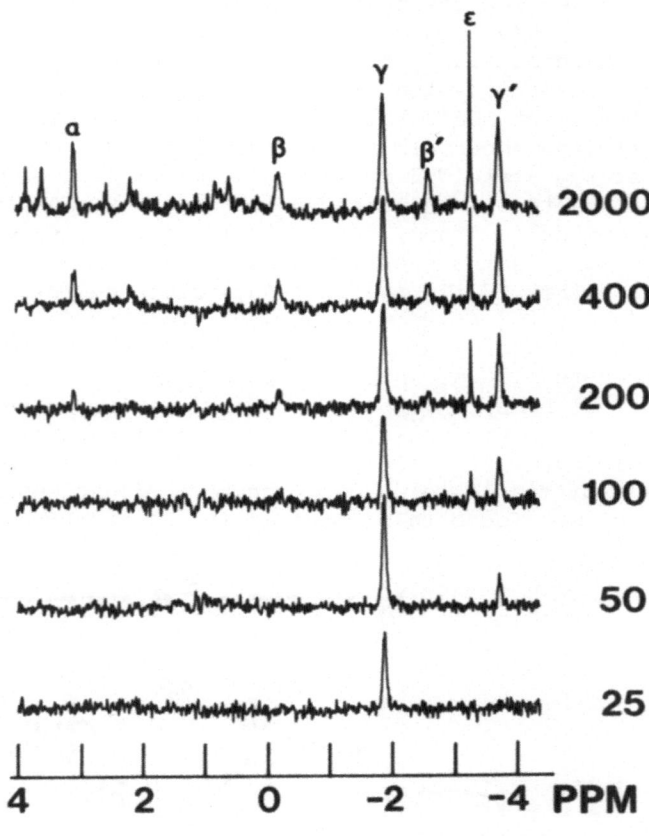

Figure 8. TOE difference spectra between -4 and 4 ppm
of ferrocytochrome c obtained with the pulse sequence
[8], where the preirradiation was on the γ-methylene
proton line of Met-80 at -1.9 ppm. The numbers on the
right indicate the irradiation times t_1 in ms. In the
top trace the α-, β-, γ- and ε-proton resonances of
Met-80 are identified.

system to recover to equilibrium, a reference spectrum without NOE is recorded. In contrast to transient NOE experiments, where no radio frequency is applied while the NOE's are built up and decay, the radio frequency field ω_A in [8] is applied over the entire time span t_1 during which NOE's in the spin system are built up. To distinguish experiments of the type [8] from transient NOE's, we use the term "radio frequency driven NOE's" or short "driven NOE's' (29). If the irradiation period t_1 in a driven NOE experiment is sufficiently long, a conventional steady state NOE difference spectrum is obtained. On the other hand, if the experiment [8] is used to record a series of "truncated driven NOE's" (TOE) with different t_1's which are all short compared to the irradiation time required for a steady state NOE, the build-up of the driven NOE's for individual resonance lines can be followed.

The use of the pulse sequence [8] is illustrated with the spectra of ferrocytochrome c in Figs. 5 and 8. The steady-state NOE difference spectra of Fig. 5 were obtained with a preirradiation time t_1 of 2 s. In Fig. 8 the build-up with time of the driven NOE's can be followed. It is seen that during the initial phase the development of the resonance intensities corresponds closely to that in the corresponding transient NOE experiment of Fig. 6. For long irradiation times the TOE difference spectra tend towards the steady state difference spectrum of Fig. 5B.

The data of both Fig. 6 and Fig. 8 are a reliable manifestation of the distances between irradiated and observed protons (28,29). Compared to transient NOE's, TOE's have important advantages in practical use. One is that the amplitude of the preirradiation field can be considerably smaller, so that more selective irradiation in crowded spectral regions may be obtained. Secondly, the sensitivity for observation of TOE's is inherently better than for observation of transient NOE's. Thirdly, from the line intensities in a single TOE difference spectrum recorded with a suitable choice of t_1, the relative distances between the preirradiated proton and several nearby protons in a macromolecule can quite reliably be estimated (29), which is considerably less time consuming than to use a transient NOE experiment for obtaining this information.

3. RECENT APPLICATIONS OF MULTIPLE PULSE EXPERIMENTS FOR STRUCTURAL STUDIES OF HEMOPROTEINS

As for other proteins (2) spin decoupling has been used extensively to identify the spin systems of aromatic rings and other amino

acid side chains in hemoproteins (31-33). Similarly the spin
systems of the heme groups were identified (Fig. 3) (34-37).
Saturation transfer studies were used to assign corresponding
resonance lines in reduced and oxidized cytochromes (23,34-37).
TOE difference spectroscopy was employed to determine nearest
neighbor relationships between the heme substituents 1-8 (Fig. 1),
which, because of the symmetry properties of the heme groups, provided individual assignments for the heme resonances (34-37).
Studies of NOE's also resulted in identifications of amino acid
side chains located near the heme (37). Overall the combined use
of the different multiple pulse experiments [5]-[8] has so far
resulted in individual assignments of the heme resonances in the
oxidized and reduced state of several cytochromes as well as the
identification of amino acids in the heme crevice (34-37). These
data present a considerable advance towards answering some of the
pertinent questions on hemoprotein structure mentioned in the
introduction. Thus the contact and pseudocontact contributions to
the hyperfine shifts (eq. [3]) can be more reliably separated on
the basis of these resonance assignments and the locations of
ring carbon atoms carrying high spin density on the heme periphery can be determined. The resulting detailed descriptions of
the electronic heme structure and the protein conformation near
the heme open new avenues to gain insights into structure-function relations and evolutionary relationships in hemoproteins.

ACKNOWLEDGEMENTS

The research projects discussed in this report were supported by
grants of the Swiss National Science Foundation (project
3.0046.76) and the ETH Zürich.

REFERENCES

(1) Wüthrich, K.: 1970, Struct. Bonding 8, pp. 53-121.
(2) Wüthrich, K.: 1976, "NMR in Biological Research: Peptides and
     Proteins", North-Holland Publ. Co., Amsterdam.
(3) Keller, R.M., Wüthrich, K., and Schejter, A.: 1977, Biochim.
     Biophys. Acta 491, pp. 409-415.
(4) Shulman, R.G., Wüthrich, K., Yamane, T., Patel, D., and
     Blumberg, W.E.: 1970, J. Mol. Biol. 53, pp. 143-157.
(5) Giessner-Prettre, C., and Pullman, B.: 1969, C. R. Acad. Sc.
     Paris 268 Série D, pp. 1115-1117.
(6) Abraham, R.J., Fell, St.C.M., and Smith, K.M.: 1977, Org.
     Magn. Res. 9, pp. 367-373.
(7) Kurland, R.J., and McGarvey, B.R.: 1970, J. Magn. Res. 2,
     pp. 286-301.

(8) McConnell, H.M.: 1956, J. Chem. Phys. 24, pp. 764-766.
(9) Solomon, I.: 1955, Phys. Rev. 99, pp. 559-565.
(10) Abragam, A.: 1961, "The Principles of Nuclear Magnetism", Oxford University Press, London.
(11) Guéron, M.: 1975, J. Magn. Res. 19, pp. 58-66.
(12) Wüthrich, K., Hochmann, J., Keller, R.M., Wagner, G., Brunori, M., and Giacometti, G.: 1975, J. Magn. Res. 19, pp. 111-113.
(13) Dwek, R.A.: 1973, "Nuclear Magnetic Resonance in Biochemistry: Applications to Enzyme Systems", Clarendon Press, Oxford.
(14) Campbell, I.D., Dobson, C.M., Williams, R.J.P., and Xavier, A.V.: 1973, J. Magn. Res. 11, pp. 172-181.
(15) De Marco, A., and Wüthrich, K.: 1976, J. Magn. Res. 24, pp. 201-204.
(16) Wagner, G., Wüthrich, K., and Tschesche, H.: 1978, Eur. J. Biochem. 86, pp. 67-76.
(17) Nagayama, K., Wüthrich, K., Bachmann, P., and Ernst, R.R.: 1977, Biochem. Biophys. Res. Commun. 78, pp. 99-105.
(18) Nagayama, K., Bachmann, P., Wüthrich, K., and Ernst, R.R.: 1978, J. Magn. Res. 31, pp. 133-148.
(19) Gibbons, W.A., Beyer, C.F., Dadok, J., Sprecher, R.F., and Wyssbrod, H.R.: 1975, Biochemistry 14, pp. 420-429.
(20) De Marco, A., Tschesche, H., Wagner, G., and Wüthrich, K.: 1977, Biophys. Struct. Mech. 3, pp. 303-315.
(21) Richarz, R., and Wüthrich, K.: 1978, J. Magn. Res. 30, pp. 147-150.
(22) Janson, T.R., and Katz, J.J.: 1972, J. Magn. Res. 6, pp. 209-220.
(23) Redfield, A.G., and Gupta, R.K.: 1971, Cold Spring Harbor Symp. Quant. Biol. 36, pp. 405-411.
(24) Noggle, J.H., and Schirmer, R.E.: 1971, "The Nuclear Overhauser Effect", Academic Press, New York.
(25) Balaram, P., Bothner-By, A.A., and Dadok, J.: 1972, J. Am. Chem. Soc. 94, pp. 4015-4017.
(26) Hull, W.E., and Sykes, B.D.: 1975, J. Chem. Phys. 1975, pp. 867-880.
(27) Kalk, A., and Berendsen, H.J.C.: 1976, J. Magn. Res. 24, pp. 343-366.
(28) Gordon, S.L., and Wüthrich, K.: 1978, J. Am. Chem. Soc. 100, pp. 7094-7096.
(29) Wagner, G., and Wüthrich, K.: 1979, J. Magn. Res. 33, pp. 675-680.
(30) Wüthrich, K.: 1969, Proc. Natl. Acad. Sci. U.S. 63, pp. 1071-1078.
(31) Dobson, C.M., Moore, G.R., and Williams, R.J.P.: 1975, FEBS Lett. 51, pp. 60-65.

(32) Moore, G.R., and Williams, R.J.P.: 1975, FEBS Lett. 53, pp. 334-338.
(33) Keller, R.M., and Wüthrich, K.: 1977, Biochim. Biophys. Acta 491, pp. 416-422.
(34) Keller, R.M., and Wüthrich, K.: 1978, Biochim. Biophys. Acta 533, pp. 195-208.
(35) Keller, R.M., and Wüthrich, K.: 1978, Biochem. Biophys. Res. Commun. 83, pp. 1132-1139.
(36) Keller, R.M., Picot, D., and Wüthrich, K.: 1979, Biochim. Biophys. Acta, in press.
(37) Keller, R.M., and Wüthrich, K.: 1979, Biochim. Biophys. Acta, submitted.

NUCLEAR MAGNETIC RESONANCE STUDIES OF PARAMAGNETIC PROTEINS

R. G. Shulman

Bell Laboratories
Murray Hill, New Jersey 07974

It is of considerable interest that in hemoglobin spectroscopic studies such as NMR represent
the forefront of our knowledge. With the crystal
structures having been determined by x-ray diffraction
the interest has shifted to what we can learn about
the energetics and the dynamics of the function. The
role of NMR in probing these functional aspects of
oxygen binding is discussed. Since much of the information about hemoglobin comes from studying the hyperfine shifted resonances, these are discussed in some
detail. In particular it is shown how contact interactions and anisotropic interactions with the unpaired
spin affect the NMR peaks.

1. INTRODUCTION

The interaction of unpaired electrons with
nuclear spins affects the nuclear magnetic resonance
in many ways. In the decade between 1955 and 1965
most of the present mechanisms and types of experiments were performed, often on non-biological materials,
and the understanding, so obtained was then applied to
paramagnetic metal ions in proteins, such as the heme
proteins.

In the earliest papers on NMR by Bloembergen,
Purcell and Pound[1] and by Block, Hansen and Packard
[2], it was shown how paramagnetic metal ions would
shorten the nuclear spin lattice relaxation times, T_1,
in solutions. This was extended to the measurement of
relaxation enhancement, where the metal ion was bound
to nucleic acids[3] in 1961, and was subsequently

135

*I. Bertini and R.S. Drago (eds.), ESR and NMR of Paramagnetic Species in Biological and
Related Systems. 419–422.*
Copyright © 1979 by D. Reidel Publishing Company

extended to proteins[4].

The first observation of NMR shifts in solids or molecules coming from transferred hyperfine interaction was made in 1956 on the ^{19}F shifts in paramagnetic MnF_2[5]. These and subsequent experiments were interpreted in terms of the unpaired spin being in a molecular orbital for which the NMR shifts defined the degree of covalency[6]. The first report of large paramagnetic NMR shifts in proteins appeared in 1965[7] and this was followed by intensive investigations of hemoglobin[8], and myoglobin[10]. In this paper the basic physical mechanisms responsible for the NMR shifts in paramagnetic molecules are reviewed. There is a book which contains a very comprehensive and authoritative discussions of these effects[11].

The magnetic moment of the electron spin is about three orders or magnitude larger than the magnetic moment of the proton. Magnetic dipole fields emanating from electrons are proportionately larger. Paramagnetic compounds contain electrons whose large magnetic fields can influence the NMR spectra of nearby nuclei by shifting the resonances or shortening the relaxation times T_1 and T_2. Under some conditions the nuclear resonances are broadened or shifted beyond detectability, while nuclei far enough away from the paramagnetic electrons may not be noticeably affected. However in crystals and molecules of paramagnetic materials many NMR lines have been observed which are shifted or broadened but still detectable. A considerable amount of attention has been paid to the observation and interpretation of NMR lines in paramagnetic compounds since it was shown that they can be interpreted to give information about the distribution of unpaired electrons[6].

In general the interactions between unpaired electron and nucleus can be divided into dipolar and hyperfine contributions. The magnetic dipole of the electron generates magnetic fields, which for isotropic dipoles like the free electron spin, have projections along the direction of the external field H_o given by

$$H_{dip} = \frac{g\beta S}{r^3} (1 - 3\cos^2\theta) \qquad (1)$$

where θ is the polar angle between, r, the radius from electron to nucleus and H_o, g is the electronic g-gactor which might be anisotropic, β the Bohr magneton and S the electron spin. Since the dipolar field has an r^{-3} dependence it is a function of distance and does not depend upon the nature of the molecular bonds in

the region between electron and nucleus. Substituting
illustrative values of $g\beta S = 0.927 \times 10^{-20}$ ergs/gauss,
for the free electron spin, and setting $(3 \cos^2\theta - 1)$
~ 1 we see that H_{dip} is $\sim 10^4$ Oersteds at $r = 1$ Å and
~ 10 Oersteds at $r = 10$ Å.

Hyperfine interactions come about between a nucleus
and an unpaired electron in the atomic orbitals associated with that nucleus. Since in this review we are
most concerned with hyperfine interactions with hydrogen
nuclei, or protons, hyperfine interactions can be considered to arise from unpaired electrons, or spin
density, in the hydrogenic 1s orbital. Because of the
spherical symmetry of the 1s orbital this hyperfine
interaction is spherically symmetrical. Because it
arises from the finite probability that the electron
and the nucleus occupy the same volume element in space,
it is sometimes refered to as a contact interaction.
The strength of this interaction for a single unpaired
electron has been measured by the hyperfine interaction
of atomic hydrogen and in terms of magnetic field at
the proton it is

$$H_{hf}^{(1s)} = 3.3 \times 10^5 \text{ Oersteds.} \qquad (2)$$

Considering that protons are generally several angstroms
away from the unpaired spin it is clear that a small
amount of spin, introduced into the hydrogenic 1s
orbital has a large effect upon the proton, compared
to electron-nuclear dipolar interactions.

Another measure of the strength of internal
magnetic fields is the nuclear-nuclear interaction,
which are usually the dominant interactions in diamagnetic compounds, and of course will also be found in
paramagentic compounds. Taking the two protons in
H_2O as a case of very close approach the dipolar field
of one at the other, neglecting corrections of a factor
of two, is

$$H_{dip}(\text{nuclear}) \sim 10 \text{ Oersteds}$$

which is very small compared to the electron dipolar
fields because of the smaller magnetic moment of the
nucleus.

2. PARAMAGNETIC SHIFTS

These internal magnetic fields which can exist at nuclei in paramagnetic compounds have large instantaneous values. In the absence of rapid motions which average these fields the nuclear resonances can be shifted very far indeed. For example in the antiferromagnetic state of MnF_2 at 4.2°K where the electron spins are ordered the NMR of the ^{19}F was observed near 160 MHz in zero external magnetic field[12]. This large shift occurs because the electrons are ordered so that each nucleus is in the full, time independent, hyperfine field. As the temperature is raised, so that the electron spins become paramagnetic and disordered, the hyperfine fields at the nuclei are time averaged. Instead of seeing the large fields from completely aligned electrons they see nonvanishing fields which are proportional to the polarization of the electron. In the simple cases which concern us this polarization is given by

$$\langle S \rangle = g^2 \beta^2 H_o \frac{S(S=1)}{3kT} \tag{3}$$

in which $\langle S \rangle$ is the expectation value of the electron spin S, g is the electron g-factor, β the Bohr magneton, k is Boltzmann's constant and T the absolute temperature. The shifts of the resonance $\Delta \nu$ arising from the hyperfine interaction is

$$\Delta \nu = A \ I \cdot \langle S \rangle \tag{4}$$

where A is the hyperfine interaction and I is the nuclear spin. The shift is proportional to the polarization of the electron in this extreme where the electron spin is averaged among its different quantum states in the external field. For a particular spin S the degree of polarization is directly proportional to the external magnetic field and inversely proportional to the temperature, and for typical values of H_o at 300°K, $\langle S \rangle$ is of the order of 10^{-2} times as large as S.

Another averaging process which occurs in solution is that the molecular rotation is usually sufficiently rapid to present an effective average of the anisotropic electron-nuclear interaction. In this event only the isotropic component of the hyperfine interaction will

be completely non-vanishing. This gives rise to the
contact shifts of the resonances. In the case of
isotropic electronic g-factors the contact term will
be the only contribution by the electron to the shift
of the nuclear resonance. This is seen from Eq.(1)
where the average over a sphere which describes the
positions of a rotating molecule gives $\langle\cos^2\theta\rangle = \frac{1}{3}$ so
that the dipolar contribution vanishes. However
if g is anisotropic the average over all values of the
angular variables does not vanish. This non-vanishing
term has been called the pseudo-contact contribution.
(13,14). Its physical origin can be illustrated by
the example of a complex which at first we consider to
have cubic symmetry. The averaged dipolar field at any
point can be considered in terms of the product of the
probability that H_o has a particular orientation (value
of θ) and the field at that point for the particular
value of θ. The most probable value of θ is 90° for
which $(3\cos^2\theta - 1) = -1$. The larger probability of
finding H_o at values of θ near 90° rather than near 0°
is compensated for by the fact that at $\theta = 0°$ the value
of $3\cos^2\theta - 1$ is twice as large and positive rather
than negative. When the g-value in and perpendicular
to the plane are the same, the average value of $3\cos^2\theta$
$- 1$ is zero. If we consider the tetragonal case we see
that if $g_\perp > g_{||}$ then the average field at a point on
the axis $^\perp$ is $^{||}$negative while for those points near
the plane the average field is positive.

Both contact and pseudocontact shifts are propor-
tional to $\langle S\rangle$ which means directly proportional to the
polarizing field H_o and inversely porportional to T the
absolute temperature.

We shall first discuss in detail the contact
shifts which are caused by the unpaired electron spin
being delocalized into the 1s atomic orbitals of the
proton measured. Although non-vanishing overlap of
this atomic orbital with orbitals· containing the electron spin will introduce spin density this mechanism
can usually be neglected compared to the transfer of
spins through covalent bonds. Furthermore, theory and
experiment have shown that the attenuation of spin
density through σ-bonds is usually much more rapid than
through π-bonds. It should be recalled that σ-bonds
include electrons which lie along the internuclear axis
and are symmetrical about it while in π-bonds the
electrons are in atomic orbitals which are perpendicular
to the internuclear axis. In planar molecules, such
are porphyrin, the σ-bonds lie in the plane and are

symmetric with respect to reflection in the plane and
the π-bonds are above and below the plane and are antisymmetric with respect to reflection. In contrast to
the attenuation of spin density in σ-bonds, spin density
in the π orbitals of porphyrin will be delocalized
through the entire conjugated system, shown by the
series of alternating single and double bonds for the
naturally occurring iron protoporphyrin IX in Fig. 1.
It is clear from this diagram that there are two kinds
of hydrogens attached to the ring which might get some
spin density from the π-electron system but that special
mechanisms for transferring the spin across the last
bond or two to the hydrogenic orbitals will be required.

    The first kinds are the meso protons at the four
methine bridge positions which lie in the plans, and
require that the spin be transferred from the π-orbital
of the methine carbon to the hydrogen 1s orbital, a
process which is forbidden by symmetry from going
through covalent bonds. However by including different
configurations it has been possible to describe how a
spin in the carbon π orbital polarizes the electrons
in the hydrogen 1s orbital and thereby produces a
contact shift of the proton resonance. One is usually
interested in the spin density in the carbon π orbital
designated as ρ_c^π. It is particularly fortunate that
the observed contact interaction A is proportional to
ρ_c^π, i.e.

$$A = Q_{CH} \rho_c^\pi \tag{5}$$

where the constant Q_{CH} for the -CH fragment has been
determined experimentally to be -6.3×10^7 Hz.[14] The
value of A can be obtained from the measured value of
$\Delta\nu_c$ via the explicit form of Eq. (4) which is

$$\Delta\nu_c = A \frac{\gamma_e}{\gamma_H} S(S+1) \frac{\nu}{3kT} \tag{6}$$

where γ_e and γ_H are the gyromagnetic ratios of electron
and proton and ν is the nuclear Larmour frequency.

    The second kind of protons are in methyl or
methylene groups attached to the porphyrin ring at one
of the eight β positions. Spin density reaches these

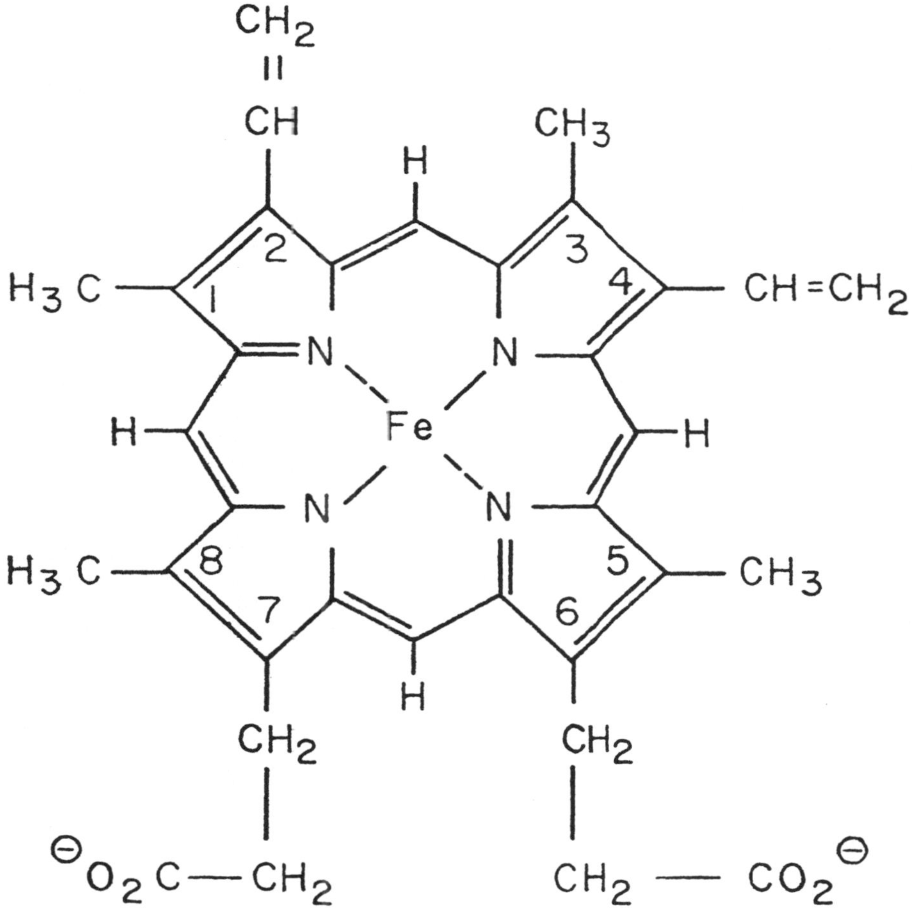

Figure 1.- Iron protoporphyrin IX.

protons by hyperconjugation which means that the π electrons are delocalized into the CH bonds. In this system the hyperfine interaction has been shown to depend upon the angle ϕ between the –CH bond and the normal to the plane of the ring by the relation

$$A = (B_0 + B_2 \cos^2\phi)\rho_c^{\pi} \qquad (7)$$

where the numerical values of $B_0 = 9\times10^6$ Hz and $B_2 = 122\times10^6$ Hz. For the methyl group, which rotates rapidly substituting the average value of $\cos^2\phi = 1/2$ we obtain

$$A = Q_{CCH_3}\rho_c^{\pi} \qquad (8)$$

where $Q_{CCH_3} \approx +70\times10^6$ Hz. The positive sign means that the spin density at the proton has the same direction of polarization as the spin in the carbon π orbital. This is to be contrasted with the negative sign of Q in the corresponding expression for the CH fragment (Eq. 5). However in contrast to the conveniently constant value of Q_{CH} the value of Q_{CCH_3} has been observed to vary by as much as 20% although the value quoted above has been observed most often. Amongst the factors which affect the value of Q_{CCH_3} is the charge on the ring with negatively charged rings delocalizing the spin onto the methyl groups more readily than positively charged rings. However in other cases the cause of the variation of Q_{CCH_3} has not been identified.

As for the other substituents on the porphyrin it is clear that the CH and CH_2 groups of the vinyl should have hyperfine coupling similar to that of CH fragments. The methylenes of the propionates separated from the heme by an intervening methylene should, as a first approximation have no contact shifts.

The expression for the pseudocontact shifts for the hemes and heme proteins where the tumbling time is long and the three g-factors may be unequal is

$$\Delta\nu_{pc}(ppm) = CK\left\{\left[g_z^2 - \frac{1}{2}\left(g_x^2 + g_y^2\right)\right]\left(1 - 3\cos^2\theta\right) + \frac{3}{2}\left(g_y^2 - g_x^2\right)\sin^2\theta\cos 2\Psi\right\} \qquad (9)$$

where θ is the polar angle between r the radius from electron spin to nucleus and Ψ is the angle between its projection in the xy plane and the x axis[16]. In this

$$C = \frac{\beta^2 S(S+1)}{9kTr^3}$$

and K is the fraction of the spin in the iron atomic orbitals where it has anisotropic g-factors. Once the g-factors and measured by EPR experiments and the structural parameters of the compound by x-ray crystallography the value of $\Delta\nu_{pc}$ for any point can be computed quite directly.

Both contact and pseudocontact shifts depend upon the state of the iron in the heme. The iron can be either in the ferrous, Fe^{2+}, or the Ferric, Fe^{3+}, state and in myoglobin and hemoglobin both oxidation states can exist either in the low spin or high spin form depending upon the sixth ligand.

The influence of electronic spin lattice relaxation times T_{1e} upon the nuclear T_1 and T_2 values depends upon a number of parameters. Here we only comment generally that the contribution of the paramagnetic electrons to the line widths of the nuclear resonance $\frac{1}{T_2}$ is proportional to T_{1e}. Hence longer values of T_{1e} tend to give broader NMR line widths. The electronic relaxation times T_{1e} have not been determined directly at room temperature for the hemes although from their effects upon the linewidths of the heme proton resonances it seems that the high spin ferric form has a much longer value of T_{1e} than the low spin ferric form.

In heme compounds the observed resonance frequency ν_{obs} can be written as

$$\nu_{obs} = \nu_{diam} + \Delta\nu_{rc} + \Delta\nu_{pc} + \Delta\nu_c \qquad (10)$$

where $\Delta\nu_{rc}$ is the ring current shift introduced by the porphyrin ring, ν_{diam} is the resonance position that the proton would have in the absence of ring current shifts and paramagnetic shifts described by $\Delta\nu_{pc}$ and $\Delta\nu_c$. In this paper our convention is that shifts to

lower fields are given negative signs and expressed
in their equivalent frequency units.   Resonance
positions are given with respect to an internal standard
of DSS (2,2-dimethyl-2-silapentane-5-sulfonate), usually
in terms of parts per million (ppm) but sometimes in
frequency units, and positions downfield from DSS are
given negative signs.

Early attempts to fit the ring current fields of
porphyrins used a Johnson-Bovey type formula.   We have
derived an empirical descriptions of the ring current
shifts[9].   We have taken the reported ring current
shifts of porphyrin and phthalocyanine derivatives
where the structure was indicated by the chemical
formulas and used them to fit a truncated dipolar
expansion with the results shown in Fig. 2.   Along the
axis this agrees very well with the other descriptions
of the ring current.   In the plane the ring currents
can be estimated by referring the resonances of the
heme substituents to analogous compounds which do not
have the porphyrin ring currents.

3. HEMES

Porphin is the simplest porphyrin with protons at
the eight β positions on the pyrrole rings labelled
a(H) in the figure and protons at the four meso positions, labelled b in Figure 3.   These have relative
intensities of two to one and in accord with the fourfold symmetry of porphin their NMR peaks are expected
to be split.   Figure 3 compares the high resolution
220 MHz spectrum of diamagnetic Zn^{2+} Porphin and paramagnetic Fe^{3+} Porphin dissolved in a d_5 pyridine, D_2O
solution (3:1) with an excess of cyanide.   The meso and
β protons are unsplit and can be identified by their
intensities.   However in the Fe^{3+} Porphin they are both
shifted to higher fields, the β protons being shifted
almost three times as far as the meso protons.   In these
spectra one can see spinning side bands of the strong
solvent resonances which are of no interest.

The spectral differences between the diamagnetic
and paramagnetic compound suggest that the shifts come
from interactions with the paramagnetic electrons.
Support for this interpretation is found in the temperature dependences of the positions of these resonances
which have been measured.   Equations 6 and 9 predict an
inverse temperature dependence for paramagnetic shifts
and unpublished experiments showed this to be approximately true for these NMR peaks.   At $\frac{1}{T} \to 0$, however,
the shifts did not extrapolate to the positions where

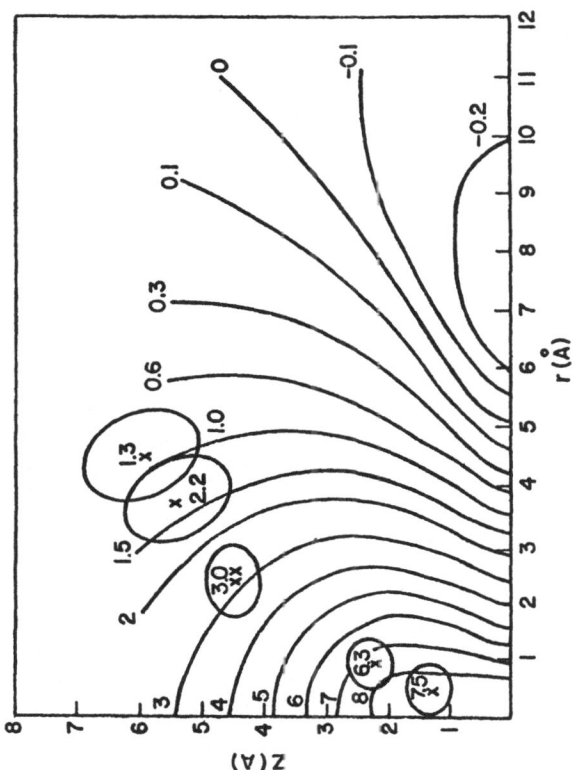

Figure 2.- A plot of the ring-current shifts measured in porphyrin and phthalocyanine complexes where the positions can be deduced from the chemical formulas. The estimated uncertainties in the positions are indicated by the ellipses and are larger than the rrrors obtained in deriving the ring-current shifts from measured positions. The coordinates Z and r refer to the distances perpendicular to the plane of the ring and outward in the plane of the ring, respectively. The lines show the contour of equal ring-current shift.

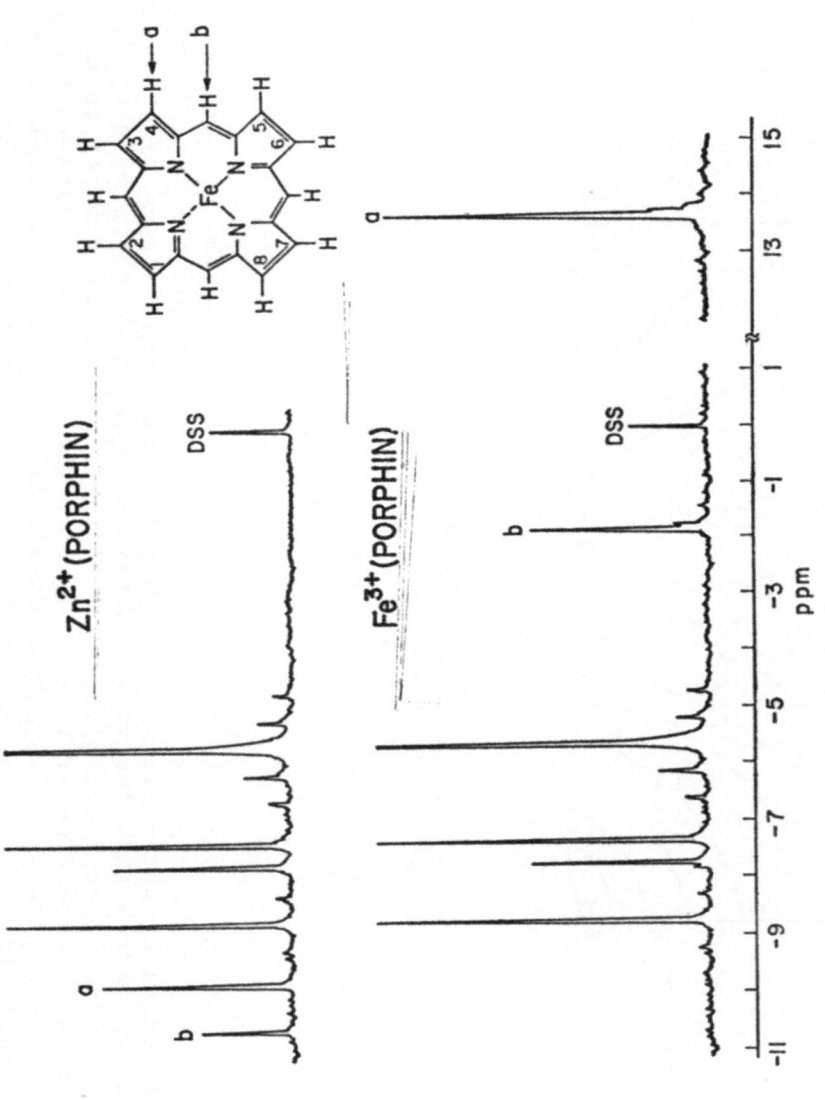

Figure 3.- A comparison of the NMR spectra at 220 MHz and 46°C of Zn^{2+} Porphin and Fe^{3+} Porphin in 75% d_5 pyridine-25% D_2O solutions.

they were observed in the corresponding diamagnetic
compound. Nonetheless it is clear that they are
paramagnetically shifted lines. These kinds of shifts
have been used to follow the states of myoglobin (9)
and hemoglobin (8,10).

These paramagnetically shifted resonances were
used (8,10) to distinguish between two models of oxygen
binding to hemoglobin. It was shown that the two state
model, originally proposed by Monod, Wyman and Changeux
(17) satisfactorily described the ligand binding. In
this model the ligand affinity is changed by the
quaternary structure switching from a low affinity (T)
structure to a high affinity (R) structure. By using
hemoglobin hybrids of the form $(\alpha^{+3}CN\beta^{+2})_2$ one could
follow very clearly the hyperfine shifted resonances
of the $\alpha^{+3}CN$ subunits (the β^{+2} peaks were broader and
not quite as indicative). In contrast to fully ligated
hemoglobin for which the R form was very stable, and
fully unligated hemoglobin, for which the T form was
very stable, the half ligated hybrids could be switched
from the R→T forms by the addition of organic phosphates
i.e. 2,3 disphosphoglycerate or inositol hexaphosphate.
These changes in quaternary structure were indicated
by small shifts in the well resolved paramagnetically
shifted heme peaks. Subsequent stopped flow experiments performed on similar solutions of these hybrids
showed that the CO binding rates changed from fast to
slow with the R→T switch. Furthermore the fast rate
was the same as the fourth ligand binding to hemoglobin while the slow rate was the same as that
observed when the first CO molecule binds to deoxyhemoglobin. Hence the CO binding rates, characteristic
of the affinities, depended completely upon the quaternary structure as revealed by the NMR spectra, and the
affinities were completely independent of the number of
ligands already bound to the hemoglobin molecule.

The NMR lines used to monitor these changes
mostly came from heme methyl groups which were shifted
to low fields by the delocalized unpaired electron of
the iron atom. In this way, they gave a measure of the
wave function of the electron, showing what its density
was at the various points around the heme. The small
changes in the NMR positions observed during the R
(high affinity) to T (low affinity) switch therefore
could be used not only to monitor the quaternary state,
as mentioned above, but also to follow the changes in
electron distribution with quaternary state. Furthermore, by comparison with model systems it was possible
to convert these NMR shifts into energy differences in

the iron-ligand bond. It was shown that the energy
differences of the ligated heme between the R and T
states were negligible compared to the 3 kcalories
differences previously reported for the binding
energy differences between high and low affinity
hemoglobin (10,18).

        Although this showed that the ligated state
did not change significantly the energy of the iron-
ligand bonds during the R→T switch it did not enable
us to say anything about the unligated or deoxy state.

        Recently a new technique for studying metal
ligand distances in metalloproteins has been developed, i.e. X-ray absorption spectroscopy using
synchrotron radiation (19), and we have used this
technique to answer these questions about binding
energies of deoxy hemoglobin in the high and low
affinity forms (19).

        The details of the experiment and the unique
aspects of the data analysis have been published
recently in the report of this work (19). Here I
shall present briefly the results and discuss them in
terms of the questions about hemoglobin stated above.

        We conclude from the EXAFS experiments that
to within the accuracy of ± 0.01Å there are no differences in the Fe-Np distances in three deoxy compounds studied. These were the deoxygenated "picket
fence" compound synthesized and characterized by
Collman (20), normal low affinity HbA and a high
affinity mutant Hb Kempsey which can be kept in the
R state even when deoxygenated. Quite simply this
means that neither adding the globin nor switching
the quaternary state affects the Fe-Np distance,
again to within the quoted accuracy. (see Table).

Table 1 (Taken from Ref. 19)

Compound	Fe-Np Distance after correction	X-Ray Diffraction
PFO_2	1.979 ± 0.01	$1.979^{a} \pm 0.01$
HbO_2	1.986 ± 0.01	
PF	2.055 ± 0.01	$2.069^{b} \pm 0.013$
Hb	2.055 ± 0.01	
Hb Kempsey	2.053 ± 0.01	

a Collman, J.P., Gagne, R.R., Reed, C.A., Robinson, W.T., Rodley, G.A.: Proc. Nat. Acad. Sci. U S 71, 1326-1329 (1974)

b Jameson, G.B., Robinson, W.T.: Private Communication.

One can estimate how much strain energy could be localized in these bonds if the strain itself is less than 0.01Å. Using a simple Morse potential, and evaluating the force constant from the symmetric stretching frequency we calculate that the strain energy is ~10^{-5} eV for 0.01Å displacement. Since the binding energy difference is 3 kcal/mole between the R and T, which corresponds to ~0.1 eV, we see that the upper limit of the energy which could be stored in the iron bonds in the deoxy state is completely negligible. Since the NMR measurements had previously shown (18,21) a similar lack of localized strain energy in the ligated form we can consider that the answer is now complete, and that the energy responsible for the change in ligand affinity during the T→R switch does not appear at the iron bonds to its neighbours.

There is a small change in the Fe-Np distance upon loss of oxygen i.e. 1.986→2.055Å. The Hoard-Perutz (22) model had predicted that the latter distance would be ~2.18Å and the structure was interpreted as showing that in deoxy hemoglobin the iron was found 0.8Å out of the plane by this long bond. Now we see that this bond is only 2.055Å and we have discussed the energetics of this proposal showing that the Fe is not forced out of the plane by this bond (19), although it may be slightly out of the plane. Our estimate was ~0.2Å. In this way we have shown that the knowledge about ligation is not transmitted to the globin by a forced iron motion coming from its long bonds. Hence, in addition to the static strain energy not being localized at the iron, these results have been used to show that the pathway is not localized in a forced iron motion in and out of the plane. As we suggested from preliminary results long ago (21), the energies in hemoglobin responsible for the cooperative oxygen binding are not localized at the heme.

REFERENCES

(1) Bloembergen, N., Purcell, E.M. and Pound, R.V. Phys. Rev. 73, 679 (1948)

(2) Bloch, F., Phys. Rev. 70, 460 (1946)
(3) Eisinger, J., Shulman, R.G. and Blumberg, W.E.
     Nature 192, 963 (1961)
(4) Cohn, M. and Leigh , J., Nature 193, 1037 (1962)
(5) Shulman, R.G. and Jaccarino, V., Phys. Rev. 103,
     1126 (1956)
(6) Sugano, S. and Shulman, R.G., Phys. Rev. 130, 517
     (1963)
(7) Kowalsky, A., J. Biol. Chem. 237, 1807 (1962)
(8) Ogawa, S. and Shulman, R.G., Biochem. Biophys.
     Res. Commun. 42, 9 (1971)
(9) Shulman, R.G., Wuthrich, K., Yamane, T., Patel,
     D.J. and Blumberg, W.E., J. Mol. Biol. 53, 143
     (1970)
(10) Shulman, R.G., Hopfield, J.J. and Ogawa, S.
     Quart. Rev. Biophysics, 8, 325 (1975)
(11) "NMR of Paramagnetic Molecules" Ed. by G.N. LaMar,
     W.D.W. Horrocks, Jr. and R.H. Holm, Academic Press
     NY and London 1973
(12) Jaccarino, V. and Shulman, R.G., Phys. Rev. 107,
     1196 (1957)
(13) McConnell, H.M. and Robertson, R.E., J. Chem.
     Phys. 29, 1361 (1958)
(14) McConnell, H.M., J. Chem. Phys. 24, 764 (1956);
     Bersohn, R., J. Chem. Phys. 24, 1066 (1956)
(15) Bolton, J.R., Carrington, A., and McLachlan, A.D.
     Mol. Phys. 5, 31 (1962)
(16) See ref 11 Chapter 1 by J.P. Jesson
(17) Monod, J., Wyman, J. and Changeux, J.P., J. Mol.
     Biol. 12, 88 (1965)
(18) Ogawa, S. and Shulman, R.G., J. Mol. Biol. 70,
     315 (1972)
(19) Eisenberger, P.M., Shulman, R.G., Kincaid, B.M.,
     Brown, G.S., Ogawa, S., Nature 274, 30-34 (1978)
(20) Collman, J.P., Accounts Chem. Res. 10, 265-272
     (1977)
(21) Shulman, R.G., Ogawa, S., Wuthrich, K., Yamane, T.,
     Peisach, J., Blumberg, W.E., Science 165, 251-257
     (1969)
(22) Perutz, M.F., Nature 237, 495-499 (1970)

PARAMAGNETIC IONS AS RELAXATION PROBES IN BIOLOGICAL SYSTEMS

Dennis R. Burton

Department of Physical Chemistry 2, University of
Lund, Lund, Sweden

The theory of nuclear relaxation in the presence of paramagnetic ions bound to macromolecules is considered in terms of a
simplified Solomon-Bloembergen-Morgan (SBM) approach. Some of
the difficulties encountered and how these can best be minimised
are discussed. The application of paramagnetic ions as relaxation probes in a number of currently controversial biological
problems is considered. These problems include the mechanism of
triggering of the complement cascade in the immune response, the
hydration number of the catalytic metal ion in carbonic anhydrase
and the identification of the catalytic metal ion and nature of
substrate binding in liver alcohol dehydrogenase.

1. INTRODUCTION

Paramagnetic ions by virtue of their unpaired electrons
possess a large electronic magnetic moment − the magnetic moment
of the electron is about 10^3 greater than that for most nuclei.
A nucleus in the vicinity of a paramagnetic ion will thus experience a large local field which will fluctuate due to electron
spin relaxation and molecular motions. This fluctuating field
will generally serve to enhance the magnetic relaxation rates of
the nucleus, giving rise to the phenomenon of paramagnetic relaxation enhancement. The magnitude of this enhancement can in principle be analysed to give the distance of the nucleus from the
paramagnetic ion and information on molecular dynamics.

In biological systems, interest generally centers around
paramagnetic ions bound to macromolecules. The paramagnetic ion
may occur naturally in association with the macromolecule e.g.

151

*I. Bertini and R.S. Drago (eds.) ESR and NMR of Paramagnetic Species in Biological and
Related Systems. 419–422.*
Copyright © 1979 by D. Reidel Publishing Company

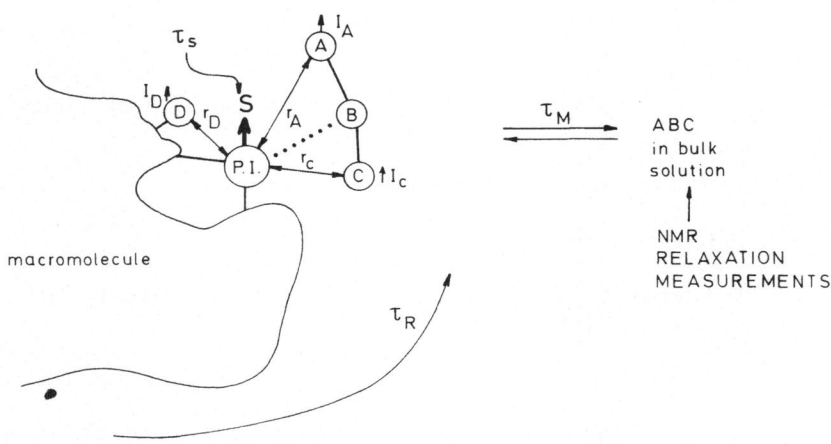

Figure 1. A schematic view of a paramagnetic ion used
as a relaxation probe in a macromolecular system. P.I.
is the paramagnetic ion with electron spin quantum
number S and electron spin longitudinal relaxation
time τ_S. The nucleus A for example has a nuclear spin
quantum number I_A and is at a distance r_A from the
paramagnetic ion. For further details see text.

Fe(III) in haemoglobin; a naturally occurring diamagnetic ion
e.g. Zn(II) or Mg(II) may be replaced by a suitable paramagnetic
ion e.g. Co(II) or Mn(II) respectively; or the paramagnetic ion
may be introduced only as a relaxation probe without biological
function e.g. Gd(III) in antibodies (see later).

A scheme showing the use of a paramagnetic ion bound to a
macromolecule as a relaxation probe is presented in Fig. 1. One
possibility is that the paramagnetic relaxation enhancement for
a nucleus on the macromolecule is observed (D in Fig 1) and analysed to yield the ion-nucleus distance (r_D). Several nuclei-
typically protons on different aminoacid residues of a protein
could be examined in this way. Combination of such studies with
the use of paramagnetic ions as shift probes which yield both
distance and angular information, could then be used to "map out"
the atoms surrounding the ion. The ultimate goal is the quantitative determination of macromolecules in solution and the

reader is referred to the extensive investigation of lysosyme (1) for detailed information. Another example of the relaxation enhancement of nuclei on a macromolecule by a paramagnetic ion is studies of the binding of Mn^{2+} (compare Mg^{2+}) to transfer RNA by ^{31}P NMR (2). The binding affinities of Mn^{2+} to the various phosphate groups of tRNA is reflected in the different paramagnetic enhancement effects observed for the ^{31}P nuclei of the phosphates. Studies of the type described above of course require that the resonance of the nucleus to be investigated can be resolved. For the most used nuclei ^1H and ^{13}C, this will generally only be possible for relatively small macromolecules (<15.000).

For larger macromolecules a useful approach is to observe the nuclei of a small molecule (ABC in Fig 1) in interaction with the macromolecule. The small molecule bound to the macromolecule is not observed directly - the problems of resolution referred to above would then be encountered. Generally the resonances of the small molecule in the bulk solution ("free sites") are observed, information on the bound site being carried over to the free site by exchange. This exchange process has been described as the "biochemical amplifier" (3). Clearly for this approach to work, exchange must be sufficiently rapid for the transfer of information to take place on the time-scale of nmr observation. We refer then to fast, intermediate and slow exchange, with the latter precluding the type of approach described.

In this article, we shall concentrate solely on the relaxation effects of paramagnetic ions on small molecules in interaction with a macromolecule. What type of information can this approach yield? The answer depends strongly on the nature of the small molecule. In the simplest case with many assumptions which will become apparent later, we can write for the longitudinal relaxation rate of say A on ABC in Fig 1:

$$R_A = \frac{Cn}{r_A^6} \cdot f(\tau_c, \omega_I) \tag{1}$$

where C is a constant related to the strength of the electron-nucleus dipolar ($S-I_A$) interaction, n is the number of A nuclei at a distance r_A from the paramagnetic ion and $f(\tau_c, \omega_I)$ is a complex function of the correlation time, τ_c, for the dipolar interaction and also (possibly) the nuclear Larmor frequency, ω_I. Consider now ABC as a substrate molecule to a metalloenzyme where we can replace the normal metal ion with a paramagnetic ion with at least some retention of activity. R_A is measured, C can generally be calculated, n is usually known (typically 1), ω is known and τ_c can possibly be determined from the frequency dependence of R_A or from PRE (see later). Hence r_A can be determined - this is the basis of the use of paramagnetic ion

relaxation probes as "the spectroscopic molecular ruler" (4).
Observation of several nuclei on a substrate molecule permits
"substrate mapping". An important feature here is the sixth
power dependence of r making it relatively insensitive to
errors in the other parameters of eqn. (1).

ABC may also be a water molecule in which case relaxation
measurements are carried out on solvent water protons, a technique referred to as Proton Relaxation Enhancement (PRE)
(reviewed (5)). The exchanging water molecule(s) is generally
assumed to be directly coordinated to the metal ion so that r
can be estimated and the unknowns in eqn. (1) now are τ_c and n/2,
the number of coordinated exchangeable water molecules. The
latter can be of vital interest as regards enzyme function e.g.
carbonic anhydrase considered later. τ_c is given by:

$$\tau_c^{-1} = \tau_S^{-1} + \tau_R^{-1} + \tau_M^{-1} \tag{2}$$

where τ_S is the electron spin relaxation time for the paramagnetic
ion, τ_R the rotational correlation time for the ABC(H_2O)-macro-
molecule complex and τ_M the lifetime of ABC(H_2O) in the bound
state. The three correlation times simply reflect three processes which cause the field at the nucleus due to the paramagnetic
ion to fluctuate. PRE is often used to attempt to determine
these correlation times for various purposes. For instance τ_R
may reveal macromolecular flexibility and if PRE shows that
$\tau_c = \tau_S$ for a metal ion with water and substrate ligands then
the PRE τ_c may be useful in substrate mapping studies. Also
since the magnitude of the PRE effect is proportional to the
amount of parmagnetic ion bound to the macromolecule, PRE can be
used as a titration indicator in metal binding studies. Of course
any molecule of type ABC could be used in this way (or indeed to
determine the details of τ_c). The advantage of PRE is the experimental ease of observing the solvent water proton signal (111M).

Having given this brief introduction to the use of paramagnetic ions as relaxation probes we shall now consider the theory
in more detail, some of the problems which may arise in applying
the theory and finally we consider the application of paramagnetic
relaxation probes in a limited number of important biological
problems.

2. THEORY

The theory of the relaxation of nuclei in the vicinity of
a paramagnetic centre is described in a set of equations generally referred to as the Solomon-Bloembergen-Morgan (SBM) equations
after the authors (6-10). The equations are complex and cumbersome
- it is possible to make some generally valid simplifications by

(a) considering the tumbling of the macromolecule-paramagnetic
ion complex as relatively slow ($\tau_R > 10^{-9}$ s), (b) considering long
τ_S ions ($\tau_S > 10^{-9}$, e.g. Mn(II), Gd(III)) separately from short τ_S
ions ($\tau_S < 10^{-10}$, e.g. Co(II)). We shall first consider the long
τ_S ions.

Theory for long τ_S ions

A simplified theory with many assumptions which are discussed to some extent below and more extensively elsewhere (5)
is presented in Scheme 1.

For substrate mapping studies, a first concern is to show
that fast exchange conditions apply to T_{1B} i.e. $T_{1M} \gg \tau_M$. This is
often taken to be the case if it is observed that $T_{2B}^{-1} \gg T_{1B}^{-1}$.
Temperature dependence studies have also been used although these
tend to be ambiguous - the best possibility being to observe the
temperature dependence of T_2 at high frequency ($\omega_I^2 \tau_c^2 \gg 1$,
intermediate exchange gives an inverse dependence of T_2^{-1} on
temperature). If fast exchange can be proved, then eqns. (3) and
(4) yield r if τ_c is known. τ_c is often determined by one of
three methods: (i) from the ratio of T_{1B}/T_{2B} at a single
frequency (compare eqns (4) and (5)) (ii) from the frequency
dependence of T_{1B} - as ω_I can appear in eqn. (4) not only explicitly but through a dependence of $\tau_c(\tau_S)$ on ω_I this can be a
difficult approach, (iii) from PRE - one assumes that at a given
field the same τ_c is valid for water protons and the nucleus
under observation.

For PRE studies, a much used approach has been to study
water proton relaxation rates, T_{1B}^{-1} and T_{2B}^{-1}, at varying temperature and frequency. From eqns. (3)-(10) one concludes that
T_{1B}^{-1}, $T_{2B}^{-1} = f(n, \tau_M, r, B, \tau_v, \tau_R, E_M, E_v, E_R, \omega, T)$ and determination of
the unkonwn parameters from variable ω- and T-studies will
require a best-fitting procedure to the T_{1B}, T_{2B} data using these
equations. The large number of parameters involved in such a
procedure can mean that it is extremely difficult to obtain a
unique set of parameters as has been pointed out (11). Furthermore the analysis requires the use of an expression for τ_S (the
Bloembergen-Morgan equation) which lacks experimental verification in a macromolecular situation.

Two methods which provide less detailed information than
the multiparameter fitting approach but which avoid the two
problems referred to above are Navon's method and the PRE/DRE
method described in Schemes 2 and 3, respectively.

Scheme 1. Simplified theory for relaxation of nucleus A in paramagnetic ion (long τ_S) · macromolecule solutions

i = 1, 2

$$\frac{1}{P_M T_{iB}} = \frac{n}{T_{iM} + \tau_M} \qquad (3)$$

$1/P_M T_{iB}$'s are the normalised paramagnetic relaxation rates which can be determined experimentally.

$P_M = $ [A bound to paramagnetic ion sites] /[total A]

N.B. Fast exchange, $T_{iM} >> \tau_M$,
$1/P_M T_{1B} = n/T_{1M}$
Slow exchange, $T_{iM} << \tau_M$,
$1/P_M T_{iB} = n/\tau_M$
Intermediate exchange, $T_{iM} \sim \tau_M$, T_{iM}'s relaxation rates in the metal ion site given by simplified Solomon-Bloembergen (SB) expressions

$$\frac{1}{T_{1M}} = \frac{C}{r^6}\left(\frac{3\tau_c}{1+\omega_I^2\tau_c^2}\right) \qquad (4)$$

$$\frac{1}{T_{2M}} = \frac{C}{2r^6}\left(4\tau_c + \frac{3\tau_c}{1+\omega_I^2\tau_c^2}\right) \qquad (5)$$

$C = 2/15 \cdot \gamma_I^2 \cdot \mu_{eff}^2$

where γ_I is the magnetogyric ratio of nucleus A and μ_{eff} the effective magnetic moment of the electron. $\mu_{eff} = g^2 S(S+1) \beta^2$ where g is the electronic g - value, S the total electron spin of the metal ion and β the Bohr magneton.

r = ion-nucleus A distance (note the sixth power dependence)

n = no. of A nuclei exchanging rapidly on the nmr time scale from the metal ion site (e.g. if observing one water molecule, n = 2).

τ_M = residence lifetime of A in the metal ion site.

ω_I = nuclear Larmor precession frequency. If $\omega_I^2\tau_c^2 > 1$ (as is generally the case for macromolecules particularly at high ω_I) terms of the type shown lead to a frequency dependence in $1/T_{1M}$ and hence in the observed T_{1B} values. Such an effect is much smaller in $1/T_{2M}$ and confined to the region $\omega_I^2\tau_c^2 \sim 1$.

τ_c is the correlation time characterising the dipole-dipole interaction between electron and nuclear spins:

$$\tau_c^{-1} = \tau_S^{-1} + \tau_R^{-1} + \tau_M^{-1} \quad (6)$$

τ_S is the longitudinal electron spin relaxation time of the metal ion which is often assumed to be given by the Bloembergen-Morgan (BM) equation:

$$\tau_S^{-1} = B\left(\frac{\tau_v}{1+\omega_S^2\tau_v^2} + \frac{4\tau_v}{1+4\omega_S^2\tau_v^2}\right) \quad (7)$$

τ_R = rotational correlation time of the macromolecule·ion· ABC complex.

τ_M occurs again (see above). Thus τ_M can contribute to the measured rates either directly as an exchange term or through τ_c.

B = a constant related to the zero-field splitting parameters of the metal ion.

τ_v = a correlation time related to modulation of the zero-field splitting of the metal ion.

ω_S = electron Larmor frequency ($\equiv 658$ x ω_I for protons). If $\omega_S^2\tau_v^2 \gtrsim 1$, τ_S will be frequency dependent. If τ_S contributes significantly to τ_c this dependence will appear in $1/T_{iM}$ and hence $1/T_{iB}$ values. Thus we have a second possible source of frequency dependence to the observed rates. (Note that τ_S may be frequency dependent independent of the validity of the BM equation).

This equation was derived for simple aqueous solutions of paramagnetic ions where τ_v describes modulation by impact of water molecules on the aquo complex. For macromolecules, it completely lacks experimental verification and its validity is in question In the literature, reference is often made to the Solomon-Bloembergen-Morgan (SBM) approach. It should be appreciated that, in our context, the SB theory is on a different theoretical footing to the BM theory. The former is a quite sound theory whereas the latter, for our purposes, is not. In our opinion, use of the BM equation should be avoided where possible.

The correlation times τ_R, τ_V (and therefore τ_S) and τ_M are temperature dependent   (8)

$$\tau_j = \tau_j^{\circ} \exp(E/_jRT)$$

$$\tau_M^{-1} = \frac{kT}{h} \exp\left(\frac{-\Delta H^{\ddagger}}{RT} + \frac{\Delta S^{\ddagger}}{R}\right) \quad (9)$$

or in another form:

$$\tau_M^{-1} = (\tau_M^{298})^{-1} \frac{T}{298} \exp\left\{\frac{\Delta H^{\ddagger}}{R}\left(\frac{T-298}{298T}\right)\right\} \quad (10)$$

Enhancement is often defined as:

$$\varepsilon_b = \frac{(T_{iB})^{-1}}{(T_{iB})^{-1} \text{ ion}} \quad (11)$$

If fast exchange conditions prevail ($T_{iM} >> \tau_M$) the temperature dependence of the observed relaxation rates will reflect the temperature dependence of τ (and therefore very roughly the dominant correlation time in τ_c). Under slow exchange ($T_{iM} < \tau_M$) the observed rates will reflect the temperature dependence of τ_M. For intermediate conditions the temperature dependence of the observed rates could be very complex.

(T_{iB}^{-1}) ion = paramagnetic relaxation rate for a solution of metal ion. $(T_{iB})^{-1}$ = rate for a solution of metal ion/macromolecule where all the metal ion (same concentration as the metal ion-only solution) is bound to the macromolecule.

j = R, v
E_j = activation energy
R^j = gas constant
T = temperature

k = Boltzmann constant
h = Planck's constant
$\Delta H^{\ddagger}$ = enthalpy of activation
$\Delta S^{\ddagger}$ = entropy of activation

$\tau_M^{298} = \tau_M$ at 298K

ε_b = characteristic enhancement of the metal-macromolecule complex

Scheme 2. Navon's method for the determination of the number of exchangeable water molecules, q, bound to a paramagnetic ion (12).

From eqns. (3)-(5) if fast exchange conditions apply to T_{1B} and T_{2B}:

$$q = \frac{n}{2} = \frac{3C}{r^6} \frac{T_{1B}/T_{2B} - 0.5}{(T_{1B}/T_{2B} - 1.17)^{1/2}} \cdot \frac{\omega_I}{P_M T_{1B}} \qquad (12)$$

and the symbols have been defined earlier.

Advantages. If a value for r can be assumed, q can be determined under fast exchange conditions from measurement of T_{1B} and T_{2B} at a single frequency. As it turns out for a reliable value of q this frequency (ω_I) should be chosen such that $\omega_1^2\tau_c^2 \sim 1$ (q is then least sensitive to error in T_{1B}/T_{2B} (13). The condition $\omega_I^2\tau_c^2 \sim 1$ corresponds roughly to the maximum normally observed in the ω_I-dependence of T_{1B}.

Limitations. It must be possible to demonstrate fast exchange conditions e.g. by temperature dependence studies. Errors in estimates of q can arise if complicating features such as internal rotation (see later) are present.

Scheme 3. The proton/deuteron relaxation enhancement (PRE/DRE) method (14).

is based on comparing proton and deuteron paramagnetic relaxation rates in a solution of macromolecule/paramagnetic ion/H_2O/D_2O.

Eqns. (3) and (4) can be rewritten as:

$$P_1 T_{1B} = \frac{\overline{C}}{q} \left(\frac{1 + \omega^2\tau_c^2}{\omega^2\tau_c}\right) \cdot B^2 + \frac{\tau_M}{q} \qquad (13)$$

where $\overline{C} = \dfrac{15\ r^6}{6\ g^2\beta^2 S(S+1)} = \gamma_I^2 r^6/3C$, B is the magnetic field and the other symbols have been defined earlier.

Eqn. (13) should also yield the paramagnetic contribution to deuteron relaxation rates if the PRE experiment is carried out in D_2O instead of H_2O. If the experiment is carried out in an H_2O/D_2O mix then for proton and deuteron paramagnetic relaxation rates at the same magnetic field (B_1) the only factor differing in eqn. (13) in the two cases is ω. Thus determination of both rates at field B_1 involves two equations and three unknowns - q, τ_c and τ_M

(r is assumed known). If it is supposed that the value of τ_c varies with magnetic field, then the determination of both proton and dueteron rates at a different field (B_2) can only introduce one new parameter - a new τ_c - while introducing two new equations. The resulting set of four simultaneous equations can then be solved to yield values for the four unknowns - q, τ_M, τ_c (B_1), τ_c(B_2).

Advantages. Simple to apply, involves exact solution of equation rather than multiparameter fitting. Does not require assumption of an equation for the frequency dependence of τ_S.

Limitations. (1) no explicit information on τ_c is obtained, although any frequency dependence is revealed; (2) if $\omega_H^2\tau_c^2 \ll 1$ for all measured ω_H (as e.g. Co(11) then method not applicable); (3) if fast exchange conditions apply only an upper limit for τ_M is obtained; (4) there may be practical problems associated with obtaining a reasonable DRE effect (depends on K_d, ε_b (14)); (5) only tested to date for Gd(111); (6) subject to errors arising from e.g. internal rotation.

Note: Method should be applicable to substrates, although not tested to date.

Theory for short τ_S ions.

For ions with $\tau_S < 10^{-10}$ e.g. Co(11) then for normal operating frequencies with a number of assumptions (5) instead of eqns. (4) and (5), we have eqns. (14) and (15):

$$\frac{1}{T_{1M}} = \frac{C}{r^6} \left(3\tau_{1S} + \frac{7\tau_{2S}}{1 + \omega_S^2\tau_{2S}^2}\right) \qquad (14)$$

$$\frac{1}{T_{2M}} = \frac{C}{2r^6} \left(7\tau_{1S} + \frac{13\tau_{2S}}{1 + \omega_S^2\tau_{2S}^2}\right) \qquad (15)$$

We should now consider the transverse electron spin relaxation time, τ_{2S}, as well as the longitudinal time, τ_{1S} (simply referred to as τ_S previously) although it is often assumed that $\tau_{1S} \sim \tau_{2S}$. A likely situation is that $\omega_S^2\tau_{2S}^2 \sim 1$ at accessible frequencies so that a frequency dependence of the relaxation rates could be observed.

One feature which can be noted by comparing eqns. (14) and (15) with the corresponding eqns. (4) and (5) for long τ_S ions is that the correlation time important in the former case ($\tau_S < 10^{-10}$ s) will lead to much smaller paramagnetic rates (T_{iM}^{-1}) than in the latter ($\tau_c \sim 10^{-9}$ s). In other words, ions with short τS

values will tend to give small paramagnetic enhancement effects as compared to the ions such as Mn(II) or Gd(III).

Complications in the use of the simplified theory.

The theory presented in Scheme 1 includes a number of simplifications and makes a number of assumptions which, loosely termed as complications, are summarized in Scheme 4. For further discussion the reader is referred to (5).

Scheme 4. Possible complications to the simplified approach presented in Scheme 1.

Type	Brief description	Likely importance
1 τ_{c2} terms	Full SB eqns. contain τ_{c2} terms where τ_{2S} replaces τ_{1S} ($= \tau_S$) in eqn. (6).	Terms only significant for short τ_S ions (cf. eqns. (14) and (15).
2 scalar interaction	Full SB eqns. contain terms allowing for ion-nucleus through-bond scalar interaction.	Not expected significant in macromolecular systems – available experimental evidence supports this belief.
3 nature of τ_S	BM eqn. not tested by independent measurement for macromolecular systems.	Introduces uncertainty to all results dependent on BM eqn.
4 ZFS approximation	SB eqns. not valid if zero-field splitting of metal ion (ZFS) $\gtrsim \omega_S$.	Problems at low nuclear frequencies and ions having large ZFS e.g. Co(II).
5 hyperfine interaction	SB eqns. not valid if ion electron-ion nuclear spin hyperfine interaction $\gtrsim \omega_S$.	Possibly problems at very low frequencies.

6 g-value aniso-tropy	SB eqns. assume iso-tropic g-value and should be modified if g-value is aniso-tropic (15).	For Mn(11), Gd(111) expect g isotropic ∿2. For Co(II) problems could arise.
7 covalency	SB eqns. use point dipole approxima-tion, delocalisa-tion of electron density could af-fect validity of eqns.	Calculations (16) sug-gest not important, at least for H_2O as ligand.
8 internal rotation	SB eqns. assume iso-tropic rotation of ion-macromolecule complex. Internal rotational motions of e.g. segment of macromolecule, ion or coordinated lig-and may also con-tribute to relaxa-tion.	Little investigated. If present could lead to incorrect estimation of magnitude of dipolar interaction or τ_c or both.
9 outer sphere relaxation	Ligand in bulk solu-tion as well as in metal ion site can be affected to some extent by paramag-netic centre (normal-ly discussed in con-text of PRE).	Generally expected to be a small effect most significant at low fre-quencies (17).

3. APPLICATIONS

Immunoglobulins (11,13,14,18-20)

    Structure and function of immunoglobulin G. An antibody is a protein synthesised by an organism in response to invasion of the organism by a foreign substance termed an antigen. The prin-cipal antibody in the serum is immunoglobulin G (IgG) which has a molecular weight of about 150,000 and a domain structure as shown in Fig. 2. The IgG molecule can be cleaved in to a number of proteolytic fragments as shown in the figure. Also indicated are the antigen and complement binding sites. In the absence of antigen, binding to the latter site is negligible. However, on the formation of specific antigen-antibody complexes, the

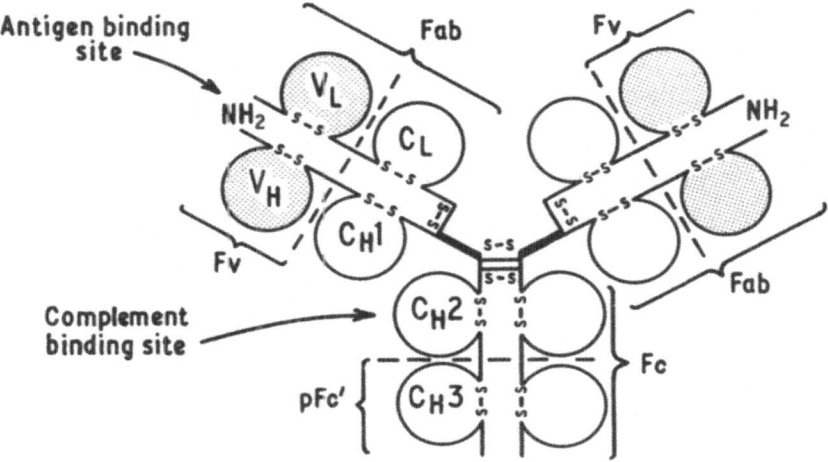

Figure 2. Structure of IgG. The different domains are shown by circled regions where V signifies variable domain, C constant domain, H belonging to the heavy chain and L belonging to the light chain. The various proteolytic fragments Fc, pFc´, Fab and Fv are also indicated. The antigen and complement binding sites are approximately 7·5 nm apart.

first protein of the complement sequence binds to this site leading to activation of other proteins in the sequence and eventual destruction of the antigen (21). We refer to complement "triggering". Much debate has centered around whether this triggering is attributable to antigen-induced conformational changes in the complement binding region or whether the aggregation of antibodies by multivalent antigens is critical (reviewed (22)).

Aim of relaxation studies. General investigation of water proton relaxation in a number of antibody and antibody fragment· Gd(III) complexes with special emphasis on factors of possible importance in relation to the mechanism of complement "triggering". PRE of antibody·Gd(III) in presence and absence of antigen : PRE as a probe to answer the question; conformational change or not an adding antigen?

Approach. The PRE technique was used as a titration indicator to study Gd(III) binding to non-immune rabbit IgG and its

proteolytic fragments. IgG was found to possess two equivalent
tight Gd(III) sites located in the C_H3 domains of the Fc region
(18,19). Other Gd(III) sites were found on the Fab region but
these were of much lower affinity. Experimental conditions could
be arranged so that for IgG and the fragments Fc and pFc´ only
effects from the same two Gd(III) sites were observed. The
three species IgG, Fc and pFc´ thus form a family of Gd(III)
tight-binding macromolecules of decreasing molecular weight:
150,000 (IgG), 50,000 (Fc) and 25,000 (pFc´). Extensive frequency and temperature dependent PRE data was obtained for the
complexes of these proteins with Gd(III). The rabbit IgG-Gd(III)
complex was further investigated by the PRE/DRE method described
earlier using data obtained at four magnetic fields.

The PRE technique as a titration indicator was also used to
show that pig IgG possesses Gd(III) sites in both Fc and Fab
regions of relatively similar affinity. The PRE effect in the
presence and absence of multi- and monovalent antigens was investigated.

Discussion. A detailed analysis of PRE data for the IgG·Gd(III
complex (11) showed that despite the large amount of data collected many of the variable parameters were ill-defined. In particular it was found that the number of exchangeable water molecules q could take values between 2 and 8 with only small effects
on the error of best-fit . The best-fit value obtained for τ_R of
9 ns was much less than the overall rotation time for IgG of approximately 170 ns given by nanosecond fluorescence spectroscopy
(23). However, the possibility of large errors in τ_R prevented
unequivocal conclusions being drawn.

Application of the PRE/DRE method to the IgG·Gd(III) complex
allowed much better definition of some of the variable parameters
involved in the analysis (14). For instance q was obtained as
4·1±0.5 - the error limits are considerably less than those
found by application of a conventional PRE approach showing the
advantage of exact solution of equations as compared to a multiparameter fitting procedure. However these limits do not include
uncertainties in q which might arise from e.g. covalency or internal rotation. The PRE/DRE method does not give directly a
value for τ_R but combination of the results from this method
with those from the conventional approach allowed τ_R to be
fixed within the limits 5-10 ns. It was suggested that an internal motion was present in the Fc portion of the IgG molecule.
This was an interesting finding in view of the crystallographic
results of Huber et al. (24) which also indicated a considerable
amount of motional freedom at the Fc position. Indeed these
authors proposed a flexible→rigid transition in the IgG molecule
as the trigger for complement activation, although recent crystallographic work (25) would seem to provide evidence against this

hypothesis. Interestingly high resolution proton nmr of the Fc and pFc′ fragments also gives evidence of considerable protein mobility (26). The spectra of these fragments appear very well resolved compared to the spectra of other proteins of similar molecular weight.

In the comparison of IgG, Fc and pFc′ (13) the observed relaxation rates at varying frequency and temperature were found to be generally very similar. This allowed fits to the three sets of data with fairly similar best-fit parameters. However, the problem of ambiguity in fitting again rises and the similarity in the observed experimental relaxation rates does not necessarily indicate that the values of the variable parameters are as similar as might intuitively be expected. Experimentally, the greatest difference between the IgG, Fc and pFc′ data was in T_1 at 84 MHz – the highest frequency used where τ_R is expected to be important in τ_c. At this frequency the values of T_{1B}^{-1} for IgG and Fc are almost identical but about 50 % smaller than for pFc′. This result can be explained by postulating that for the former larger molecules only internal rotation is important whereas for the smaller pFc′ the overall tumbling time of the molecule contributes to relaxation.

Experimental considerations lead to the use of pig anti-Dnp antibodies (as opposed to non-immune rabbit IgG discussed above) being used to investigate the effects of antigen on a Gd(III) PRE probe (20). With Gd(III) bound to both Fc and Fab regions of a non-precipitating pig anti-Dnp antibody, no change in PRE behaviour was found on addition of a multivalent 80,000 M.W. Dnp-dextran as antigen. The large differences in experimental conditions for complement activation studies and the PRE experiment mean that care should be exercised in interpreting data. Nevertheless the PRE results would seem to support the growing body of evidence which argues against the conformational theory of complement activation.

Carbonic anhydrase (27-31)

Enzyme function. Carbonic anhydrase catalyses the reversible hydration of CO_2 as well as other reactions such as the hydration of aldehydes and the hydrolysis of esters. It has a molecular weight of 30,000 and contains one zinc atom per protein molecule essential for activity. Lindskog et al. (32) have suggested that a basic form of a group closely linked to the metal ion and having a pK_a of approximately 7 is critically involved in catalysis. Despite many efforts the chemical nature of the catalytic group has not been conclusively identified. Four principal possibilities have been proposed (33) : (1) a metal-coordinated water molecule ionising to OH^-; (2) a titratable imidazole group associated with the metal ion via one or

more water molecules; (3) a neutral imidazole group ionising to
a metal-coordinated imidazolate anion; (4) a partially buried
carboxyl group connected to a metal-coordinated water molecule
via a hydrogen bond system. Undoubtedly the simplest and most
favoured theory has been (1) which is distinguished from the
other theories by its prediction that, at high pH, a hydroxyl
ion and not a water molecule should be coordinated to the metal
ion.

Aim of relaxation studies. Determination at high pH of the
metal ion coordination number in carbonic anhydrase.

Approach. A relatively early observation was that the PRE
of the Co(II) enzyme was greater at high pH and titrated with a
pK_a of roughly 7. This correlated well with the pK_a for activity
of the enzyme so that it was natural to suggest that an $OH^{\ominus}$ ion
was bound to the metal at high pH and either a slowly exchanging
H_2O molecule or no H_2O molecule at all at low pH (see theory (1)
above). Of course PRE should in principle be able to distinguish
a water molecule (n = 2 in eqn. (3)) from a hydroxyl ion (n = 1)
and quantitative studies have been carried out to this end on
the Co(II) bovine enzyme (27) and on the Mn(II) bovine enzyme
(28,29). More recently qualitative PRE studies in the absence
of buffer (30,31) have proved very interesting as described
below.

Discussion. For the Co(II) enzyme, we have found that the
PRE data in (27) can be fit equally well assuming either a H_2O
molecule or a $OH^{\ominus}$ ion is coordinated to the metal ion at high pH
(5). There is furthermore the problem that for most of the data
in (27), the $ZFS \gtrsim \omega_S$ and so the SB eqns. may be invalid (see
Scheme 4).

For the Mn(II) enzyme, which has a low activity compared to
the Zn(II) and Co(II) enzymes, it was originally claimed (28,29)
that the PRE data demonstrated the coordination of one water
molecule to the metal ion at high pH. In a reexamination, we
found this to be the best explanation of the data although we
also pointed out the possibility of there being more than one
water molecule (3to4) coordinated (5). The latter has incidentally
recently been proposed on the basis of a Monte-Carlo simulation
of the water structure in the active cleft (34).

Recently Bertini et al. (30,31) have described pH titrations
of Co(II) bovine carbonic anhydrase B in the presence and absence
of buffer (tris/sulphate). In the presence of buffer they observed the normal titration curve with a pK_a of very roughly 7.
In the absence of buffer no titration occurred, the PRE effect
remaining approximately constant in the pH region 5.8 to 9.
Bertini et al. proposed than an interaction between sulphate and

the metal ion was responsible for their observations and this was supported by electronic spectra data. Furthermore, as a $OH^\ominus$ ion is not expected on the metal ion at a pH as low as 5.8 and as no pH dependence of the PRE was observed, the authors concluded that a water molecule must be bound to the metal even at high pH.

Other techniques have been used in support of a metal-bound hydroxyl ion at high pH (e.g. (35)) and the mechanism of action of carbonic anhydrase remains even today a matter of considerable controversy.

Alcohol Dehydrogenase (34-39)

Enzyme function. Alcohol dehydrogenase catalyses the inter-conversion of alcohol and aldehyde. The enzyme from liver (LADH) is a dimer of molecular weight 80,000. It has a requirement of four zinc atoms per molecule. Two of these atoms, one per sub-unit, are known to be intimately involved in catalytic activity ("catalytic zinc atoms") and two, again one per subunit , seem to lack direct catalytic importance ("non-catalytic zinc atoms"). Enzyme activity is also dependent on the presence of one NAD^+ coenzyme molecule per subunit.

Aim of relaxation studies. Elucidation of the role of cata-lytic metal ion in enzyme action.

Approach. Zn(II) can be replaced in two stages by Co(II) with full retention of enzymic activity. The order of replace-ment of the two classes of Zn(II) ion (catalytic and non-catalytic) is expected to be apparent from comparison of the paramagnetic effects of the Co_2Zn_2 and Co_4 LADH enzymes on sub-strate and inhibitor molecules since only catalytic Co is ex-pected to show significant effects. The magnitude of the ob-served paramagnetic enhancement effects of substrate/inhibitor nuclei in a suitable coenzyme complex, can then hopefully be analysed in terms of the SBM equations to give distance infor-mation. This was essentially the approach of the three groups referred to in the above references. The perturbed nuclei were generally methyl and/or methylene protons on the substrate ethanol and the inhibitor isobutyramide. Of course the reactive ternary enzyme·ethanol·NAD^+ complex could not be studied but instead the abortive NADH ternary complex, which is thought to have a very similar structure, was used.

Discussion. Isotope studies combined with activity measure-ments and optical spectroscopy (40) and EPR (41) indicate that it is the non-catalytic pair of Zn atoms which exchange first with Co (40). From paramagnetic relaxation effects, this con-clusion has been reached by two of three groups (38,39) although the third (36,37) arrived at the opposite conclusion. It seems perhaps that the small magnitude of the paramagnetic effects from Co(II) (see earlier) may have contributed to the confusion here. In any case the weight of evidence available favours exchange of first non-catalytic and then catalytic Zn(II) by Co(II).

As regards distance information, Boccalon et al. (38) found paramagnetic effects on the CoLADH·ethanol·NADH complex more consistent with a direct coordination of ethanol to the metal ion than a "second sphere" type of coordination with for example an intervening water molecule. Andersson et al. (39) found the reverse. Young and Mildvan have consistently argued from paramagnetic relaxation effects, for a second sphere substrate coordination in LADH (36,37).

Interestingly, crystallographic studies indicate direct coordination of the competitive inhibitors imidazole and dimethylsulphoxide (DMSO) to the catalytic metal ion (42,43). Optical studies similarly indicate direct coordination of the inhibitor trans-4-N, N-dimethylaminocinnamaldehyde (DCA) to the metal ion (44). Thus there exists a possibility that the SBM theory may have broken down in this case (i.e. derived distances are too large). Further studies should clarify this point.

4. CONCLUSIONS

The examples considered above demonstrate the versatility of paramagnetic ions as qualitative relaxation probes in biology. For the problems outlined – the mechanism of complement triggering, the hydration state of carbonic anhydrase and the identification of the catalytic metal ions in LADH – the qualitative results make a valuable contribution to their respective fields. The quantitative application of paramagnetic ions as relaxation probes is seen to pose greater problems. Here great care should probably be exercised. As other techniques are applied to the many problems hereto tackled by use of paramagnetic relaxation probes, we shall learn more about the reliability of such probes used quantitatively and understand more clearly where the principal problems lie.

Acknowledgements

I should like to thank Professor Sture Forsén and Dr Raymond A. Dwek for their continuous encouragement and advice.

References

1   Dobson, C.M. 1977, in NMR in Biology, ed. R.A. Dwek, I.D.
    Campbell, R.E. Richards and R.J.P. Williams, Academic Press,
    New York, p. 63.
2   Leroy, J.L., and Guéron, M.: 1978, poster C17, VIIIth
    International Conference on Magnetic Resonance in Biological
    Systems, Nara, Japan.
3   Morris, A.T., and Dwek, R.A.: 1977, Quart. Rev. Biophys. 10,
    p. 421.
4   Marshall, A.G.: 1978, Biophysical Chemistry, John Wiley and
    Sons, New York, p. 544.
5   Burton, D.R., Forsén, S., Karlström, G., and Dwek, R.A.: 1979,
    Progress in Nuclear Mag. Res. Spectroscopy, Pergamon Press,
    Oxford, in press.
6   Solomon, I.: 1955, Phys. Rev. 99, p. 59.
7   Solomon, I., and Bloembergen, N.: 1956. J. Chem. Phys. 25,
    p. 261.
8   Bloembergen, N.: 1957, J. Chem. Phys. 27, p. 572.
9   Bloembergen, N., and Morgan, L.O.: 1961, J. Chem. Phys. 34,
    p. 842.
10 Dwek, R.A.: 1973, Nuclear Magnetic Resonance in Biochemistry,
    Applications to Enzyme Systems, Clarendon Press, Oxford.
11 Burton, D.R., Forsén, S., Karlström, G., Dwek, R.A.,
    McLaughlin, A.C., and Wain-Hobson, S.: 1976, Eur. J. Biochem.
    71, p. 519.
12 Navon, G.: 1970, Chem. Phys. Lett. 7, p. 390.
13 Burton, D.R., Forsén, S., Karlström, G., Dwek, R.A.,
    McLaughlin, A.C., and Wain-Hobson, S.: 1977, Eur. J. Biochem.
    75, p. 444.
14 Burton, D.R., Dwek, R.A., Forsén, S., and Karlström, G.: 1977,
    16, p. 250.
15 Sternlicht, H.: 1965, J. Chem. Phys. 42, p.2250.
16 Waysbort, D., and Navon, G.: 1978, J. Chem. Phys. 68, p.3074.
17 Koenig, S.H., Brown, R.D., and Studebaker, J.: 1971, Cold
    Spring Harbour Symp. Quant. Biol. 36, p. 551.
18 Dower, S.K., Dwek, R.A., McLaughlin, A.C., Mole, L.E., Press,
    E.M., and Sunderland, C.A.: 1975, Biochem. J. 149, p. 73.
19 Burton, D.R., Forsén, S., Karlström, G., Willan, K.J.,
    Wright, C., and Dwek, R.A.: 1978, in Rare Earths in Modern
    Science and Technology, p. 115. Plenum Press, New York.
20 Burton, D.R., Forsén, S., Franek, F., and Novotny, J.: 1979,
    FEBS Letts, in press.
21 See e.g. Mayer, M.: in "Immunology: readings from Scientific
    American" ed. F.M. Burnet., 1976, W.H. Freeman, San Francisco.
22 Metzger, H.: 1978, Contemporary Topics in Immunology, in press.
23 Yguerabide, J., Epstein, H.F., and Stryer, L.: 1970, J. Mol,
    Biol. 51, p. 573.
24 Huber, R., Deisenhofer, J., Colman, P.M., Matsushima, M., and

Palm, W.: 1976, Nature, (London) 264, p. 415.

25 Silverton, E.W., Navia, M.A., and Davies, D.R.: Proc.
Nat. Acad. Sci. USA in press.

26 Boyd, J., Easterbrook-Smith, S.B., Zavodsky, P., Mountford-
Wright, C., and Dwek, R.A.: 1979, submitted for publication.

27 Fabry, M.H., Koenig, S.H., and Schillinger, W.E.: 1970,
J. Biol. Chem. 245, p. 4256.

28 Lanir, A., Gradsztajn, S., and Navon, G.: 1973, FEBS Lett.
30, p. 351.

29 Lanir, A., Gradsztajn, S., and Navon, G.: 1975, Biochemistry,
14, p. 242.

30 Bertini, I., Canti, G., Luchinat, C., and Scozzafava, A.:
1977, Biochem. Biophys. Res. Commun. 78, p. 158.

31 Bertini, I., Canti, G., Luchinat, C., and Scozzafava, A.:
1978, J. Amer. Chem. Soc. 100, p. 4873.

32 Lindskog, S., Henderson, L.E., Kannan, K., Lhjas, A.,
Nymand, P.O., and Strandberg, B.: 1971, Enzymes. 3rd edn.
5, p. 587.

33 Campbell, I.D., Lindskog, S., and White, A.I.: 1977, Biochim.
Biophys. Acta 48, p. 443 and references therein.

34 Clementi, E., Corongin, G., Jönsson, B., and Romano, S.:
FEBS Lett., in press.

35 Bauer, R., Linkilde, P., and Johansen, J.T.: 1976, Biochemistry, 15, p. 334.

36 Sloan, D.L., Young, J.M., and Mildvan, A.S.: 1975, Biochemistry 14, p. 1998.

37 Young, J.M., and Mildvan, A.S.: 1977, in Alcohol and Aldehyde
Metabolising Systems, Academic Press, New York.

38 Boccalon, G., Grillo, G., Baroncelli, V., Renzie, P., and
Parretta, A.: 1978, J. Mol. Catalysis 4, p. 307.

39 Andersson, I., Burton, D.R., Drakenberg, T., Maret, W., and
Zeppezauer, M.: to be published.

40 Sytkowski, A.J. and Vallee, B.L.: 1975, Biochem. Biophys.
Res. Commun. 67, p. 1488.

41 Drott, H.R.: 1977, in Alcohol and Aldehyde Metabolizing Systems, Academic Press, New York.

42 Boiwe, T., and Brändén, C.-I.: 1977, Eur. J. Biochem. 77,
p. 173.

43 C.-I. Brändén, private communication.

44 Dietrich, H., Andersson, I., Maret, W., and Zeppezauer, M.:
submitted for publication.

THEORY OF ELECTRON SPIN RESONANCE

BRUCE R. McGARVEY

Department of Chemistry, University of Windsor
Windsor, Ontario, Canada

In this chapter we shall treat only the theory of transition
metals and their complexes. Much of the material is directly
applicable to organic free radicals or triplet state molecules
except that the spin-orbit interaction is less important for
s and p electrons. The treatment for rare earth metals is
similar in principle but the fact that the crystal field is much
smaller than the spin-orbit interaction for rare earths makes
many of the details different.

It is assumed here that the reader is familiar with crystal
and ligand field theories for transition metal ions such as
given by Ballhausen (1), Figgis (2) or similar texts. A
familiarity with quantum mechanics equivalent to that given in
introductory courses is also assumed.

1. S=1/2 SYSTEMS

1.1. g-Tensor and its relation to electronic structure.

For transition metal ions the relative ordering by energy
is: (electron repulsion, crystal field interactions) > spin-
orbit interactions > Zeeman interactions. The normal theoretical
treatment is then to assume we have good wave functions for the
ground and excited d levels in a given crystal field and then
calculate the effect of the spin-orbit interaction using first
and second order perturbation theory. If the resulting ground
state is still degenerate we then do a degenerate perturbation
treatment using the Zeeman operator to discover how the degener-
ancy is lifted by the magnetic field.

171

*I. Bertini and R.S. Drago (eds.), ESR and NMR of Paramagnetic Species in Biological and
Related Systems. 419–422.*
Copyright © 1979 by D. Reidel Publishing Company

For our purposes we will assume that we can approximate the the electronic wave function for the ion with a Slater determinant made up of atomic orbitals. The symbol $d(m)^{\pm}$ will be used to represent a d orbital which obeys the following important operator relations:

$$\hat{\ell}_z d(m) = m d(m)$$

$$\hat{\ell}_{\pm} d(m) = (\hat{\ell}_x + i\hat{\ell}_y) d(m) = [6-m(m\pm1)]^{1/2} d(m\pm1)$$

$$\hat{s}_z d(m)^{\pm} = \pm(1/2) d(m)^{\pm}$$

$$\hat{s}_+ d(m)^- = d(m)^+ \qquad\qquad\qquad (1.1)$$

$$\hat{s}_- d(m)^+ = d(m)^-$$

$$\hat{s}_+ d(m)^+ = s_- d(m)^- = 0$$

where $\hat{\ell}$ and $\hat{s}$ are orbital and spin moment operators. From crystal field theory we know that in octahedral or tetrahedral fields the 5d orbitals separate into two degenerate sets:

$$t_2(0) = -(i/\sqrt{2})\,[d(2) - d(-2)] = d(xy)$$

$$t_2(\pm1) = d(\pm1) \qquad\qquad\qquad (1.2)$$

$$e(1) = d(0) = d(z^2)$$

$$e(2) = (1/\sqrt{2})\,[d(2) + d(-2)] = d(x^2-y^2) \qquad (1.3)$$

1.1.A. d^1 configuration. In an octahedral field the ground state is $(t_2)^1$. In many instances there are distortions from octahedral symmetry and to illustrate the effect of such distortions we will assume a tetragonal distortion producing an energy level diagram of the type given in Figure 1.

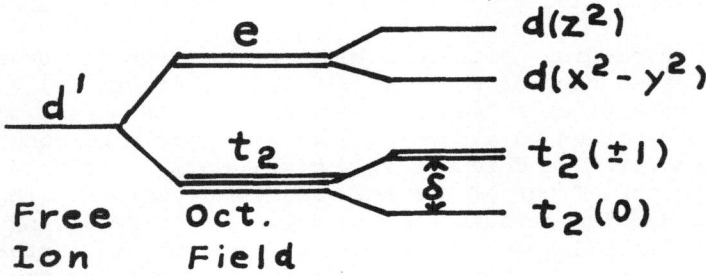

Figure 1. Energy levels for a tetragonally distorted octahedral field.

Thus in an octahedral field the ground state is $^2T_{2g}$. For a tetrahedral field the above energy levels are inverted and the ground state becomes ^2E.

We shall first consider a $^2T_{2g}$ ground state and since we wish to allow for the possibility that the tetragonal distortion and the spin-orbit interaction are of the same order of magnitude we shall assume a perturbation operator of the form.

$$\hat{H}' = \hat{V}_T + \hat{H}_{LS} = \hat{V}_T + \xi \underset{\sim}{\ell} \cdot \underset{\sim}{s} \qquad (1.4)$$

where $\hat{V}_T$ is the tetragonal component of the crystal field and is defined such that

$$<t_2(\pm 1)|\hat{V}_T|t_2(\pm 1)> = \delta \;; \quad <t_2(0)|\hat{V}_T|t_2(0)> = 0 \qquad (1.5)$$
$$<t_2(\pm 1)|\hat{V}_T\,t_2(\mp 1)> = 0$$

$\hat{H}_{LS}$ is the spin-orbit operator which can also be written as:

$$\hat{H}_{LS} = \xi \hat{\ell}_z \hat{s}_z + (1/2)\, \xi(\hat{\ell}_+ \hat{s}_- + \hat{\ell}_- \hat{s}_+) \qquad (1.6)$$

Perturbation theory requires that we solve the determinant: $\det|H_{ij} - \delta_{ij}E| = 0$ where $H_{ij} = <\phi_i|\hat{H}'|\phi_j>$ and δ_{ij} is Kronecker's delta. Using Eq. (1.1),(1.2), (1.5) and (1.6) the complete determinant can be written as:

	$t_2(0)^+$	$t_2(-1)^-$	$t_2(0)^-$	$t_2(+1)^+$	$t_2(+1)^-$	$t_2(-1)^+$
$t_2(0)^+$	$-E$	$-(i/\sqrt{2})\xi$	0	0	0	0
$t_2(-1)^-$	$(i/\sqrt{2})\xi$	$(\delta+1/2\xi-E)$	0	0	0	0
$t_2(0)^-$	0	0	$-E$	$(i/\sqrt{2})\xi$	0	0
$t_2(+1)^+$	0	0	$-(i/\sqrt{2})\xi$	$(\delta+1/2\xi-E)$	0	0
$t_2(+1)^-$	0	0	0	0	$(\delta-1/2\xi-E)$	0
$t_2(-1)^+$	0	0	0	0	0	$(\delta-1/2\xi-E)$

For $t_2(+1)^-$ and $t_2(-1)^+$: $E = \delta-(1/2)\xi$

For the two (2 x 2) determinants: $E = (\delta/2)+(\xi/4)$
$$\pm(1/2)[\delta^2+\xi\delta+(9\xi^2/4)]^{1/2}$$

Thus after application of a tetragonal distortion plus spin-orbit interaction the $^2T_{2g}$ state becomes three doubly degenerate states which are often referred to as Kramer's doublets. To better see which state is the ground state we will express energy in units of ξ

$$\epsilon = E/\xi \quad \text{and} \quad \eta = \delta/\xi \tag{1.7}$$

$$\epsilon = \eta - (1/2) \quad ; \quad \{\eta + 1/2 \pm [\eta^2 + \eta + (9/4)]^{1/2}\}/2 \tag{1.8}$$

A plot of these three values of ϵ vs. η reveals that the ground state has $\epsilon = \eta - 1/2$ for $\eta < 0$ and $\epsilon = \{\eta + 1/2 - [\eta^2 + \eta + (9/4)]^{1/2}\}/2$ for $\eta > 0$. When $\eta = 0$ the ground state would be fourfold degenerate with $\epsilon = -1/2$. The Jahn-Teller theorem would predict that a distortion would always take place so that $\eta \neq 0$.

The preceding results are very similar to that found for a ^2P state in the H atom and it has been shown that T_1 and T_2 states can be treated as P states provided that the angular momentum operator $\hat{L}_z$ is replaced by $\alpha \hat{L}_z$ and $\hat{L}_x$ and $\hat{L}_y$ by $\alpha'\hat{L}_x$ and $\alpha'\hat{L}_y$ in the spin-orbit and Zeeman operators. The values of α and α' depend on the particular configuration being treated. In our example $\alpha = 1$ and $\alpha' = -i$. This "p formalism" is often used in the ESR literature.

We are now ready to determine the effect of the magnetic field or Zeeman operator on the ground state doublet but before we can proceed we need to know the wave functions for this doublet. For $\eta < 0$ the wave functions are $t_2(+1)^-$ and $t_2(-1)^+$. The Zeeman operator is:

$$\hat{H}_Z = \mu_B (\hat{\ell} + g_e \hat{s}) \cdot \hat{B}$$

$$= \mu_B (\hat{\ell}_z + g_e \hat{s}_z) B_z + (1/2)(\hat{\ell}_+ + g_e \hat{s}_+) B_- + (1/2)(\hat{\ell}_- + g_e \hat{s}_-) B_+ \tag{1.9}$$

$$B_\pm = B_x \pm iB_y$$

where μ_B is the Bohr magneton, $g_e = 2.0023$, and B is the magnetic induction (often called the magnetic field in the literature). If we apply degenerate perturbation theory to the ground state doublet using H_Z as the perturbation operator we obtain the following determinant:

$$
\begin{array}{c|cc}
 & t_2(-1)^+ & t_2(+1)^- \\
\hline
t_2(-1)^+ & (1/2)(g_e-2)\,\mu_B B_z - E & 0 \\
t_2(+1)^- & 0 & -(1/2)(g_e-2)\,\mu_B B_z - E
\end{array} = 0
$$

Since $g_e - 2 \cong 0$ we find that this Kramer's doublet is not split by the magnetic field and the ground state is equivalent to an $S = 1/2$ state in which $g = 0$. This means no ESR would be detectable for this ground state.

For $\eta > 0$ the Kramer's doublets are mixtures of $t_2(0)^+$ plus $t_2(-1)^-$ and of $t_2(0)^-$ plus $t_2(+1)^+$ which can be written as:

$$\psi_+ = a\ t_2(0)^+ - ibt_2(-1)^-$$
$$\psi_- = a\ t_2(0)^- + ibt_2(+1)^+$$

(1.10)

The mixing coefficients a and b can be found for ψ_+ in the following manner using matrix elements from the determinant:

$$a(-E) - ib[-(i/\sqrt{2})\xi] = 0$$

or $b = -\sqrt{2}\ \varepsilon a$

$$a^2 + b^2 = 1 = a^2(1 + 2\varepsilon^2)$$

$$a^2 = 1/(1 + 2\varepsilon^2) = (1/2)\{1 + (\eta + 1/2)/[\eta^2 + \eta + (9/4)]^{1/2}\}$$ (1.11)

$$a = (2)^{-1/2}\{1 + [\eta + (1/2)]/[\eta^2 + \eta + (9/4)]^{1/2}\}^{1/2}$$

$$b = (2)^{-1/2}\{1 - [\eta + (1/2)]/[\eta^2 + \eta + (9/4)]^{1/2}\}^{1/2}$$

a goes from $\sqrt{2/3}$ to 1 as η goes from 0 to ∞ while b goes from $1/\sqrt{2}$ to 0. Using ψ_+ and ψ_- as our two degenerate wave functions, perturbation theory gives the following determinant for the energy:

$$\begin{array}{c} \\ \psi_+ \\ \psi_- \end{array} \begin{vmatrix} (1/2)[g_e(a^2-b^2)-2b^2]\mu_B B_z - E & (1/2)[g_e a^2 - 2\sqrt{2}\ ab]\mu_B B_- \\ (1/2)[g_e a^2 - 2\sqrt{2}ab]\mu_B B_+ & -(1/2)[g_e(a^2-b^2)-2b^2]\mu_B B_z - E \end{vmatrix} = 0$$

This determinant is identical to that obtained by using a fictitious spin of $S = 1/2$ with the spin Hamiltonian:

$$\hat{H} = g_\parallel \mu_B \hat{S}_z B_z + g_\perp \mu_B (\hat{S}_x B_x + \hat{S}_y B_y)$$

(1.12)

$$
\begin{array}{c|cc}
 & |1/2\rangle & |-1/2\rangle \\
\hline
|\ 1/2\rangle & (1/2)g_\parallel \mu_B B_z - E & (1/2)\ g_\perp \mu_B B_- \\
|-1/2\rangle & (1/2)g_\perp \mu_B B_+ & -(1/2)g_\parallel \mu_B B_z - E
\end{array} = 0
$$

where $g_\parallel = g_e(a^2-b^2)-2b^2$ (1.13)

$\qquad g_\perp = g_e a^2 - 2\sqrt{2}ab$

Thus $g_\parallel$ and $g_\perp$ both vary from 0 to g_e as η varies from 0 to ∞. For large values of η, $g_\parallel$ and $g_\perp$ become approximately:

$\qquad g_\parallel \stackrel{\sim}{=} g_e$

$\qquad g_\perp \stackrel{\sim}{=} g_e -(2/\eta) = g_e-(2\xi/\delta)$

For large η we must include higher order perturbation terms involving the e orbitals.   Doing this gives

$\qquad g_\parallel = g_e - 8\xi/[E(x^2-y^2) - E(xy)]$

$\qquad g_\perp = g_e - (2\xi/\delta)$ (1.15)

The preceding demonstrates that the spin-orbit interaction produces g values smaller than the free spin value. It also produces an anisotropy in the g values. We thus see that g values depend on the nature of the ground state making it possible to distinguish ground states by the experimentally determined g values.

A trigonal distortion from octahedral symmetry will also produce an axially symmetric g tensor. In this case the energy splitting pattern is given by Fig. 2.

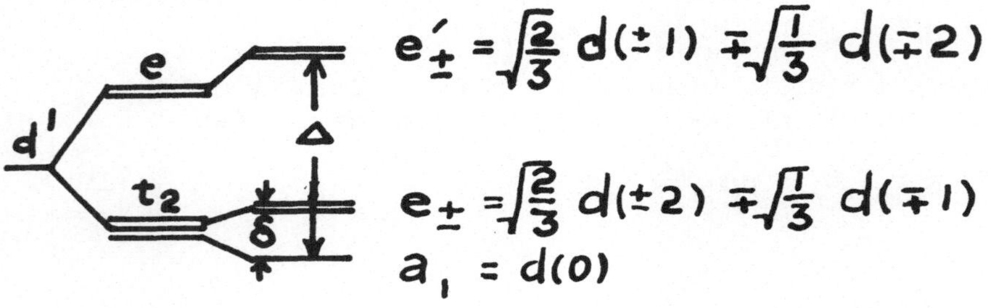

$$e'_\pm = \sqrt{\tfrac{2}{3}}\,d(\pm 1) \mp \sqrt{\tfrac{1}{3}}\,d(\mp 2)$$

$$e_\pm = \sqrt{\tfrac{2}{3}}\,d(\pm 2) \mp \sqrt{\tfrac{1}{3}}\,d(\mp 1)$$

$$a_1 = d(0)$$

Figure 2. Energy levels for a trigonally distorted octahedral field.

If we do a degenerate perturbation treatment as before for the t_2 orbitals we get the same equations for $g_\parallel$ and $g_\perp$ we obtained for the tetragonal distortion. The second order contribution from the e_f^- orbitals, however, is different. For a large distortion we get

$$g_\parallel = g_e$$
$$g_\perp = g_e - (2\xi/\delta) - (4\xi/\Delta) \qquad (1.16)$$

Thus for large distortions in the crystal field we can, in principle, distinguish a trigonal distortion from a tetragonal distortion of the $^2T_{2g}$ ground state.

For tetrahedral crystal fields the energy level diagram in Fig. 1 is inverted giving a ^2E state as ground state. A tetragonal distortion will leave either the $d(z^2)$ or $d(x^2-y^2)$ orbital lower in energy. The spin-orbit interaction produces no mixing of these two orbitals even for small distortions (as was the case for the $^2T_{2g}$ state) because all matrix elements are zero. Thus large deviations of g from g_e will not be expected for the ^2E state even for small distortions. Any orbital contributions will only come from second order interactions involving states of much higher energy. In a situation such as this we can derive a general equation for $g_\parallel$ and $g_\perp$.

Let $\phi_n^\pm$ be wave functions for a set of states in the absence of the spin-orbit interaction with ϕ_o^- being designated as the ground state. Each state has a spin of $\pm 1/2$. From perturbation theory we find that introduction of the spin-orbit interaction produces a ground state Kramer's doublet with the following wave functions:

$$\psi_\pm = \phi_o^\pm - (\xi/2)\Sigma[<\phi_n|\hat{\ell}_z|\phi_o>/(E_n - E_o)]\phi_n^\pm$$
$$- (\xi/2)\Sigma[<\phi_n|\hat{\ell}_\pm|\phi_o>/(E_n - E_o)]\phi_n^\mp \qquad (1.17)$$

From our handling of the d^1 problem in $^2T_2$ we see that g values are obtained in the following manner from $\psi_\pm$.

$$g_{zz} = 2<\psi_+|\hat{\ell}_z+g_e\hat{s}_z|\psi_+> = g_e-2\xi\Sigma\Sigma<\phi_o|\hat{\ell}_z|\phi_n><\phi_n|\hat{\ell}_z|\phi_o>/(E_n-E_o) \quad (1.18)$$

We have kept only terms of order $\xi/\Delta E$ in obtaining Eq. (1.18). Similarily

$$g_{xx} = 2<\psi_+|\hat{\ell}_x+g_e\hat{s}_x|\psi_-> = g_e-2\xi\Sigma\Sigma<\phi_o|\hat{\ell}_x|\phi_n><\phi_n|\hat{\ell}_x|\phi_o>/(E_n-E_o) \qquad (1.19)$$

$$g_{yy} = 2i\langle\psi_+|\hat{\ell}_y + g_e\hat{s}_y|\psi_-\rangle = g_e - 2\xi\Sigma\langle\phi_o|\hat{\ell}_y|\phi_n\rangle\langle\phi_n|\hat{\ell}_y|\phi_o\rangle/(E_n - E_o) \quad (1.20)$$

For a $d(x^2-y^2)$ ground state these equations give

$$g_{zz} = g_\| = g_e - 8\xi/[E(xy) - E(x^2-y^2)]$$

$$g_{xx} = g_\perp = g_e - 2\xi/[E(\pm1) - E(x^2-y^2)] \quad (1.21)$$

and for a $d(z^2)$ ground state

$$g_{zz} = g_\| = g_e$$

$$g_{zz} = g_\perp = g_e - 6\xi/[E(\pm1) - E(z^2)] \quad (1.22)$$

Again we see a distinct connection between the g tensor and the ground state wave function.

1.1.B. Other d^n configurations. The d^9 configuration is readily treated as a positive hole d^1 configuration so that we can apply preceding equations with minor modifications. The d^9 configuration occurs mostly in Cu^{2+} which is most often found in a tetragonally distorted octahedral field with the unpaired electron in the $d(x^2-y^2)$ orbital. The major change in the "hole formalization" from that of a simple one electron system is that the spin-orbit coupling constant ξ must be taken as negative. Therefore the g values for Cu^{2+} are given by Eq. (1.21) except that ξ is a negative number making $g_\|$ and $g_\perp$ greater than g_e.

For d^5 in a strong field all five electrons are in the t_2 orbitals in an octahedral field and, therefore, we have a $(t_2)^1$ hole configuration. Thus we can use the preceding treatment for $(t_2)^1$ except we change the sign of ξ in the appropriate equations. This must be done with some care to avoid ambiguities creeping into the problem due to the presence of square roots.

The best way to arrive at the correct equations is to let $\hat{H}_{LS} = -\xi\hat{\ell}\cdot\hat{s}$. The constant ξ is then kept as a positive parameter. Doing this and following through the earlier calculation we find for a tetragonal distortion the energies of the three Kramer's doublets to be:

$$E = \delta + (1/2)\xi \quad ; \quad (\delta/2) - (\xi/4) \pm (1/2)[\delta^2 - \delta\xi + (9\xi^2/4)]^{1/2}$$

$$\text{or} \quad \epsilon = E/\xi \quad ; \quad \eta = \delta/\xi \quad (1.23)$$

$$\varepsilon = \eta + (1/2) \quad , \quad (1/2)\{\eta^2 - (1/2)\pm [\eta^2 - \eta + (9/4)]^{1/2}\}$$

Plotting these values of ε from $\eta = -\infty$ to $+\infty$ reveals that the lowest energy is always

$$\varepsilon = (1/2)\ \{\eta - (1/2) - [\eta^2 - \eta + (9/4)]^{1/2}\}$$

unlike the d^1 case in which two energy levels crossed at $\eta = 0$. The two Kramer's doublets become

$$\psi_\pm = a\ t_2(0)^\pm \pm i\ b\ t_2(\pm 1)^\mp$$

$$a = (1/\sqrt{2})\ \{1 + (\eta - 1/2)/[\eta^2 - \eta + (9/4)]^{1/2}\}^{1/2}$$

$$b = (1/\sqrt{2}\ \{1 - (\eta - 1/2)/[\eta^2 - \eta + (9/4)]^{1/2}\}^{1/2} \quad (1.24)$$

$$g_\parallel = g_e\ (a^2 - b^2) - 2b^2$$

$$g_\perp = g_e\ a^2 + 2\sqrt{2}\ ab$$

A plot of $|g_\parallel|$ and $g_\perp|$ as a function of η is given in Fig. 3

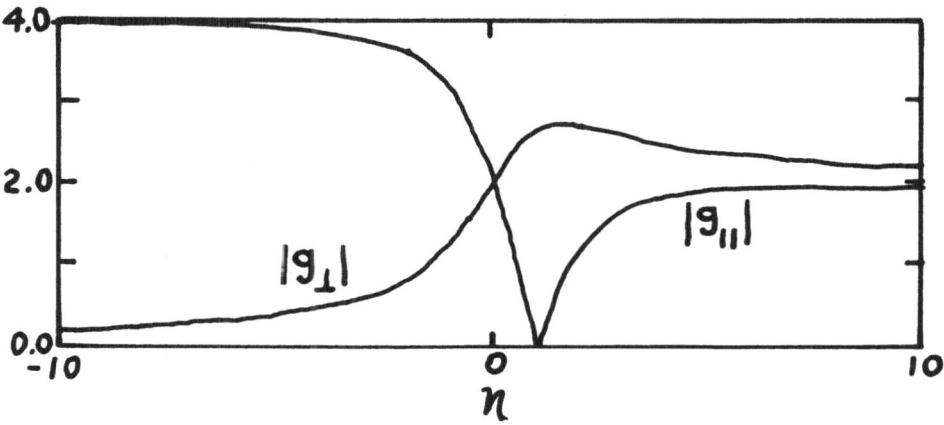

Figure 3. $|\ g_\parallel\ |$ and $|\ g_\perp\ |$ vs. η for strong field d^5.

Absolute values are plotted because the negative values of $g_\parallel$ for $\eta < 1$ are not measureable in the normal ESR experiment. It will be noted that these g values are much different from those obtained for the d^1 case and are not the values that would be obtained if one were careless and simply went to Eq. (1.11) and

(1.13) and simply changed the sign of ξ.

Care must be taken in trying to calculate any contribution to g from the e orbitals because any excited state produced by promotion of an electron from a t_2 to e orbital is no longer a single electron hole state.

The d^7 low spin system of Co^{2+} in square planar complexes is also an S = 1/2 system but its treatment is a little more complex because it can't be reduced to a one electron problem as were d^5 and d^9. Earlier treatments of this system assumed that only spin-orbit interactions between the ground state doublet and excited doublet states were important. It has been shown (3) that spin-orbit interactions between the ground state doublet and certain excited quartet states can also make significant contributions to the g tensor.

1.1.C. g tensor and molecular orbitals. Application of preceding equations to experimental data leads to difficulties. For example, DeSimone and Drago (4) found $g_\parallel$ = 1.61 and $g_\perp$ = 2.61 for $Fe(bipy)_3(PF_6)_3$. This is a strong field d^5 system in a trigonally distorted octahedral field. As for the d^1 configuration the d^5 configuration gives the same equations for both the trigonal and tetragonal distortions if only the t_2 orbitals are considered. Since $g_\parallel$ and $g_\perp$ both depend on one parameter η, we can calculate η from $g_\parallel$ and $g_\perp$ and compare to see how well the theory works. Rearranging Eq. (1.13) and using the fact that $b^2=(1-a^2)$ we get

$$a^2 = (g_\parallel + g_e + 2)/(2g_e + 2)$$

for $g_\parallel$ = +1.61   ; a^2 = 0.9347 or η = 2.988

for $g_\parallel$ = -1.61   ; a^2 = 0.3984 or η = 0.2065

For g there are also two solutions: a^2 = 0.8792, η = 2.146 or a^2 = 0.3711 or η = 0.1226. We see then that theory does not quantitatively account for $g_\parallel$ and $g_\perp$ with one parameter. One approach used to modify the equations is to propose that covalency interactions reduce the orbital momentum integrals from that for a pure atomic orbital by the factor k where k is a fraction between 0 and 1.0. The correct equations for g then become:

$$g_\parallel = 2\langle\psi_+|g_e\hat{s}_z + k\hat{\ell}_z|\psi_+\rangle = g_e(a^2-b^2)-2kb^2$$

$$(1.25)$$

$$g_\perp = 2\langle\psi_+|g_e\hat{s}_x + k\hat{\ell}_x|\psi_-\rangle = g_e a^2 + 2\sqrt{2}k\ ab$$

Solving these two for η and k in our example gives

For g $= +1.61$; $k = 1.063$; $a^2 = 0.9360$; $\eta = 3.019$

  g $= -1.61$; $k = 1.230$; $a^2 = 0.4411$; $\eta = 0.3324$

Although simple, this method of accounting for covalency is less than satisfactory and sometimes leads to absurd values for k.

    A better but naturally more complex approach is to use molecular orbitals instead of atomic orbitals in evaluating appropriate matrix integrals. All attempts to do this have used the LCAO-MO approach. To illustrate what is involved, we will consider a square planar complex of Cu^{2+} (d^9) using the coordinate system given in Fig. 4.

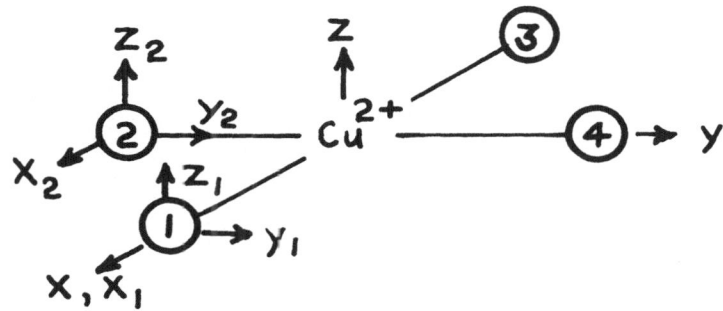

Figure 4. Coordinate system for copper square planar complex.

The ligand atoms are presumed to lie on the x and y axes and the coordinate systems for the ligand atomic orbitals are centered at each ligand nucleus and are parallel to the central Cu^{2+} coordinate system.   The appropriate orbitals are antibonding MO's and can be written as follows:

$$|x^2-y^2> = \alpha d(x^2-y^2) + (\alpha'/2)[p(\sigma)^1+p^2(\sigma)-p^3(\sigma)-p^4(\sigma)]$$

$$|xy> = \beta d(xy) + (\beta'/2)[-p(y)^1 +p(x)^2 +p(y)^3-p(x)^4]$$

$$|z^2> = \delta d(z^2) + (\delta'/2)[-p(\sigma)^1 +p(\sigma)^2 + p(\sigma)^3 - p(\sigma)^4] \quad (1.26)$$

$$|xz> = \gamma d(xz) + (\gamma'/\sqrt{2})[-p(z)^1 + p(z)^3]$$

$$|yz> = \gamma d(yz) + (\gamma'/\sqrt{2})[p(z)^2-p(z)^4]$$

$p(\sigma)^1$ is an sp hybrid orbital using $p(x)^1$ and s^1, etc.

Earlier we have seen that g for Cu^{2+} can be calculated from the equation

$$g_{\parallel} = g_e - 2<x^2-y^2|\hat{\ell}_z|xy> <xy|\hat{H}_{LS}(z)|x^2-y^2>/[E(xy)-E(x^2-y^2)] \tag{1.27}$$

For a multicentered MO, $\hat{H}_{LS}$ has the form

$$\hat{H}_{LS} = \Sigma \xi_i(r_i)\hat{\underset{\sim}{\ell}}_i \cdot \underset{\sim}{\hat{s}}$$

where $\hat{\ell}_i$ and r_i are defined in terms of the coordinate system centered at nucleus i. $\xi_i(r_i)$ is a function that varies approximately as r_i^{-3}. In our example we must introduce a negative sign in $\hat{H}_{LS}$ because we are using the "hole formalism".

$$<xy|\hat{H}_{LS}(z)|x^2-y^2> = <xy|\Sigma\xi_i(r_i)\hat{\ell}_{zi}|x^2-y^2>$$

If we ignore overlap terms by reason of the r_i^{-3} dependence of $\xi_i(r_i)$, we obtain

$$<xy|\hat{H}_{LS}(z)|x^2-y^2> = -2i\xi_M\alpha\beta+(\alpha'\beta'/4)\{<p(y)^1|\xi_1(r_1)\hat{\ell}_{z1}|p(\sigma)^1>$$

$$-<p(x)^2|\xi_2(r_2)\hat{\ell}_{z2}|p(\sigma)^2>+<p(y)^3|\xi_3(r_3)\hat{\ell}_{z3}|p(\sigma)^3>$$

$$-<p(x)^4|\xi_4(r_4)\hat{\ell}_{z4}|p(\sigma)^4>\}$$

If we let $p(\sigma)^1 = ns^1 + \sqrt{1-n^2}p(x)^1$, $p(\sigma)^2 = ns^2 + \sqrt{1-n^2}p(y)^2$, etc.

$$<xy|\hat{H}_{LS}(z)|x^2-y^2> = -2i\alpha\beta\xi_M+i\alpha'\beta'\sqrt{1-n^2}\xi_L$$

where ξ_M is spin-orbit constant for a d orbital in Cu^{2+} and ξ_L is spin-orbit constant for p orbital in ligand atom. If we had used a pure d orbital the integral would have been $-2i\xi_M$. It is instructive to compare the above MO integral to that in the pure atomic case

$$<\hat{H}_{LS}>(MO)/<\hat{H}_{LS}>(atomic) = \alpha\beta - (\alpha'\beta'/2)\sqrt{1-n^2}(\xi_L/\xi_M) \tag{1.29}$$

This ratio is less than one and could even be negative if $\xi_L >> \xi_M$. For first row atoms in the periodic table $\xi_L << \xi_M$ so that the ratio is $\alpha\beta$ which will be less than one for extensive covalency. For ligand atoms in the second or third row, ξ_L is similar in

magnitude or even larger than ξ_M so that the ratio is generally
very small or even negative when there is extensive covalency.

Evaluation of $\langle x^2-y^2 | \hat{\ell}_z | xy \rangle$ is more complex and will not be
done here but similar results are obtained in that the integral
is much smaller than the atomic case when there is extensive
covalency. The general result found is that g values computed
from the atomic model are always further from the spin only value
of g_e than the g values found experimentally. In favourable
situations, experimental g values have been used to estimate the
MO parameters α, β, γ, etc. To do so, usually requires additional
data such as $E(x^2-y^2)$, ξ_M, etc. be obtained from other spectroscopic measurements.

Such calculations are useful and qualitative relationships
can be established using the parameters obtained but do not put
too much reliance in the actual numbers quoted for these parameters
because many approximations have generally been made to get them.
Attempts have been made in the literature to account for every
small overlap term including those between ligand orbitals on
different ligands but these efforts are a waste of time, in my
opinion, because they are trying to overrefine a very approximate
model. The LCAO-MO functions we start with are, at best, a poor
approximation to the real wave functions. They are only better
than using pure atomic orbitals in the calculations and they
demonstrate where and why the simple atomic calculations fail.

1.2. Hyperfine Tensor

1.2.A. <u>Terms from metal ion nucleus</u>. The terms in the
Hamiltonian that involve interaction between the nuclear spin
and the electron's spin or orbital motion can be written in the
form:

$$\hat{P}_z \hat{I}_z + \hat{P}_x \hat{I}_x + \hat{P}_y \hat{I}_y \qquad (1.30)$$

$$\hat{P}_z = P'\{(8\pi/3)\ \delta(r)\ \hat{s}_z + (3\cos^2\theta - 1)\ r^{-3}\hat{s}_z$$

$$+\ (3/2)\ \sin\theta\ \cos\theta\ r^{-3}(e^{-i\phi}\hat{s}_+ + e^{i\phi}\hat{s}_-) + r^{-3}\hat{\ell}_z\} \qquad (1.31)$$

$$\hat{P}_x = P'\{(8\pi/3)\ \delta(r)\hat{s}_x - (1/2)(3\cos^2\theta - 1)r^{-3}\hat{s}_x$$

$$+(3/4)\sin^2\theta\ r^{-3}(e^{-2i\phi}\hat{s}_+ + e^{2i\phi}\hat{s}_-) + 3\sin\theta\ \cos\phi\ r^{-3}\hat{s}_z + r^{-3}\hat{\ell}_x\}$$

$$\qquad (1.32)$$

$$\hat{P}_y = P'\{(8\pi/3)\ \delta(r)s_y - (1/2)(3\ \cos^2 -1)r^{-3}s_y \tag{1.33}$$

$$+(3/4)i\ \sin^2\theta\ r^{-3}(e^{-2i\phi}\hat{s}_+ + e^{2i\phi}\hat{s}_-) + 3\ \sin\theta\cos\theta\sin\phi r^{-3}\hat{s}_z + r^{-3}\hat{\ell}_y\}$$

$$P' = g_e g_N \mu_B \mu_N \tag{1.34}$$

where r,θ,ϕ are the polar coordinates of the electron, g_N is the nuclear g factor, and μ_N is the nuclear magneton and $\underset{\sim}{I}$ is the nuclear spin operator. The first term in Eq. (1.31 - 1.33) is the Fermi contact term for s electrons. The last term is a nuclear spin-electron orbit term and the terms in between are dipolar interaction terms appropriate to p,d and f electrons.

To include the effect of these terms we must add these terms for the Zeeman operator when we do a degenerate perturbation treatment on the Kramer's doublet of ψ_+ and ψ_-. The results can be represented by adding the term

$$\underset{\sim}{\hat{S}} \cdot \underset{\approx}{A} \cdot \underset{\sim}{\hat{I}} \tag{1.35}$$

to the spin Hamiltonian. In a diagonal representation using principal axes x,y and z this becomes

$$A_{xx}\hat{S}_x\hat{I}_x + A_{yy}\hat{S}_y\hat{I}_y + A_{xx}\hat{S}_z\hat{I}_z \tag{1.36}$$

Using the procedures developed for the g tensor we can calculate these paramaters from ψ_+ and ψ_- by use of the equations

$$A_{zz} = 2<\psi_+|\hat{P}_z|\psi_+>$$

$$A_{xx} = 2<\psi_+|\hat{P}_x|\psi_-> \tag{1.37}$$

$$A_{yy} = 2i<\psi_+|\hat{P}_y|\psi_->$$

as an example we apply these equations to the d^1 system with a tetragonal distortion for the case of $\eta>0$.

$$A_\| = A_{zz} = 2P'<\psi_+|(8\pi/3)\delta(r)\hat{s}_z|\psi_+> + P'a^2<t_2(0)|(3\cos^2\theta-1)r^{-3}|t_2(0)>$$

$$-P'b^2<t_2(-1)|(3\cos^2\theta-1)r^{-3}|t_2(-1)> \tag{1.38}$$

$$-6iabP^{\prime}<t_2(0)|\sin\theta\cos\theta\ e^{-i\phi}r^{-3}|t_2(-1)>+2P^{\prime}<\psi_+|r^{-3}\hat{\ell}_z|\psi_+>$$

The first term (Fermi contact term) gives zero using our wave function because d orbitals have a node at the nucleus. Experimental and extended theoretical analysis have shown, however, that a contribution from this term is always found due to a small polarization of spin in inner 1s, 2s and 3s orbitals of the ion. Although this polarization is small, the sensitivity of the Fermi term to even small densities of spin in the s orbitals produces a measureable effect. It is customary to represent this term by a parameter K which will be determined by experiment. K is normally found to be negative unless there is sufficient hybridization of 4s with 3d in the transition metal orbital which would give a positive K. Such hybridization is normally not permitted in complexes of high symmetries.

$$P^{\prime}<t_2(0)|(3\cos^2\theta-1)r^{-3}|t_2(0)> = -(4/7)P^{\prime}<r^{-3}> = -(4/7)P$$

$$P^{\prime}<t_2(-1)|(3\cos^2\theta-1)r^{-3}|t_2(-1)> = (2/7)P \qquad (1.39)$$

$$P^{\prime}<t_2(0)|\sin\theta\cos\theta\ e^{-i\phi}r^{-3}|t_2(-1)> = (\sqrt{2}/7)iP$$

$$P^{\prime}<\psi_+|r^{-3}\hat{\ell}_z|\psi_+> = -b^2P^{\prime}<r^{-3}> = -b^2P$$

Therefore:

$$A_{\parallel} = K + P\{-(4/7)\ a^2-(2/7)b^2+(6\sqrt{2}/7)\ ab-2b^2\} \qquad (1.40)$$

In a similar manner $A_{\perp}= A_{xx}$ can be obtained as:

$$A_{\perp} = K + P\{(2/7)a^2-(2/7)b^2-(3\sqrt{2}/7)ab-2\sqrt{2}ab\} \qquad (1.41)$$

If the distortion from octahedral symmetry is large we should include additional second order terms from the e orbitals. It should be obvious that the hyperfine tensor is also strongly determined by the nature of the ground state, particularly the dipolar part.

For the case of ^2E ground state where a tetragonal distortion gives either the $d(x^2-y^2)$ or $d(z^2)$ orbital as the ground state the following expressions are obtained for $A_{\parallel}$ and $A_{\perp}$: for the $d(x^2-y^2)$ ground state

$$A_{\parallel} = K + P\{-(4/7) + (g_{\parallel} - g_e) + (3/7)(g_{\perp} - g_e)\}$$

$$A_\perp = K + P\{(2/7) + (11/4)(g_\perp - g_e)\} \tag{1.42}$$

and for the $d(z^2)$ ground state

$$A_\parallel = K = P\{(4/7) - (1/7)(g_\perp - g_e)\}$$

$$A_\perp = K + P\{-(2/7) + (15/14)(g_\perp - g_e)\} \tag{1.43}$$

Covalency manifests itself in these equations primarily in the value of P. Extensive covalency will give smaller values of P since the electron spends more time further away from the nucleus.

An MO analysis similar to that for g tensors can be done. Most MO analyses have been done on systems having both a g and A tensor since the additional experimental parameters allow the determination of more MO coefficients.

1.2.B. <u>Ligand hyperfine terms</u>. The observation of hyperfine terms from ligand nuclei requires an MO type of analysis to explain its existance and its observation is direct evidence for covalency in the metal ion–ligand bond. The theoretical analysis is similar to that for the hyperfine interaction in the metal nucleus. One important difference occurs in the orientation of the principal axes of the tensor. Normally the principal axes for the metal nuclear interaction coincide with that for the g tensor but this is not generally true for the ligand terms.

Consider our earlier example of a square planar complex of Cu^{2+}. The dipolar part of the hyperfine interaction for nucleus 1 comes from the electron in a $p(\sigma)^1$ orbital pointed along the x axis. The hyperfine interaction is thus found to be axial but the axis of symmetry is the x axis not the z axis which is a symmetry axis for the complex. Further for nucleus 2 the symmetry axis is the y axis. Thus even if nucleus 1 and 2 are chemically equivalent their hyperfine interactions are not spacially equivalent for all orientations of the magnetic field.

For our example, it can be shown that

$$A_\parallel(L) = (\alpha')^2 n^2 (2\pi/3)|s(0)|^2 P' + (\alpha')^2(1-n^2)P_p/5 + g_\perp P_M \alpha^2$$

$$A_\perp(L) = (\alpha')^2 n^2 (2\pi/3)|s(0)|^2 P' - (\alpha')^2(1-n^2)P_p/10 - g_\parallel P_M \alpha^2/2$$

$$\tag{1.44}$$

$$P_p = P'\langle r^{-3}\rangle_p \quad ; \quad P_M = P'/R^3$$

where R is the metal ion-ligand atom distance. Since values of $\langle r^{-3}\rangle_p$ and $|s(0)|^2$ are available for some ligand atoms, measurements of $A_{||}$ (L) and $A_{\perp}$ (L) provide a direct means of obtaining values of α' and n.

1.3 Experimental determination of g and A tensor.

   1.3.A. <u>Solution of Spin Hamiltonian</u>. For S=1/2 our spin Hamilton is

$$\hat{H} = \mu_B [g_x B_x \hat{S}_x + g_y B_y \hat{S}_y + g_z B_z \hat{S}_z] + A_x \hat{S}_x \hat{I}_x + A_y \hat{S}_y \hat{I}_y + A_z \hat{S}_z \hat{I}_z$$

$$-g_N \mu_N \underset{\sim}{B} \cdot \underset{\sim}{\hat{I}} \tag{1.45}$$

If $g_N \mu_N B < A_x, A_y, A_z < g\mu_B B$ the spin Hamiltonian can be solved by first order perturbation theory to give

$$E(M_s, M_I) = g\mu_B B M_s + K M_s M_I - g_N \mu_N B M_I \tag{1.46}$$

$$g = (g_x^2 \sin^2\theta\cos^2\phi + g_y^2\sin^2\theta\sin^2\phi + g_z^2\cos^2\theta)^{1/2} \tag{1.47}$$

$$K = (g_x^2 A_x^2 \sin^2\theta\cos^2\phi + g_y^2 A_y^2 \sin^2\theta\sin^2\phi + g_z^2 A_z^2\cos^2\theta)^{1/2}/g \tag{1.48}$$

where θ, ϕ are polar angles for B in principal axes. Allowed transitions for ESR are $\Delta M_S = \pm 1$, $\Delta M_I = 0$. Frequencies for allowed transitions are then

$$h\nu(M_I) = g\mu_B B + K M_I \tag{1.50}$$

Normally the ESR spectrometer uses a constant frequency ν_0, and measures absorption for different values of the magnetic filed B. The magnetic fields for allowed transitions are then

$$B(M_I) = (h\nu_0/g\mu_B) - (K/g\mu_B)M_I \tag{1.50}$$

This gives (2I + 1) lines evenly spaced with a separation of $(K/g\mu_B)$. The spacing and center change with changes in the orientation of the magnetic field.

   For larger values of A_x, A_y and A_z we must go to second order perturbation theory which gives for axial symmetry the following equation (5)

$$h\nu(M_I) = g\mu_B B + KM_I + (A^2/4g\mu_B B)[(A^2+K^2)/K^2][I(I+1)-M_I^2]$$

$$+ [(A_\parallel^2 - A_\perp^2)^2/2g\mu_B BK^2](g_\parallel g_\perp/g^2)^2 M_I^2 \sin^2\theta\cos^2\theta \qquad (1.51)$$

Since the second order terms depend on M_I^2 they produce an uneven spacing of the hyperfine lines. The second order terms very often must be included in the analysis of metal ion nuclear hyperfine interactions. In the first transition series they are found to be important for ^{51}V, ^{55}Mn, ^{59}Co and ^{63,65}Cu. When the hyperfine terms become comparable in magnitude to the Zeeman terms only a complete solution of the spin-Hamiltonian will give a correct fit to the experimental data. This is not too difficult these days using computers.

One important thing to notice in the above equations is that both the g and A tensor elements are all squared so that the sign of either cannot be determined from the allowed transitions. Further, hyperfine lines of different M_I do not cross at any orientation of the magnetic field. The presence of a nuclear quadrupole moment for $I>1/2$ can in some orientations of the magnetic field relax the selection rules allowing transitions in which both the M_S and M_I change simultaneously. The relative signs of the hyperfine terms can sometimes be determined from these "forbidden transitions".

1.3.B. Solutions. In solutions the molecules or ions tumble rapidly giving rise to an average g and A so that the spin-Hamiltonian becomes

$$\hat{H} = \bar{g}\mu_B B_z \hat{I}_z + \bar{A}\hat{S}\cdot\hat{I} - g_N\mu_N B_z \hat{I}_z \qquad (1.52)$$

$$\bar{g} = (g_x+g_y+g_z)/3 \;;\; \bar{A} = (A_x+A_y+A_z)/3$$

If $\bar{g}\mu_B B \gg \bar{A}$ then

$$E(M_s,M_I) = \bar{g}\mu_B B M_s + \bar{A}M_s M_I - g_N\mu_N B M_I$$

$$h\nu(M_I) = \bar{g}\mu_B B + \bar{A}M_I \qquad (1.53)$$

$$B(M_I) = (h\nu_0/\bar{g}\mu_B) - (\bar{A}/\bar{g}\mu_B)M_I$$

There is a tendency to report hyperfine parameters in magnetic

field units ($\overline{A}/\overline{g}\mu_B$) because this is what is measured. This should
be discouraged, however, because it includes the $\overline{g}$ value and hence
is not directly comparable to other systems in which the $\overline{g}$ value
is different. Energy or frequency units such as MHz or cm^{-1} are
much preferred.

1.3.C. Single crystal studies. Many crystal systems have
more than one site in the unit cell for the magnetic ion. These
sites are often equivalent in that one can be generated from the
other by appropriate crystal symmetry operations. In general
the principal axes of the g and A tensors for one site are
oriented in different spatial directions from the second site.
Thus for a general direction of the magnetic field we obtain
separate resonances for the two sites. We say that the two
crystallographically equivalent sites are magnetically inequivalent.
Thus depending on the crystal system we might detect 1,2,3,4,6,8,
12 or 24 separate resonances from the same entity. Often the
number of resonances seen can be reduced if the magnetic field is
oriented in special directions in the crystal.

It is often impossible to determine g and A tensors for
single crystals of pure magnetic materials. The large magnetic
fields of neighboring unpaired spins in the crystal broadens the
ESR lines to such an extent that little information can be obtained.
In some instances the magnetic ions are so close together that an
exchange interaction takes place between the ions which narrows
the observed lines but destroys magnetic inequivalence and hyperfine
interactions due to the spins being rapidly averaged over the
whole crystal lattice. To avoid these effects the magnetic ions
must be kept far apart. This occurs naturally in some large
complexes such as the porphyrins. The most common technique is to
grow an isomorphic crystal with a diamagnetic ion and dope it with
a small amount of the magnetic ion as an impurity. We then have
a crystal of a dilute solid solution. For example we might grow
a crystal of $ZnSO_4$ with a small percentage of $CuSO_4$ to study the
Cu^{2+} resonance.

In general for a given magnetic site in the crystal we must
determine six quantities for both the g and A tensors. We wish
to know the three principal values of the tensor (g_x, g_y and g_z
for the g tensor) and the orientation of the principal axes in
relation to the crystal lattice (generally the three Euler angles).
If the site symmetry is axial only four quantities must be
determined. In principle, spectra for six different orientations
of the magnetic field would suffice to give all the data needed.
In practice, however, the problem of identifying which resonance
belongs to which magnetic site makes it necessary to do a rotation
study in which the spectra are taken at incremental angles so
that the orientation behavior of each resonance can be plotted.

The physical setup of most spectrometers makes it convenient to rotate the crystal or the magnet such that the magnetic field is kept in one crystallographic plane.

The choice of planes for rotation depends on what is known about the crystal structure. If the tensor has all three principal values different (rhombic symmetry!), three mutually orthogonal rotations will give sufficient data for a complete determination.

1.3.D. <u>Frozen solutions and powders</u>. It is possible to extract information about the g and A tensor from a frozen solution or powder spectrum when single crystals are not available or difficult to grow. First consider a simple system with hyperfine interaction and axial symmetry. From Eq. (1.50) we get

$$B = (h\nu_o/\mu_B) (g_\parallel^2 \cos^2\theta + g_\perp^2 \sin^2\theta)^{-1/2} \qquad (1.54)$$

In a powder system where all values of θ are equally possible we expect a broad absorption line since resonances will be detected for all field values between $(h\nu_o/g_\parallel\mu_B)$ and $h\nu_o/g_\perp\mu_B)$. The shape will not be symmetrical, however, and an analysis of the shape will enable us to determine $g_\parallel$ and $g_\perp$ from the shape of the spectrum. If we let dN be the number of spins whose axis of symmetry is between θ and $(\theta + d\theta)$ we can write

$$N^{-1}dN = \sin\theta \; d\theta \; = \sin\theta \; (d\theta/dB) \; dB = A(B) \; dB$$

where we assume the total range of θ is 0 to $\pi/2$. A(B) is fraction of spins with resonances in the interval B and (B + dB) and is the normalized shape function for the powder resonance line.

$$(dB/d\theta) = [(h\nu_o/\mu_B)(g_\parallel^2 - g_\perp^2)/g^{3/2}]\sin\theta\cos\theta$$

$$A(B) = \sin\theta \; (dB/d\theta)^{-1} = (g^{3/2}\mu_B/h\nu_o)(g_\parallel^2 - g_\perp^2)^{-1}\cos^{-1}\theta \; (1.55)$$

at $\theta = \pi/2$ where $B = (h\nu_o/g_\perp\mu_B)$, A(B) becomes infinite. In Fig.5 A(B) is plotted along with the more normally encountered shapes when each resonance has a finite line width rather than the very narrow line widths assumed in this derivation.

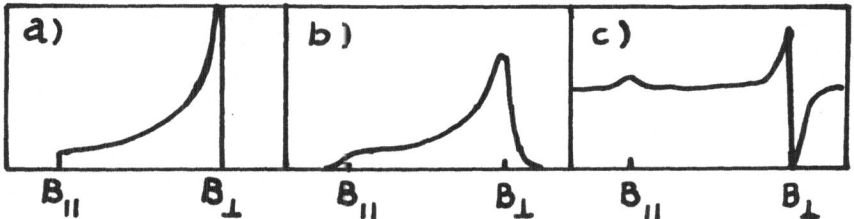

Figure 5. a) Axial powder spectrum for individually sharp resonances, b) same for finite line widths, c) first derivative of spectrum b).

For systems having hyperfine interactions we obtain (2I+1) lines similar to above one for each hyperfine line. For I=3/2 (Cu^{2+} example) we expect what is pictured in Fig. 6.

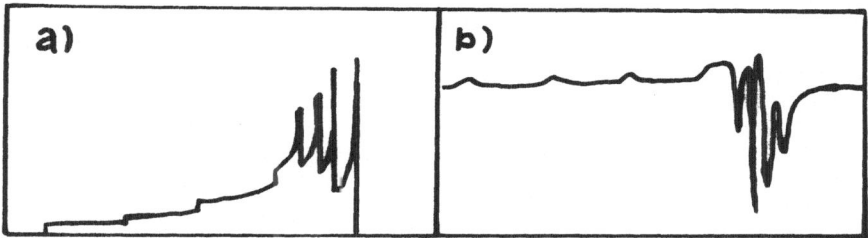

Figure 6. a) Axial powder spectrum for I=3/2 for individually sharp resonances, b) first derivative for finite line widths.

For a system with three principal g values (g_1, g_2, g_3) we would expect A(B) to have the shape shown in Fig. 7 when there is no hyperfine interaction.

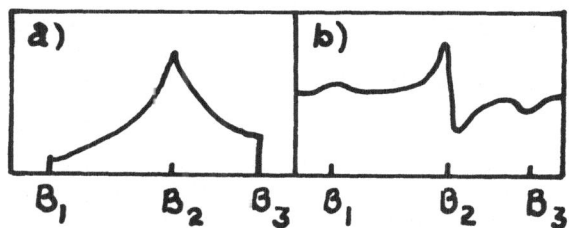

Figure 7. a) Powder spectrum for systems with three different g values, b) first derivative spectrum of same when individual resonances have finite widths.

There are reports in the literature of ESR studies of biological systems in which spectra similar to Fig. 7.b. are reported but mistakenly identified as resonances from three distinct substances. This is to be avoided, of course.

There are certain pitfalls and problems to be recognized when trying to obtain g and A tensors from frozen solution and powder spectra. It is not possible to assign a principal g or A value to a given molecular direction. This can be done for axial systems but even here one can be fooled. For example the square planar complex

has g values of 2.66, 2.01, 1.98. If this were obtained from a frozen solution spectrum alone, one would be tempted to say that the distortion from four fold symmetry is small and that the 2.66 value belongs to the direction perpendicular to the plane of the complex. Single crystal work (6), however, shows that g = 2.66 is associated with an axis in the plane of the complex.

Another problem is extraneous peaks. In systems with hyper-fine interaction, Eq. (1.50) gives

$$(dB/d\theta) = \{h\nu_o(g_{\parallel}^2 - g_{\perp}^2) - M_I[2K(g_{\parallel}^2 - g_{\perp}^2) - K^{-1}(g_{\parallel}^2 A_{\parallel}^2 + g_{\perp}^2 A_{\perp}^2)]\}$$

$$\times \quad g^{-3}\mu_B^{-1} \sin\theta\cos\theta \qquad\qquad (1.56)$$

In some instances the term in brackets can become zero at some angle between 0 and $\pi/2$ for one of the M_I values. This leads to an extra peak in the perpendicular region of the spectrum. This has been observed for copper and vanadyl complexes.

Normally line widths are fairly large and it becomes a problem to determine the correct values of B for the principal axes. To overcome this problem many workers calculate theoretical powder patterns assuming a line width and shape for the resonance from each molecule. They then adjust the parameters to get the best fit between experiment and theory. Often the fit is not as good as one could hope (very often in the literature the calculated and experimental curves are not shown and one suspects that this is due to the poor agreement between them). This poor fit is due to several assumptions made in the calculation. For convenience it is assumed that the line width and intensity of resonance lines does not depend on orientation or M_I and this is not strictly true.

We have found it necessary (7) to assume a different line width for each hyperfine line before we could begin to reproduce powder spectra of low spin Co(II) complexes. Another source of error is in neglect of weakly allowed transitions such as $\Delta M_s = \pm 1$, $\Delta M_I = \pm 1$ lines that become weakly allowed at certain orientations when there is a strong interaction between the quadrupole moment of the nucleus and the electric field gradient.

Before we leave the subject of frozen solutions, mention should be made of the use of frozen nematic solvents in which the molecules have been partially oriented by an electric or magnetic field. In favorable cases this will allow the assignment of particular principal g and A values to a particular molecular axis.

2. S = 3/2 AND 5/2 SYSTEMS

2.1. D Tensor

We will discuss here only the S = 3/2 and S = 5/2 systems in which there is no orbital degeneracy as exists in T_1 or T_2 states. These are d^3 in octahedral fields and d^7 in tetrahedral fields for S = 3/2 and d^5 systems in either field for S = 5/2. For these systems the g and A tensors are close to isotropic and hold little interest for chemical studies. The spin-spin interaction represented by the D tensor dominates the behavior of the ESR resonance in these systems.

When S > 1/2 an additional term must be added to the spin-Hamiltonian

$$\hat{\underset{\sim}{S}} \cdot \underset{\sim}{D} \cdot \hat{\underset{\sim}{S}}$$

to account for the interaction between electron spins. If x,y and z are the principal axes for this D tensor, there will be three terms, D_x, D_y, D_z. In true octahedral or tetrahedral symmetry, these terms would be equal and the spin-spin term becomes

$$D \, \hat{\underset{\sim}{S}} \cdot \hat{\underset{\sim}{S}}$$

Since all spin functions are eigenfunctions of $\hat{\underset{\sim}{S}} \cdot \hat{\underset{\sim}{S}}$ this term adds a constant energy of DS(S + 1) to every state and would not be detected in an ESR experiment which measures only energy differences. Thus we can only detect differences in the three D terms in the tensor and this occurs only for lower symmetries. If we define

$$D = D_z - (D_x + D_y)/2$$

$$E = D_x - D_y \qquad (2.1)$$

then we can write

$$\hat{\underset{\sim}{S}} \cdot \underset{\sim}{D} \cdot \hat{\underset{\sim}{S}} = D\hat{S}_z^2 + E(\hat{S}_x^2 - \hat{S}_y^2) + (1/2)(D_x + D_y)\,\hat{\underset{\sim}{S}} \cdot \hat{\underset{\sim}{S}}$$

The last term is not measureable and can be dropped. Often a
term $(1/3)DS(S+1)$ is substracted from the spin–Hamiltonian to get
a form that transforms more neatly in coordinate rotations. In
general the spin–Hamiltonian for S = 3/2 and 5/2 can be written as

$$\hat{H} = \mu_B \underset{\sim}{B} \cdot \underset{\sim}{g} \cdot \hat{\underset{\sim}{S}} + D[\hat{S}_z^2 - (1/3)S(S+1)] + E(\hat{S}_x^2 - \hat{S}_y^2) + \hat{\underset{\sim}{S}} \cdot \underset{\sim}{A} \cdot \hat{\underset{\sim}{I}} \qquad (2.2)$$

Theoretically the D and E terms come from two interactions in the
molecule. One is the normal dipolar interaction between two
magnetic spins and the second is an indirect mechanism involving
the spin–orbit interaction. The first term dominates in organic
triplet molecules where the spin-orbit interaction is weak and
distortions from a high symmetry is large. In transition metal
complexes the main contribution comes from the spin–orbit coupling
which is very large in these ions. Values of D and E are useful
in determining symmetries of electric fields about the metal ion
but have proven of little value for determining the nature of
bonding.

     Note that it is customary to designate the axes such that
$|D| > |E|$.

2.2. $E(M_s)$ vs B for S = 3/2 and 5/2.

     To illustrate the general features of the spectra and its
complexity we shall consider a S = 3/2 system with isotropic g
and E = 0. First consider case of $\underset{\sim}{B}$ along the z axis. In this
case the solution of Eq. (2.2) is readily shown to be

$$E(M_S) = g\mu_B B M_S + D[M_S^2 - (5/4)] \qquad (2.3)$$

allowed transitions are for $\Delta M_S = \pm 1$. These transitions are shown
in Fig. 8 for two situations: 1) $h\nu_o \ll D$ and 2) $h\nu_o \gg D$.

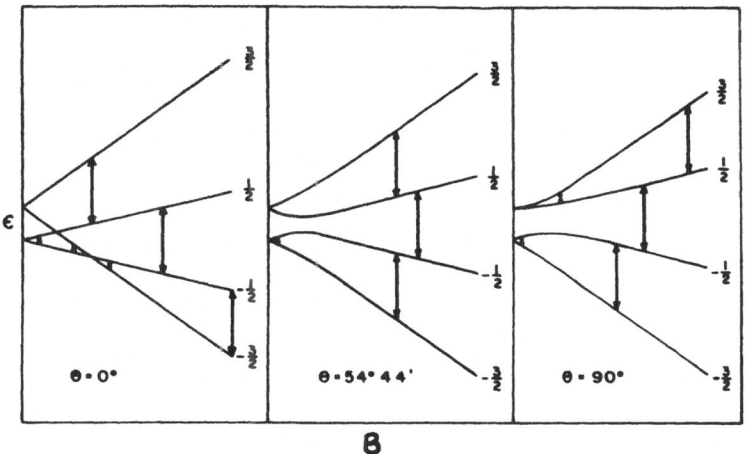

Figure 8. $E(M_S)$ vs B for S = 3/2 for three orientations of B to the z axis. Large arrows are for case when $h\nu_O \gg D$ and small ones for $h\nu_O \ll D$.

Diagrams for two other orientations of the magnetic field are also included. When $h\nu_O \gg D$ three lines are seen at all orientations but for the opposite case this is not true. In many cases when $h\nu_O \ll D$ only the one line at lowest fields is detected because the spectrometer's magnet is incapable of reaching the higher resonant fields. D values of 1 cm^{-1} are not uncommon for Cr^{3+} complexes and are even larger for second and third transition series metals.

For S = 5/2 similar behavior is found except for more lines. When $h\nu_O \gg D$ five lines are always detected but when $h\nu_O \ll D$ often only one line is found in the magnetic field range available on most spectrometers. D values of 1 cm^{-1} to 10 cm^{-1} are not uncommon in some Fe^{3+} systems.

2.3 Experimental Determination of Spin-Hamiltonian Parameters.

2.3.A. In single crystals. When $g\mu_B B \gg D$ and E = 0 we can use second order perturbation theory to obtain the equation

$$h\nu(M_S) = g\mu_B B + (D/2)[3(g_\|/g)^2 \cos^2\theta -1](2M_S-1)$$

$$-(D^2/2g\mu_B B)(g_\| g_\perp/g^2)^2 \sin^2\theta\cos^2\theta[4S(S+1)-24M_S(M_S-1)-9] \quad (2.4)$$

$$+(D^2/8g\mu_B B)(g_\perp/g)^4 \sin^4\theta[2S(S+1)-6M_S(M_S-1)-3]$$

There are (2S+1) lines equally spaced to first order with a spacing of $D[3(g_\parallel/g)^2\cos^2\theta-1]$. Maximum splitting of these fine structure lines is found when B is parallel to the z axis.

When $g\mu_B B \sim D$ we must solve complete Hamiltonian and adjust all parameters for best fit with experimental data. This is best done by computer.

When $g\mu_B B \ll D$ we normally can see only one resonance. In this case the spin system is quantized with respect to the molecular field irrespective of the orientation of the magnetic field. In this case we can only see the transition between (+1/2) and (-1/2) spin states whose energy is $-D$ for S = 3/2 when B=0 and $-8D/3$ for S = 5/2. If we consider $g\mu_B B \cdot S$ as a small perturbation and apply degenerate perturbation treatment to the $\pm 1/2$ pair of spin states we get for S = 3/2 system and axial symmetry the determinant

$$\begin{array}{c|cc} & 1/2 & -1/2 \\ \hline 1/2 & (1/2)\,g_\parallel\mu_B B_z - E & g_\perp\mu_B B_- \\ -1/2 & g_\perp\mu_B B_+ & -(1/2)\,g_\parallel\mu_B B_z - E \end{array} = 0$$

This same determinant can be obtained from the spin-Hamiltonian

$$\hat{H} = g'_\parallel\,\mu_B B_z \hat{S}_z + g'_\perp\,\mu_B(B_x\hat{S}_x + B_y\hat{S}_y) \tag{2.5}$$

solved for a S = 1/2 system if $g'_\parallel = g_\parallel$ and $g'_\perp = 2g$. Since $g_\parallel, g_\perp \sim 2.0$ for S = 3/2 systems considered here, we typically get a single resonance which appears to be an S = 1/2 system with $g_\parallel \cong 2$ and $g_\perp \cong 4$.

A similar behavior is found for S = 5/2 when $D \gg g\mu_B B$ except that $g_\parallel \cong 2$ and $g_\perp \cong 6$. If $E \neq 0$ the single resonance can be treated as an S = 1/2 system with three separate g values.

2.3.B. _Powders and frozen solutions._ When $D \ll g\mu_B B$ and $g_\parallel = g_\perp$ we can write to first order

$$B(M_s) = (h\nu_o/g\mu_B) - (D/2g\mu_B)(3\cos^2\theta-1)(2M_s-1) \tag{2.6}$$

For S = 3/2 there are three lines

$$B(-1/2) = (h\nu_o/g\mu_B) + (D/g\mu_B)(3\cos^2\theta-1)$$

$$B(1/2) = (h\nu_o/g\mu_B) \qquad\qquad (2.7)$$

$$B(3/2) = (h\nu_o/g\mu_B) - (D/g\mu_B)(3\cos^2\theta - 1)$$

The powder pattern should look like that shown in Fig. 9

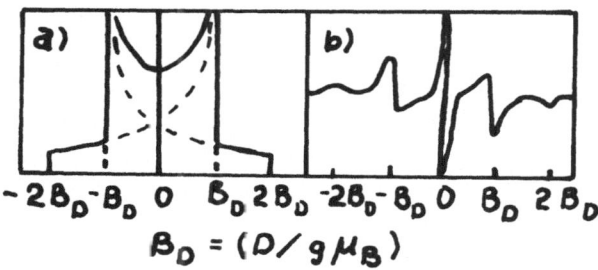

$$B_D = (D/g\mu_B)$$

Figure 9. a) Powder pattern for S = 3/2 when D << $g\mu_B B$,
b) First derivative of same when individual lines have
finite line widths.

When D ~ $g\mu_B B$ one must do a computer calculation of line
shapes and adjust parameters to get best fit for computed vs.
experimental curve. This is not easy and can use up large amounts
of computer time (particularly when there is a hyperfine interaction as in Mn^{2+}). Often researchers treat only the (-1/2 ⟷ 1/2)
transition which is narrower and centered more in the g = 2 region
of the spectrum.

For D >> $g\mu_B B$ we see only the (-1/2 ⟷ 1/2) transition which
appears similar to the S = 1/2 with axial or rhombic g values
except $g_\perp$ ~ 4 for S = 3/2 and $g_\perp$ ~ 6 for S = 5/2. In these systems
it is often difficult to see the $g_\parallel$ portion of the spectrum and
experimenters report instead an isotropic resonance at g ~ 4 or 6.
A g ~ 6 resonance is seen in soft glass due to Fe^{3+} impurities.

At values of θ between 0° and 90° the D and E terms in the
spin-Hamiltonian cause mixing of states (M_s±1) and (M_s±2) with
the M_s spin state. Thus at orientations away from θ = 0° and
θ = 90° the ΔM_s = ±2,±3 transitions become allowed to some extent
and are sometimes detected in powder patterns at lower magnetic
fields. For example if D < $g\mu_B B$ the (-3/2 ⟷ 3/2) transition is
to first order independent of θ and occurs at B = $(h\nu_o/3g\mu_B)$.
Thus this line although weak at all orientations will appear
abnormally strong in a powder pattern because it produces a
narrow line in a powder.

3. PSEUDO S = 1/2 SYSTEMS

We have seen for S = 3/2 and 5/2 that when D >> $g\mu_B B$ we see only one transition that can be interpreted by a S = 1/2 spin-Hamiltonian. These are what I would call pseudo S = 1/2 systems. The same type of pseudo S = 1/2 behavior is found for ions with T_1 or T_2 ground states and an odd number of electrons. We will consider the best known and studied example of Co^{2+} in an octahedral field.

Co^{2+} has a ^4F ground state which in the presence of an octahedral field splits to give a $^4T_{1g}$ ground state. The $^4T_{1g}$ functions can be written as

$$|0,M_s> \quad ; \quad (5/8)^{1/2}|\pm 3,M_s> + (3/8)^{1/2}|\pm 1,M_s> = e_{\pm}(M_s) \quad (3.1)$$

where $|M_L, M_s>$ is a ^4F function of eigenvalues M_L and M_s for L_z and S_z. Since d^7 is a configuration in which the d shell is more than half filled, λ is negative. We will use for H_{LS} the spin-orbit operator

$$\hat{H}_{LS} = -\lambda \hat{\underline{L}} \cdot \hat{\underline{S}} = -\lambda \hat{L}_z \hat{S}_z - (\lambda/2)(\hat{L}_+ \hat{S}_- + \hat{L}_- \hat{S}_+) \quad (3.2)$$

where λ will be taken as positive. Applying degenerate perturbation theory using Eq. (3.2) as the perturbation operator and the twelve wave functions of Eq. (3.1) we get a 12 x 12 determinant which can be factored into two 2 x 2 determinants of the form.

	$\|0,\pm 3/2>$	$e_{\mp}(\pm 1/2)$	
$\|0,\pm 3/2>$	$-E$	$-3(3/8)^{1/2}\lambda$	
$e_{\mp}(\pm 1/2)$	$-3(3/8)^{1/2}\lambda$	$(3/4)\lambda - E$	$= 0$

and into two 3 x 3 determinants of the form

	$\|0,\pm 1/2>$	$e_{\pm}(\pm 3/2)$	$e_{\mp}(\mp 1/2)$	
$\|0,\pm 1/2>$	$-E$	$-3(3/8)^{1/2}\lambda$	$-(3/\sqrt{2})\lambda$	
$e_{\pm}(\pm 3/2)$	$-3(3/8)^{1/2}\lambda$	$-(9/4)\lambda - E$	0	
$e_{\mp}(\mp 1/2)$	$-(3/\sqrt{2})\lambda$	0	$-(3/4)\lambda - E$	$= 0$

The $e_{\pm}(\mp 3/2)$ states factor out immediately with $E = (9/4)\lambda$. The
two 2 x 2 determinants have solutions of $E = (9/4)\lambda$ and $-(3/2)\lambda$
while the 3 x 3 determinants give $E = (9/4)\lambda$, $-(3/2)\lambda$ and $-(15/4)\lambda$.
Thus the spin-orbit interaction splits the $^4T_{1g}$ states into three
states of $E = (9/4)\lambda$, $-(3/2)\lambda$ and $-(15/4)\lambda$ with degenerancies of
6,4 and 2 respectively. The lowest state of $E = -(15/4)\lambda$ has a
wave function of

$$\psi_{\pm} = (3)^{-1/2}|0,\pm 1/2> - (2)^{-1/2}e_{\pm}(\pm 3/2) + (6)^{-1/2}e_{-}(\mp 1/2) \quad (3.3)$$

Applying the same procedure used in Sect. 1.1.A. we obtain by
application of the Zeeman operator

$$E = \pm(1/2) \ g\mu_B B$$

$$g = 13/3 = 4.3333 \qquad\qquad\qquad\qquad\qquad\qquad (3.4)$$

Thus the Kramer's doublet ground state in $^4T_{1g}$ behaves in a
magnetic field like an $S = 1/2$ system with $g = 4.3333$ even though
the system is in reality an $S = 3/2$ system. This is what meant
by referring to it as a "pseudo spin 1/2" system.

    The above behavior is often observed for Kramer's ions with
ground states of T_1 or T_2 in the transition metals and with rare
earth Kramer's ions. Non Kramer's ions (even number of electrons)
give ground states that behave similar to $S = 0, 1$, etc. systems.

4.   REFERENCES

1.   Ballhausen, C.J. "Introduction to Ligand Field Theory",
(McGraw-Hill, New York, 1962).
2.   Figgis, B.N. "Introduction to Ligand Fields" (Interscience
Pub. New York, 1966).
3.   McGarvey, B.R. 1975, Can. J. Chem. 53, pp. 2498-2511.
4.   DeSimone, R.E. and Drago, R.S. 1970, J. Am. Chem. Soc. 92,
pp. 2343-2352.
5.   Bleaney, B. 1951, Phil. Mag. 42, pp. 441-
6.   Malatesta, V. and McGarvey, B.R. 1975, Can. J. Chem. 53,
pp. 3791-3800.
7.   Reuveni, A., Malatesta, V. and McGarvey, B.R. 1977, Can. J.
Chem. 55, pp. 70-75.

THE ELECTRONIC GROUND STATE OF 3d METAL IONS WITH RESPECT TO THE
ESR AND NMR EXPERIMENT.

Ivano Bertini

Istituto di Chimica Generale ed Inorganica della
Facoltà di Farmacia, University of Florence, Italy.

1. SURVEY OF GENERAL THEORY

1.1. Paramagnetism in metal containing compounds

        Coordination compounds may be regarded in a rough approxima-
tion as compounds in which the d orbitals are not deeply involved
in the bonding. With this in mind the configuration of the free
ion can be a good starting point. The splitting of the five d orbitals may be considered as arising either from the electrostatic
repulsion due to the donor atoms or from their different antibonding character due to the coordination bond. The metal ions in
a coordination compound have a number of unpaired electrons which
depend on the d^n configuration and on the overall symmetry of the
molecule. When n is odd, of course there will be at least one
unpaired electron; however, in order to know the number of
unpaired electrons, the relative energies of the d orbitals
should be known and the n electrons should be placed in these
orbitals according to the Aufbau principle. The relative energies
of the d orbitals depend on the metal-donor bond strength and on
the symmetry around the metal ion (Fig. 1 and Table 1). Our
interest in this chapter will be limited only to systems with net
unpaired electrons.
        In the case of d^1 or d^9 configurations one electron will be
unpaired under any circumstance. For the d^2 configuration two
unpaired electrons will result if two orbitals at the lowest
energy are degenerate or are separated by an energy smaller than
the difference in repulsion energy. The same holds for the d^8
configuration if one keeps in mind that eight negatively charged
electrons over five d orbitals correspond to two positively
charged positrons (hole formalism). In the latter case the energy

201

I. Bertini and R.S. Drago (eds.), ESR and NMR of Paramagnetic Species in Biological and
Related Systems, 201–223.

	Octahedral	Tetrahedral	Trigonal Bipyramidal	Square Planar	Square Pyramidal
$d(x^2-y^2)$	7680	-3413	300	10320	9000
$d(z^2)$	7680	-3413	7700	-80	3800
$d(xy)$	-5120	2275	300	-2480	-3800
$d(xz)$	-5120	2275	-4150	-3880	-4500
$d(yz)$	-5120	2275	-4150	-3880	-4500

Table 1. Energies (cm^{-1}) of the d orbitals in various symmetry fields calculated through the Angular Overlap parameterization with $e_\sigma = 5200$ cm^{-1}, $e_\pi = 700$ cm^{-1} corresponding to a 10 Dq of 12800 cm^{-1}. For the transfer from the various parameterizations used in the literature see Ref. 1.

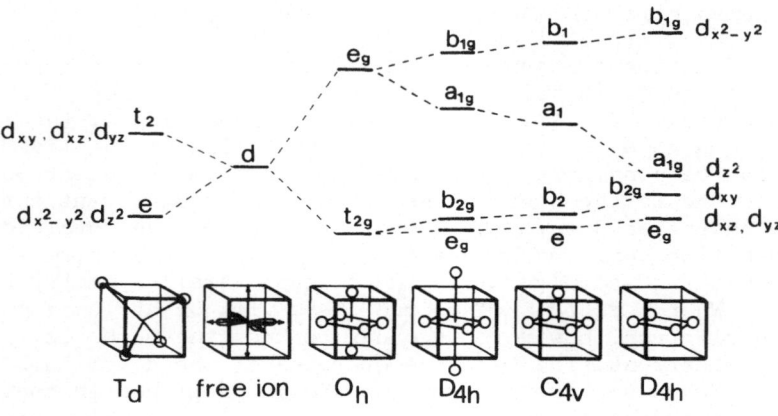

Figure 1. Splitting of d orbitals in various symmetry fields.

scale of the orbitals is reversed. With the same procedure the distribution of electrons in the split d orbitals can be obtained for the systems d^3-d^7, d^4-d^6, and d^5.

In the free ion the d^n configurations, upon the introduction of the interelectronic repulsion, give rise to free ion terms (Russell–Saunders coupling scheme) which are reported in Table 2. When the free ion is surrounded by a potential of a given symmetry, the orbital degeneracy of these terms is (partially) removed (Table 3). The number and symmetry of energy levels arising from the splitting of the free ion terms biunivocally correspond to those resulting from the interelectronic repulsion of the ground and excited configurations arising from the electron population of the split d orbitals (Fig. 2).

d^1, d^9	$^2(D)$				
d^2, d^8 $^1(S,D,G)$		$^3(P,F)$			
d^3, d^7	$^2(D)$		$^2(P,D,F,G,H)$	$^4(P,F)$	
d^4, d^6 $^1(S,D,G)$		$^3(P,F)$ $^1(S,D,F,G,I)$	$^3(P,D,F,G,H)$	$^5(D)$	
d^5	$^2(D)$		$^2(P,D,F,G,H)$	$^4(P,F)$ $^2(S,D,F,G,I)$	$^4(D,G)$ $^6(S)$

Table 2. Russell-Saunders terms for the d^n configurations. The ground state term is that with maximum spin multiplicity, and among those with equal spin multiplicity that with maximum overall multiplicity.

	O_h	T_d		O_h	T_d
$d^1, {}^2D$	^2E_g $^2T_{2g}$	$^2T_2$ ^2E	$d^6, {}^5D$	^5E_g $^5T_{2g}$	$^5T_2$ ^5E
$d^2 {}^3F$	$^3A_{2g}$ $^3T_{2g}$ $^3T_{1g}$	$^3T_1$ $^3T_2$ $^3A_2$	$d^7, {}^4F$	$^4A_{2g}$ $^4T_{2g}$ $^4T_{1g}$	$^4T_1$ $^4T_2$ $^4A_2$
$d^3, {}^4F$	$^4T_{1g}$ $^4T_{2g}$ $^4A_{2g}$	$^4A_2$ $^4T_2$ $^4T_1$	$d^8 {}^3F$	$^3T_{1g}$ $^3T_{2g}$ $^3A_{2g}$	$^3A_2$ $^3T_2$ $^3T_1$
$d^4, {}^5D$	$^5T_{2g}$ ^5E_g	^5E $^5T_2$	$d^9, {}^2D$	$^2T_{2g}$ ^2E_g	^2E $^2T_2$

Table 3. Splitting of the ground terms of free ions (d^1- d^9) in T_d and O_h symmetries.

Quantitative data on the energy levels at this stage are obtained through the Hamiltonian

$$\hat{H}_1 = \Sigma_i \, \nabla_i{}^2 - \Sigma_i \, \frac{Ze^2}{r_i} + \Sigma_{i>j} \, \frac{e^2}{r_{ij}} + V \qquad (1)$$

For 3d metal ions in a first order perturbation treatment the last two perturbations, i.e. the interelectronic repulsion and the ligand field V, are considered simultaneously.

1.2. The zero field splitting

Before the ESR experiment is taken into consideration, the energies involved in the coupling between spin and orbital angular momenta should be evaluated. For 3d metal ions this is generally a perturbation of the above energy levels which leads to a further splitting of the overall multiplicity. Besides this splitting, the Jahn Teller theorem establishes that distortions will occur in such a way to remove any orbital degeneracy through vibrational coupling. However, coordination compounds generally have very little symmetry, either because of Jahn Teller effects or since it is the most probable event.

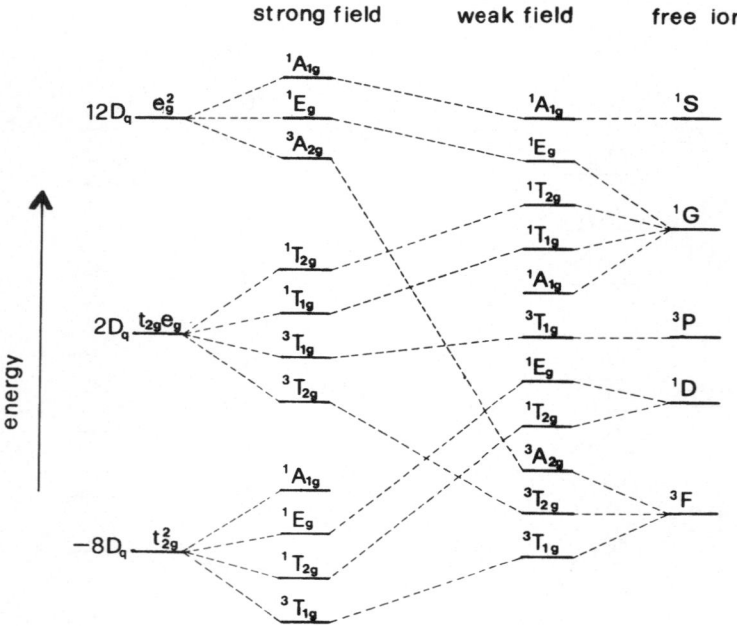

Figure 2. Correlation diagram for d^2 configuration in O_h symmetry.

In the case of systems with total spin S equal to an
integer, distortions in general remove every degeneracy. For
example in a six coordinate d^2 system the free ion term is a ^3F
which under a ligand field of O_h symmetry splits into three
levels (Fig. 2). When spin-orbit coupling and low symmetry
components are introduced (e. g. D_{4h} symmetry), the ground state
results non degenerate (vide infra). On the contrary, when the
total spin momentum is a half integer (i.e. 1/2, 3/2, 5/2 etc.)
every possible distortion leaves a double degeneracy (Kramers'
doublets). The same result is obtained from symmetry considerations
alone, since in the double point groups the system with a total
quantum number half-integer invariably belongs to an at least
doubly degenerate symmetry species. Therefore a d^1 system with ^2D
free ion term splits into two levels in O_h symmetry (Fig. 3). Spin-
orbit coupling and tetragonal elongation leaves a $^2B_{2g}$ ground
state with total multiplicity 2 given by $M_S = 1/2$. No geometrical

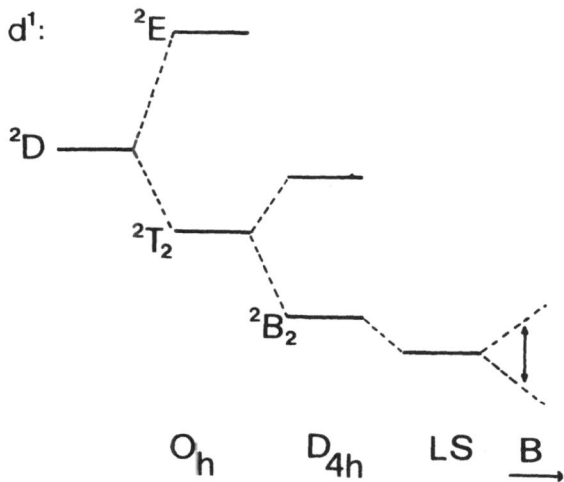

Figure 3. Splitting of a ^2D term deriving from a d^1 configuration
in D_{4h} symmetry.

distortion can remove this degeneracy.
    The splitting of the energy levels at this stage is called
zero field splitting, i.e. the splitting at zero magnetic field,
which is the starting point for any discussion on the ESR
experiment. In terms of the Hamiltonian the spin-orbit coupling
and the geometrical perturbation $\hat{V}'$, which are of the same
magnitude, are:

$$\hat{H}_2 = \Sigma_i \, \zeta \, \hat{\underline{l}}_i \cdot \hat{\underline{s}}_i + \hat{V}' \qquad\qquad (2)$$

where ζ is the monoelectronic spin-orbit coupling constant for
d orbitals. Alternatively, if the perturbation is limited to
the levels arising from a single free ion term, the Hamiltonian
can be written as

$$\hat{H}_2 = \lambda \, \hat{\underline{L}} \cdot \hat{\underline{S}} + \hat{V}' \tag{3}$$

where $\hat{\underline{L}}$ and $\hat{\underline{S}}$ characterize that free ion term and λ is the
polyelectronic spin-orbit coupling equal $\pm\zeta/2S$ for the
fundamental term.

1.3. The energies involved in the ESR experiment

When an external magnetic field is applied to a system, a
new term is added to the Hamiltonian and the energies and wavefunctions are accordingly modified. In particular, even the
Kramers' doublets degeneracy is removed.

The Hamiltonian relative to the interaction of the magnetic
field with the electron spin momentum (Zeeman effect) is given
by:

$$\hat{H}_{Zeeman} = \mu_B(K\hat{\underline{L}} + 2\hat{\underline{S}}) \cdot \underline{B} \tag{4}$$

In general the energy calculations due to Zeeman operator are
tedious since they require that all the previous perturbations
be calculated. However, this is the correct procedure in order
to obtain meaningful information from the experimental data.
Alternatively for a first glance information, it may be convenient
to define the "spin Hamiltonian": it is a new operator in which
$\underline{L}$ is considered pragmatically as affecting some parameters which
are introduced for this purpose depending on the complexity of
the system. The operator

$$\hat{H} = \mu_B \hat{\underline{S}} \cdot \underline{\underline{g}} \cdot \underline{B} + \hat{\underline{S}} \cdot \underline{\underline{D}} \cdot \hat{\underline{S}} \tag{5}$$

acts only on the spin functions and $\underline{\underline{g}}$ and $\underline{\underline{D}}$ are parameters
corresponding to the magnetogyric ratio and to the zero field
splitting. It should be remembered even if not considered here in
detail, that the total Hamiltonian contains other terms, among
which higher order S-S interaction terms for systems with $S \geq 2$,
and terms relative to the interaction between the nuclear and the
electron spin angular momenta.

The ESR experiment induces electromagnetic transitions of

low energy between levels arising from the Zeeman perturbation. The probability of a transition to occur, which is related to the intensity of the energy absorption, is different from zero at a first approximation when ΔM_S of the levels involved in the transition is ± 1.

1.4. Systems with one unpaired electron

These systems arise from the d^1, d^9, and d^7 configurations. The last configurations give rise to $S = 1/2$ only in the low spin case, i.e. when the maximum spin pairing takes place. No low spin complex is known with d^3 configuration. A d^1 complex in octahedral symmetry gives rise to three fold orbital degeneracy from which under tetragonal compression the d(xy) orbital becomes the ground state (Fig. 4). The wavefunctions of the corresponding Kramers' doublets are $d(xy)$ $(+1/2)$ ($|xy \uparrow\rangle$) and $d(xy)$ $(-1/2)$ ($|xy \downarrow\rangle$), or in the m_1, m_s formalism, $-\frac{i}{\sqrt{2}}(2-\bar{2})$ $(+1/2)$ and $-\frac{i}{\sqrt{2}}(2-\bar{2})$ $(-1/2)$. The spin operator, which is written as $\Sigma_i(\hat{1}_z \times \hat{s}_z + \hat{1}_x \times \hat{s}_x + \hat{1}_y \times \hat{s}_y) = \Sigma_i (\hat{1}_z \times \hat{s}_z + \hat{1}^+ \times \hat{s}^- + \hat{1}^- \times \hat{s}^+)$ mixes into the ground state wavefunctions with $m_1 = \pm 1$ ($d(xy)$, $d(xz)$) and $m_1 = \pm 2$ ($d(x^2-y^2)$). The resulting wavefunctions in the Dirac formalism are:

$$|\psi \uparrow\rangle = |xy \uparrow\rangle - \frac{i\zeta}{\Delta_{x^2-y^2}} |x^2-y^2 \uparrow\rangle + \frac{i\zeta}{2} \frac{1}{\Delta_{xz,yz}} (|xz \rangle -i|yz \rangle)$$

$$|\psi \rangle = |xy \rangle + \frac{i\zeta}{\Delta_{x^2-y^2}} |x^2-y^2 \uparrow\rangle + \frac{i\zeta}{2} \frac{1}{\Delta_{xz,yz}} (|xz \uparrow\rangle +i|yz \uparrow\rangle) \quad (6)$$

When the Zeeman Hamiltonian described in Eq. 4 is applied to the above functions, the corresponding energies are obtained. Now, instead of performing all these calculations, the orbital contribution to the total angular moment could have been parametrically included in the spin Hamiltonian which in this case contains only the first term of Eq. 5

$$\hat{H} = \mu_B \underset{\approx}{B} \cdot \underset{\approx}{g} \cdot \hat{\underset{\sim}{S}} \qquad (7)$$

The Hamiltonian applies to states with $M_S = \pm 1/2$. For a tetragonal symmetry when $\underset{\approx}{B}$ is parellel to the tetragonal axis

$$\hat{H} = \mu_B S_{zz} g_{\parallel} B_{\parallel} + S_\perp g_\perp B_\perp \qquad (8)$$

and the energies are $1/2\mu_B g_{\parallel}B$ and $-1/2\mu_B g_{\parallel}B$ (first term) and the transition energy $\mu_B g_{\parallel}B$. For the perpendicular case the energies are $1/2\mu_B g_\perp B$ and $-1/2\mu_B g_\perp B$ and the transition occurs at $\mu_B g_\perp B$.

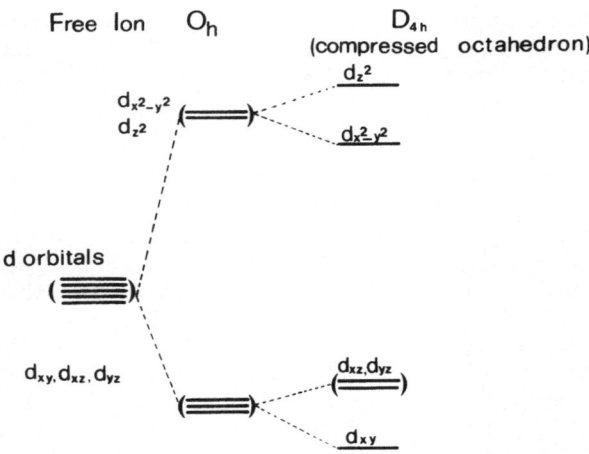

Figure 4. Splitting of d orbitals in D_{4h} (compressed octahedral) symmetry.

The values of $g_{\parallel}$ and $g_{\perp}$ are experimentally obtained; they may be empirically related to the ground state, to the coefficients of the molecular orbital of this state, to the actual value of the spin-orbit coupling constant, and so on. From the above ligand field treatment the following relations are found

$$g_{\parallel} = 2.0023 - \frac{8\lambda}{\Delta_{x^2-y^2}}$$

and   (9)

$$g_{\perp} = 2.0023 - \frac{2\lambda}{\Delta_{xz}}$$

An orbitally twice degenerate ground level occurs under the following conditions:

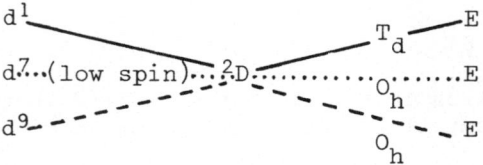

Tetragonal distortions may give rise either to $d(x^2-y^2)$ or $d(z^2)$ ground orbitals (either in the electron or in the hole formalism) and accordingly the g values are given by

$$d(x^2-y^2) \qquad\qquad d(z^2)$$

$$g_{\|} = 2.0023 - \frac{8\lambda}{\Delta_{xy}} \qquad\qquad g_{\|} = 2.0023$$

$$g_{\perp} = 2.0023 - \frac{2\lambda}{\Delta_{xz}} \qquad\qquad g_{\perp} = 2.0023 - \frac{6\lambda}{\Delta_{xz}}$$

Some typical values are reported in Table 4.

	Ground state	Symmetry	$g_{\|}$	$g_{\perp}$	Ref.
$Cu(NH_3)_4(SCN)_2$	$d(x^2-y^2)$	D_{4h}	2.257	2.056	(2)
$Cu(dien)_2(NO_3)_2$	$d(z^2)$	C_2	2.052	2.151	(2)
Cs_2CuCl_4	$d(xy)$	D_{2d}	2.384	2.093	(2)
$Cu(Me_6tren)Br_2$	$d(z^2)$	D_{3h}	1.956	2.182	(3)

Table 4. ESR parameters of some copper(II) complexes

1.5. Systems with two unpaired electrons

These systems arise from d^2 and d^8 configurations. The free ion term is a ^3F which in the d^8 case splits in O_h symmetry and under following perturbations according to the diagrams of Figure 5. In absence of zero field splitting and assuming that only the T_2 level mixes with the ground level, two degenerate transitions with $|\Delta M_S| = 1$ occur (Figure 5, dotted lines), the experimental g values being equal to $g_e - 8\lambda/\Delta$, where Δ is the energy separation of the T_2 level. The zero field splitting, in the approximation that only the first excited level is contributing, is given by $D = -4\lambda^2\delta/\Delta^2$ and the g values by

$$g_{\|} = g_e - \frac{8\lambda}{\Delta_{\|}} \qquad \text{and} \qquad g_{\perp} = g_e - \frac{8\lambda}{\Delta_{\perp}}$$

The g and D values are extracted from the experimental transitions through the spin Hamiltonian (axial case)

$$\hat{H} = \mu_B \hat{S}_{zz} g_{\|} B_{\|} + \mu_B \hat{S}_{xx} g_{\perp} B_{\perp} + D (\hat{S}_z^2 - 2/3) \qquad (10)$$

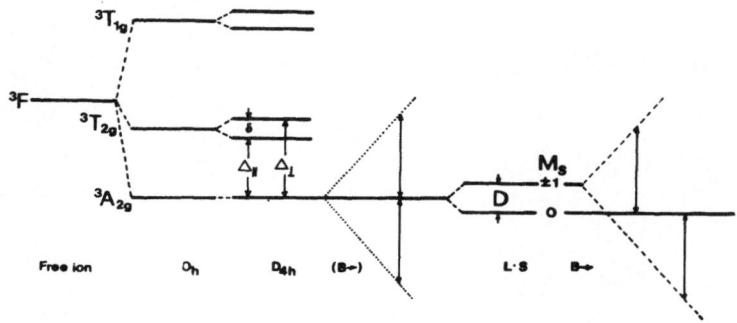

Figure 5. Splitting of the states of a free d^8 ion in an octahedral field with an added tetragonal distortion.

1.6. Systems with higher spin multiplicity

    Systems with three unpaired electrons.- d^3 and d^7 configura-
tions may give rise to systems with S = 3/2 and ^{4}F free ion ground
terms. In the case of sixcoordinate d^3 under tetragonal distortion
the $^4A_2$ state gives rise to two Kramers' doublets (Fig. 6). In
the case only one signal is obtained, it can be interpreted in
terms of the fictitious spin S = 1/2, just like the transition
always occurred between the Kramers' doublets M_S = 1/2, otherwise a spin Hamiltonian with S = 3/2 of the type of that of the
nickel can be used. In these systems the spin-orbit coupling is
strongly operative especially if a T ground state occurs in the
idealized symmetry, and small geometrical distortions dramatically
affect the shape of the spectra as well as the magnitude of the
spin Hamiltonian parameters.
    Systems with four unpaired electrons.- Such systems, which
derive from d^4 and d^6 configurations, are quite unknown owing to
the difficulty to observe the signal. If the spin-orbit operator
does not mix too hevily all the spin states, only one transition
is expected when the triply degenerate level is the ground one.
    Systems with five unpaired electrons.- Manganese(II) and
Iron(III) often display ground states with spin multiplicity
arising from the free ion term ^{6}S. Higher order spin-orbit
coupling perturbations together with a deviation from the cubic
symmetry splits the energy levels into three Kramers' doublets.
The ESR transition between the levels with $M_S = \pm1/2$ falls in the
range of g = 2. The spin Hamiltonian is of the form:

$$\hat{H} = \mu_B \hat{\underset{\approx}{S}} \cdot \underset{\approx}{g} \cdot \underset{\approx}{B} + D(\hat{S}_z^2 - \frac{35}{12}) + E(\hat{S}_x^2 - \hat{S}_y^2) + (\hat{\underset{\approx}{I}} \cdot \underset{\approx}{A} \cdot \hat{\underset{\approx}{S}}) +$$

$$+ \frac{1}{6} a (\hat{S}_x^4 + \hat{S}_y^4 + \hat{S}_z^4 - \frac{707}{16}) + \frac{1}{180} F (35 \hat{S}_z^4 - \frac{475}{2} \hat{S}_z^2 + \frac{3255}{16}) \quad (11)$$

Whereas D can easily be determined from experiments, quite accurate data are necessary for the determination of a, which corresponds to higher order spin-orbit perturbation. For a determination of g and D values as they depend on the geometry of the metal ion, see chapter VIII.

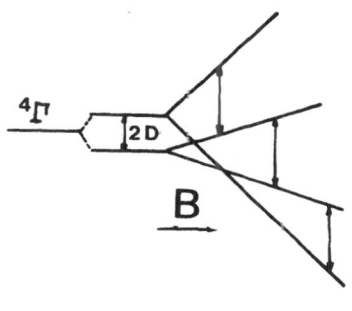

Fig. 6. ESR transitions for a
d^3 ion in tetragonal
symmetry

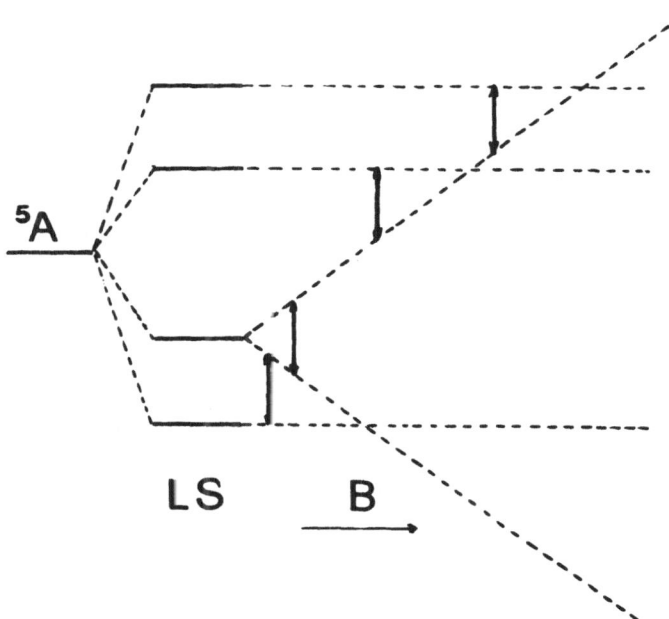

Figure 7. ESR transitions for a d^4 ion in tetragonal symmetry

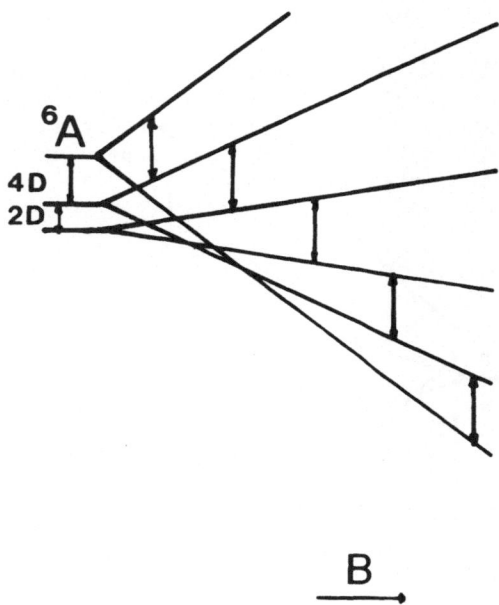

Figure 8. ESR transitions for a d^5 ion in tetragonal symmetry

2. THE ESR EXPERIMENT

2.1. Conditions of detectability of the ESR signal

     Before entering the problems of the electron relaxation
times it should be kept in mind that the investigability through
the ESR experiment is limited by the energy of the microwave
(commercially available frequencies are 2.7GHz (S-band), 9.1GHz
(X-band), 18.2GHz (K-band), 34.2GHz (Q-band)) and by the intensity
of the field which generally may be as high as 1.5 T and anyhow
can reach a few T with superconducting magnets.
     Therefore the ESR signal is generally detectable when the
transition occurs within a Kramers' doublet.
     In case of S integer the observability of the signal depends
on the zero field splitting: if this is large for example in low
symmetry nickel(II) (Fig. 5) no signal is observed, whereas two
signals can be observed when D is small (4). In the case of high
spin manganese(II) several transitions can be observed besides
the one arising from the Kramers' doublet with M_s = ±1/2 (5). It
should be remembered that in principle a transition is observed

when $\Delta M_S = \pm 1$; however, the spin orbit coupling operator mixes levels of different M_S in such a way that the rule may not be valid any more.

2.2. Factors affecting the electronic relaxation times

The possibility of detecting a resonance signal depends on the width of the signal itself which on its turn depends on what is called transversal electronic relaxation rate T_{2e}^{-1}. This rate is given by

$$T_{2e}^{-1} = T_{1e}^{-1} + T_e^{-1}$$

the longitudinal relaxation rate T_{1e}^{-1}, which corresponds to the rate constant of the first order process for an assembly of electron spins to reach the thermal equilibrium in a magnetic field, and T_e^{-1} the rate with which the spins loose their phase coherency as due to local non homogeneus fields. Whereas T_e^{-1} mostly depends on the experimental conditions, i.e. concentration of the paramagnetic species, inhomogeneties arising from librations, tumbling of the molecules etc., T_{1e}^{-1} depends largely on the intrinsic possibility for a spin to relax through mechanisms which involve energy dissipation through the spin environment. For this reason T_{1e} is called spin lattice relaxation time.

The efficiency of this relaxation mechanism is bound to the coupling of electronic and vibrational wavefunctions. In the direct (single phonon) process, the vibrational quantum must have the same energy as the Zeeman energy, which is much smaller than the average phonon energy given by the Boltzmann distribution, unless the temperature is very close to 0 K. A further mechanism (two phonon process) is therefore needed to yield the relaxation rates observed at higher temperatures. The process involves the absorption of a thermal phonon (which is easily available) by the spin system, and the emission of a slightly more energetic phonon, the difference in energy being again the Zeeman energy. The spin system is passing through an excited state, which can be a real (resonance) or a virtual (Raman) level. The existence of low-laying energy levels favours the relaxation mechanism, either causing a more efficient mixing of ground and excited levels or allowing a "resonance" process which is far more efficient than the "Raman" one. It may be taken as a general rule that in the presence of a large number of states close to each other at zero field the electronic relaxation times are short. Just to give a rough idea of such times, organic radicals display T_{1e} values of the order of 10^{-7}s, manganese(II) compounds 10^{-8}s, cobalt(II) compounds from 10^{-11} to 10^{-13}s. In this respect the symmetry plays an important role: tetrahedral cobalt(II) complexes have a $^4A_2$ ground state and a $^4T_1$ at about 4000 cm^{-1}, whose wavefunctions are in general well mixed. As a result large zero field splittings occur as well as large magnetic susceptibility anisotropy. The electronic relaxation times are estimated to be about 10^{-11}s. When high spin cobalt(II) is in an octahedral field the ground state is $^4T_1$ with an overall degeneracy of 12. Zero field splittings are here more dramatic and accordingly the electronic relaxation times shorten up to 10^{-13}s. Since T_2 is related to the half height linewidth through the relation $T_{2e}^{-1} \simeq \pi \Delta \nu$, a T_{2e} value of 10^{-11} s gives a half height linewidth of about 5000G, which is hardly observable. By freezing the system the electronic relaxation times increase and the signals sharpen up. Therefore at liquid helium temperature the signal is often sharp enough to be observed. However, it may be still so broad to mask hyperfine interactions and therefore to limit the number of observables.

When copper(II) is taken into consideration one should remind that in general Jahn-Teller effects are so strong that the deviations from the cubic symmetries are very large. Therefore in this case there never is any zero field splitting except that due to major coordination distortions and therefore the electronic relaxation times are less dramatically affected by the coordination number or coordination geometry.

Some estimated electronic relaxation times in dilute solutions at room temperature are reported in Table 5.

		ref.			ref.
V^{2+}		$10^{-9}-10^{-10}$ (6)	Fe^{2+}	$\sim 10^{-12}$	(9)
Cr^{3+}		$10^{-9}-10^{-10}$ (6)	Co^{2+}	$10^{-11}-5\times10^{-13}$	(10,11)
Mn^{2+} (high spin)	$10^{-8}-10^{-9}$ (6)	Ni^{2+}	$10^{-10}-10^{-12}$	(6,12)
Fe^{3+} (high spin)	$10^{-10}-10^{-11}$ (7)	Cu^{2+}	$10^{-8}-10^{-9}$	(6,13)
Fe^{3+} (low spin)	$<10^{-12}$ (8)			

Table 5. Common electronic relaxation times (s) for some metal ions.

2.3. The hyperfine coupling constant A.

When the signal is sharp enough some fine structures can be observed due to coupling between electron and nuclear spins. Generally the coupling with the nuclear spin of the metal itself (hyperfine structure) is larger than that with the other nuclei (superhyperfine structure); sometimes coupling with the nuclear quadrupole is also observed. Of particular interest is the coupling between the electron spin and the metal nuclear spin. The extent of this coupling depends on the Fermi contact and on the dipolar interaction between the electronic magnetic moment in its spin and orbital components and the nuclear magnetic moment. All of these components and in particular the contact contribution and the contribution due to the orbital magnetic moment strongly depend on the symmetry and nature of the donor atoms. Ligand Field

calculations tend today to reproduce with the same set of
parameters the energies of the d-d transitions, the g values, the
A values, as well as their main directions. In Table 6 the

	$A_{\parallel}$ (cm^{-1}x10^{-4})	$A_{\perp}$ (cm^{-1}x10^{-4})
D_{3h} (trigonal bipyramidal)	60-100	60-100
C_{4v} (square pyramidal)	120-150	<40
D_{4h} (four or six coordinate)	160-200	<30
D_{2d} (flattened tetrahedral)	The A values average to about 50 as the T_d symmetry is approached.	

Table 6. - Range of A values for copper complexes in different
geometries.

range of A values for copper(II) complexes in different geometries
are reported. Studies have been performed also on geometries
intermediate between trigonal bipyramidal and square pyramidal,
and between square planar and tetrahedral (14). All the experimental
values have been fully accounted for by using negative contact
contributions of about 50-100 x 10^{-4} cm^{-1}.

Of course the values of Table 6 as well as their intermediate
values allowed to provide structural information on copper
substituted enzymes. For example copper thermolysin, copper
carbonic anhydrase, and copper carboxypeptidase (15), whose ESR
spectra are shown in Fig. 9, display $A_{\parallel}$ values which indicate
deviation from planarity in the above order.

All the above considerations are valid for dilute systems in
a solid matrix (doped crystals or frozen solutions). When averaging
problems arise due to polycrystalline samples or to rotating systems
for example in solution other parameters should be taken into
consideration.

2.4. ESR in solution

In a dilute solution of a paramagnetic complex it can be
assumed that the average distance between two paramagnetic centers
is large enough to vanish dipole-dipole and spin exchange interactions, just like in a doped crystal. However, an additional
factor is still present which can influence the shape of an ESR
spectrum in solution. This factor is bound to the anisotropy of
the g values and hyperfine or spin-spin interactions, and to the
tumbling or other motion frequencies of the complex. If the
motional rates are rapid enough to average out all the anisotropies,
the signal widths are again determined by the intrinsic electronic
relaxation times. If these rates are very slow, or quenched as in
frozen solution, all the anisotropies are conserved and sometimes

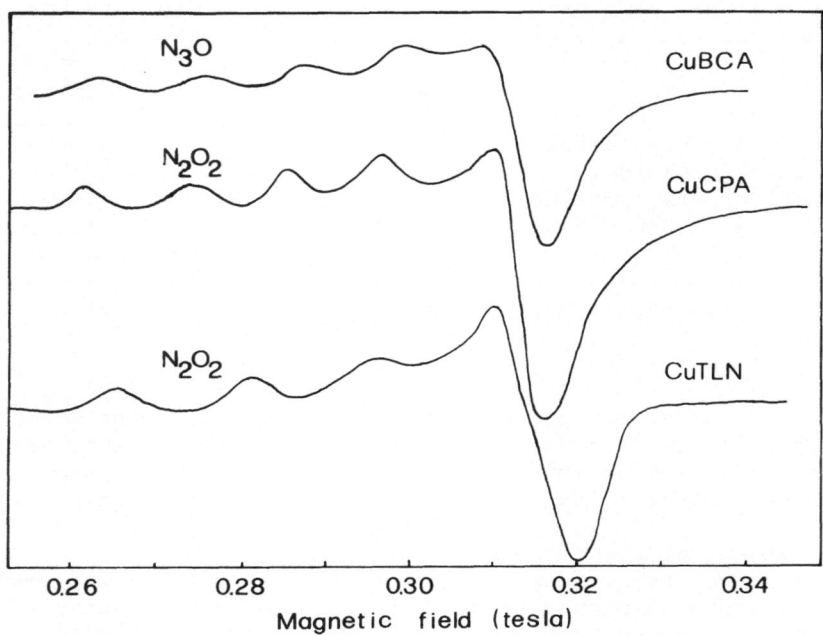

Figure 9. ESR spectra and donor atoms sets of copper in carbonic anhydrase (CuBCA), carboxypeptidase (CuCPA) and thermolysin (CuTLN). The $A_{||}$ values are 131, 124, and 163 x 10^{-4} cm^{-1}, respectively.

it is still possible to obtain the parameters of the spin Hamiltonian although they cannot be related to the coordinate system of the molecule. For S=1/2 systems two main mechanisms are operative (16) with opposite temperature dependence; the first one is a spin rotational mechanism due to the coupling between the spin angular momentum and the rotational angular momentum of the molecule, which is field and I independent (I being the nuclear spin multiplicity) and yields a positive temperature dependence of the linewidth; the second one is due to the modulation of g and A anisotropic tensors caused by the change in orientation of the molecule with respect to the external field, is field and I dependent and shows a negative temperature dependence. For S>1/2 systems the main mechanism arises from zero field splitting modulations, and the linewidths are directly proportional to D^2 and inversely proportional to the temperature.
    Figure 10 illustrates the temperature dependence of the

$Cu(H_2O)_6^{2+}$ ESR signal in aqueous solutions adsorbed on a silica gel (17): at room temperature only one average signal is observed, while at -70 °C an axial spectrum is apparent which also clearly shows hyperfine structure in the parallel region. The linewidth decreases and then increases again owing to the two opposite mechanisms discussed before; at intermediate temperature the signal is sharp enough to observe the isotropic hyperfine splitting.

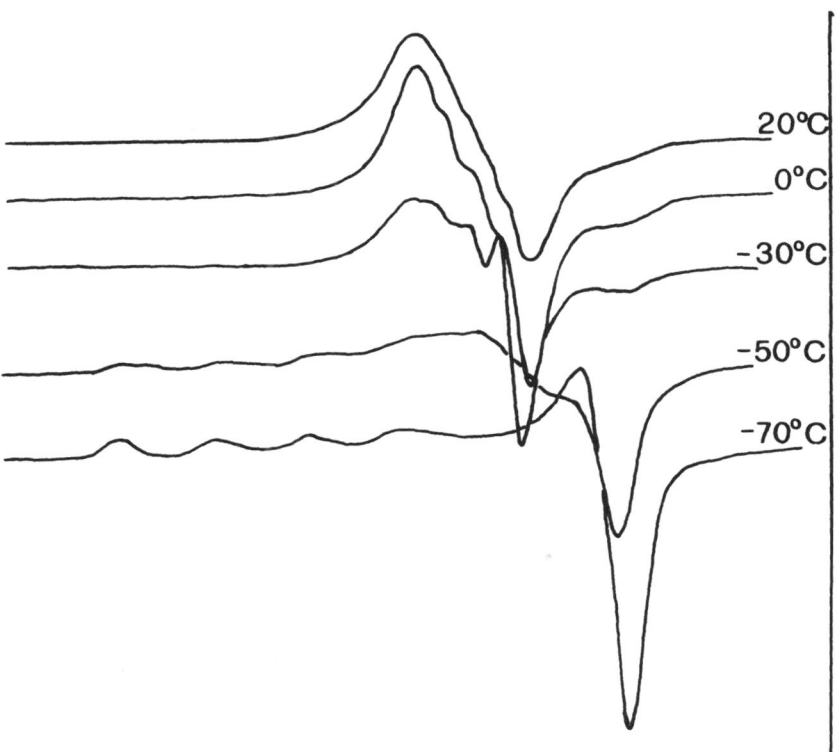

Figure 10. ESR spectra of $Cu(H_2O)_6^{2+}$ aqueous solution adsorbed on a silica gel S4, as a function of temperature.

3. THE NMR EXPERIMENT

3.1. The electronic relaxation times and the NMR experiment

The NMR parameters of a nucleus somehow interacting with

unpaired electrons are quite different from those in absence of
such interaction. In particular an additional shift is observed
which is called isotropic shift and which is due to a contact
contribution and to a dipolar contribution. The NMR experiment
which we will consider here is performed in solution and in this
case the dipolar contribution is that due to the anisotropy of
the magnetic tensor. In terms of the Hamiltonian this is just
the same term $\hat{I} \cdot \underset{\sim}{A} \cdot \hat{S}$ discussed for the electron. The other parameters which we will mostly be interested to are the longitudinal
relaxation rate, T_1^{-1}, and the transverse relaxation rate, T_2^{-1},
which are defined in a parallel fashion as for the electron. T_2^{-1}
can be directly obtained from linewidth measurements ($T_2^{-1} = \pi \Delta \nu$).
As discussed in chapters IV and VII, the increase in relaxation
rates due to a paramagnetic species can be described by the
Solomon-Bloembergen and Morgan equations:

$$\frac{1}{T_{1M}} = \frac{2}{15} \frac{S(S+1)\gamma_I^2 g^2 \mu_B^2}{r^6} \left(\frac{3\tau_{c1}}{1+\omega_I^2 \tau_{c1}^2} + \frac{7\tau_{c2}}{1+\omega_S^2 \tau_{c2}^2} \right)$$

$$+ \frac{2}{3} \frac{S(S+1)A^2}{\hbar^2} \frac{\tau_e}{1+\omega_S^2 \tau_e^2} + \frac{1}{T_{1LC}} \qquad (12)$$

$$\frac{1}{T_{2M}} = \frac{1}{15} \frac{S(S+1)\gamma_I^2 g^2 \mu_B^2}{r^6} \left(4\tau_{c1} + \frac{3\tau_{c1}}{1+\omega_I^2 \tau_{c1}^2} + \frac{13\tau_{c2}}{1+\omega_S^2 \tau_{c2}^2} \right)$$

$$+ \frac{1}{3} \frac{S(S+1)A^2}{\hbar^2} \left(\tau_e + \frac{\tau_e}{1+\omega_S^2 \tau_e^2} \right) + \frac{1}{T_{2LC}} \qquad (13)$$

    The first term is the dipolar coupling between the unpaired
electrons residing on the metal and the resonating nucleus, the
second term is due to the contact term and the third term is due
to a dipolar coupling between the electrons delocalized onto the
ligand and the resonating nucleus. Whereas all the symbols have
the usual meaning, τ_{ci} deserves a further comment. It is the time
constant for reorientation of the coupled electron and nuclear
spins, which is given by

$$\tau_{ci}^{-1} = T_{ie}^{-1} + \tau_r^{-1} + \tau_e^{-1} \qquad (14)$$

where T_{ie} corresponds to the electronic relaxation times discussed before, τ_r is the rotational time of the coupled system, and τ_e is the electron exchange rate. Whichever is the shortest time it determines τ_c. Rotational times of small molecules may range between 10^{-10} and 10^{-12}s whereas for proteins such value is even larger than 10^{-8}s. Therefore for macromolecules containing 3d metal ions T_{ie} is usually shorter than τ_r and it determines the value of τ_c. Under these conditions tetrahedral nickel complexes with orbitally triply degenerate ground states (Table 5) give relatively sharp signals. For example they provided a powerful tool to detect subtle structural differences in proton continuous wave NMR owing to the large isotropic shifts observed without a dramatic line broadening. Analogous results were obtained with octahedral vanadium(III), iron(II) and several other metal complexes.

Sometimes it is possible to shorten up the electronic relaxation times of a given metal ion through spin-spin coupling with other metal ions having short electronic relaxation times. For example $Fe(CN)_6^{3-}$ shows a broad ESR signal only at liquid helium temperature on account of its short electronic relaxation times. It binds planar copper(II) complexes giving rise to five coordinate adducts.

The ESR signal of the copper(II) ion disapperars at room and liquid nitrogen temperature as a result of the coupling with iron(III). Contemporarily the NMR signals of the ligand nuclei of the copper complexes dramatically sharpen up (18).

3.2. The NMR experiment under chemical exchange conditions

When recording the NMR of diamagnetic moieties chemically exchanging with a paramagnetic center the correlation rates τ_{ci}^{-1} (Eq. 14) should include also the reciprocal of the residence time τ_M, or ligand exchange rate. However, τ_M is often orders of magnitude longer than either T_{ie}, τ_c or τ_r and we will not consider its effect on τ_c, here. On the other hand, it still dramatically influences the width of the experimental lines through an exchange broadening mechanism. The molecule containing the nuclei under consideration can reside in two different environments, the paramagnetic site and the diamagnetic one, usually the bulk solvent site. If τ_M is slow enough compared to $(\Delta\omega)^{-1}$, $\Delta\omega$ being difference in chemical shift between the two sites, two signals are in principle observable, although the paramagnetic one can be sometimes undetectable if too broad, or too far in chemical shift, or both. If τ_M is shorter than $(\Delta\omega)^{-1}$ only one signal is observed, which partly retains the information on the paramagnetic site.

In this case the observed T_{2obs}^{-1} ($=\pi\Delta\nu$) must diminished by the term T_{2dia}^{-1} obtained in the absence of paramagnetic species, to yield the paramagnetic contribution T_{2p}^{-1},

$$T_{2p}^{-1} = T_{2obs}^{-1} - T_{2dia}^{-1} \qquad (15)$$

which on its turn must be normalized to the molar fraction of ligand bound to the paramagnetic center, f. fT_{2p} is now a function of $\Delta\omega$, τ_M and T_{2M}

$$fT_{2p}^{-1} = \tau_M^{-1} \frac{T_{2M}^{-2} + (T_{2M}\tau_M)^{-1} + (\Delta\omega)^2}{(T_{2M}^{-1} + \tau_M^{-1})^2 + (\Delta\omega)^2} \qquad (16)$$

When τ_M is shorter than both $(\Delta\omega)^{-1}$ and T_{2M} the NMR signal sits at the weighted average position between the two sites, and fT_{2p} is equal to T_{2M} as defined in Eq. 13. When τ_M is slowed down for example by lowering the temperature, towards the values of T_{2M} (quasi fast region) the signal gets broadened, fT_{2p}^{-1} approximating $(\Delta\omega)^2\tau_M$. By further increasing τ_M, the signal coalesces and then separates into two signals; this is the quasi slow exchange region, in which fT_{2p} measured on the diamagnetic signal is equal to τ_M. Finally, when τ_M is small enough, fT_{2p} is not measurable anymore, since $T_{2obs} \simeq T_{2dia}$ in Eq. 15, and the natural diamagnetic linewidth is obtained. In the case of the quasi slow exchange region, if the detatchment of a molecule from the paramagnetic site is thought to be a first order kinetic process, then τ_M^{-1} is related to $\Delta G^{\ddagger}$ through the equation

$$\tau_M^{-1} = \frac{KT}{h} \ \exp{(\Delta G^{\ddagger} / RT)} \qquad (17)$$

which also shows that τ_M decreases with increasing temperature. From the slope of the temperature dependence of fT_{2p} and τ_M also the other parameters $\Delta H^{\ddagger}$ and $\Delta S^{\ddagger}$ are determined. In the case of longitudinal relaxation rates the measured value, T_{1p}, should similarly be normalized to the ligand fraction bound to the paramagnetic center, f. The dependence of fT_{1p} on the residence time τ_M is

$$fT_{1p} = T_{1M} + \tau_M \qquad (18)$$

where T_{1M}^{-1} is given by Eq. 12. The limiting cases are $fT_{1p} = T_{1M}$ (fast exchange) and $fT_{1p}=\tau_M$ (slow exchange). The fT_{1p}^{-1} dependence on τ_M is similar to that of fT_{2p}^{-1} when the isotropic shift ($\Delta\omega$) is zero. Therefore under rapid exchange conditions the relaxation times are related to f, τ_c, and to the distance of the resonating nucleus from the paramagnetic center.

In the case of the cobalt substituted metalloenzyme carbonic anhydrase we could conclude, supported also from other experimental data, that a water molecule is always bound to the metal at any pH, giving rise to a tetrahedral geometry with three histidine nitrogens (19). This information was obtained through ^1H relaxation study of water solutions containing the metalloenzyme. As previously mentioned tetrahedral cobalt(II) complexes have electronic relaxation times longer up to two orders of magnitude than the corresponding octahedral complexes. Accordingly, T_{1p}^{-1} measurements gave larger values for the cobalt enzyme than for $Co(OH_2)_6^{2+}$. These measurements allowed also to establish that five coordinate trigonal bipyramidal cobalt complexes have intermediate electronic relaxation times since T_{1p}^{-1} is close to that of $Co(OH_2)_6^{2+}$ whereas the latter complex contributes to the relaxivity with six water molecules. This is consistent with our hypothesis that sometimes inhibited cobalt carbonic anhydrase may be five coordinate, still with a water molecule but with smaller T_{1p}^{-1} values. Of course this is related to the always short electronic relaxation times of cobalt. With the copper(II) derivative of carbonic anhydrase the electronic relaxation times are longer, and larger is the proton relaxivity (20). In Fig. 11 ^1H $(fT_{1p})^{-1}$ values are reported for several metallosubstituted carbonic anhydrases, (19-23), together with the estimated electronic relaxation times. It should be noted that although VO^{2+} displays the shortest electronic relaxation times its relaxing capability is smaller than that, for example, of copper(II). If we look at eq. 12 the efficiency of the dipolar coupling results to depend on a non linear function of τ_c (Fig. 12). The function reaches a maximum for τ_c values equal to the reciprocal of the nuclear Larmor precession frequency, ω_I, and then decreases again. For a proton 80 MHz spectrometer, $(\omega_I)^{-1}=2\times10^{-9}$ s.

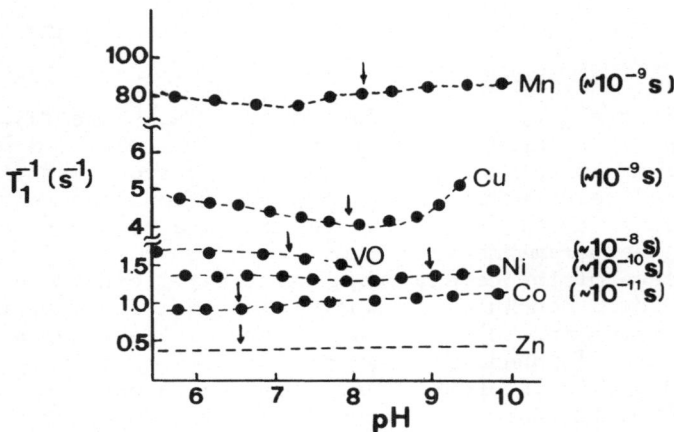

Figure 11. Water proton relaxation rates for 10^{-3}M solutions of
metallocarbonic anhydrases. In parentheses the electronic
relaxation times. The arrows indicate the pK$_a$ of the metal-linked
acid-base equilibria.

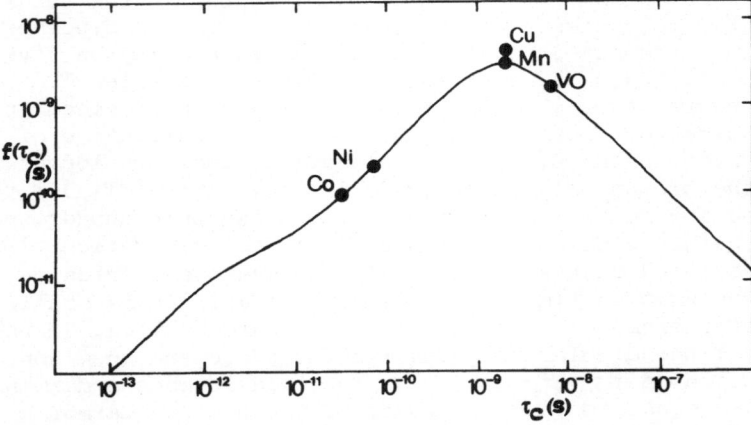

Figure 12. $f(\tau_c)$ part of the Solomon-Bloembergen equation (Eq. 12)
as a function of τ_c. The dots represent the $f(\tau_c)$ values
calculated through eq. 12 using the relaxation data for cobalt(II),
nickel(II), copper(II), manganese(II), and oxovanadium(IV)
carbonic anhydrase derivatives and taking r=280 pm.

REFERENCES

1) Bertini, I., Gatteschi, D., and Scozzafava, A.: 1977, Isr. J. Chem. 15, p. 189.
2) Hathaway, B. J., and Billing, D. E.: 1970, Coord. Chem. Rev. 5, p. 143.
3) Barbucci, R., Bencini, A., and Gatteschi, D.: 1977, Inorg. Chem. 16, p. 2117.
4) Reedijk, J., and Nieuwenhuise, 1972, Rec. Trav. Chim., 91, p. 533.
5) Abragam, A., and Bleaney, B., "Electron Pramagnetic Resonance of Transition Ions", Clarendon Press, Oxford, 1970, p. 436.
6) Bloembergen, N., and Morgan, L. O.: 1961, J. Chem. Phys. 34, p. 842.
7) Wishnia, A.: 1960, J. Chem. Phys. 32, p. 871.
8) Baker, J. M., Bleaney, E., and Bowers, K. D.: 1956, Proc. Phys. Soc. B, p. 1205.
9) Eisinger, J., Shulman, F.G., and Szymanski, B. M.: 1962, J. Chem. Phys. 36, p. 1721.
10) Luz, Z., and Meiboom, S.: 1964, J. Chem. Ohys. 40, p. 1508.
11) Fabry, M. E., Koenig, S. H., and Shillinger, W. E.: 1970, J. Biol. Chem. 245, p. 4526.
12) Luz, Z., and Meiboom, S.: 1964, J. Chem. Phys. 40, p. 1066.
13) McConnell, H.: 1956, J. Chem. Phys. 25, p. 709.
14) Bencini, A., Bertini, I., Gatteschi, D., and Scozzafava, A.: 1978, Inorg. Chem. 17, p. 3194.
15) Bertini, I., Canti, G., Kozlowski, H., and Scozzafava, A.: J. C. S. Dalton, in press
16) Lewis, W. B., and Morgan, L. O.: 1968, Trans. Metal Chemistry 4, p. 33.
17) Bassetti, V., Burlamacchi, L., and Martini, G.: J. Amer. Chem. Soc., in press
18) Bertini, I., Luchinat, C., Mani, F., and Scozzafava, A.: unpublished results.
19) Bertini, I., Canti, G., Luchinat, C., and Scozzafava, A.: 1978 J. Amer. Chem. Soc. 100, p. 4873
20) Bertini, I., Canti, G., Luchinat, C., and Scozzafava, A.: 1978, J. C. S. Dalton, p. 1269.
21) Bertini, I., Luchinat, C., and Scozzafava, A.: 1978, FEBS Letters 87, p. 92.
22) Bertini, I. Borghi, E., and Luchinat, C.: 1978, Bioinorg. Chem. 9, p. 495.
23) Bertini, I., Canti, G., Luchinat, C., and Scozzafava, A.: Inorg. Chim. Acta, in press.

E.P.R. OF IRON IN BIOLOGICAL SYSTEMS

John F. Gibson

Imperial College,
London, SW7 2AY, U.K.

        The theory of low-spin ferric and high-spin ferric are
discussed with particular reference to the effect of ligand field
strength and symmetry on the latter. The e.p.r. signal is very
dependent on the magnitudes of the zero-field splitting parameters
D and E. Very small D is easily distinguished from very large
which itself is easily recognizable if either near axial (e.g.,
haemoglobin) or near-rhombic (e.g., transferrin). Some geometries
responsible for the rhombic e.p.r. signal are discussed.
Symmetries intermediate between these extremes or having only
moderate D values give broad spectra whose interpretation may be
approached through the use of D/B graphs and more fully understood by computer fitting. Ferredoxins were the first proteins
fully to be isolated with the aid of a unique e.p.r. iron signal
at g = 1.94. The explanation of this signal for the 2Fe2S type
of cluster, in terms of anti-ferromagnetic coupling between the
iron atoms is given.

INTRODUCTION

        Iron occurs widely in biological material in both aerobic
and anaerobic situations. Since it is a typical transition metal
ion it is therefore likely to be found in more than one oxidation
state. The atoms most likely to be available in proteins for
binding to iron are oxygen, nitrogen and sulphur and these occur
in this order in the spectrochemical series. The latter is a
ligand strong enough to cause spin pairing of the metal ion
while the former two are likely to leave it spin-free. Add to
this the possibility that the coordination number of the iron is
likely to vary between three and seven and we can see that a

225

*I. Bertini and R.S. Drago (eds.), ESR and NMR of Paramagnetic Species in Biological and
Related Systems, 225–253.*
Copyright © 1979 by D. Reidel Publishing Company

variety of situations are likely to be found for this important
element. Viewed in this light, it is fortunate that e.p.r. is a
technique sensitive to both the strength and symmetry of the
ligand field; however, there are limitations which we will now
enumerate. (i) Though tetrahedral and octahedral geometries
are common for iron, no ligand field is likely to be strong enough
to cause a tetrahedral ferrous ion, or even a ferric ion, to go
low-spin and octahedral low-spin ferrous is diamagnetic so these
three ions are eliminated from our interests. (ii) The high-
spin ferrous ion, whatever its geometry, would always have an
even number of electrons; thus it is a non-Kramers ion so there
may be no zero-field spin degeneracy. Put another way, there is
no guarantee ever of seeing a resonance because two things militate
against it. Firstly, the zero-field splitting is likely to be
large enough to ensure that no two spin levels are close enough
to each other to be spanned by the quantum of microwave energy
and secondly if they were, nearby excited states would speed up
the relaxation so much that helium temperatures would be required
to see the resonance. (iii) The high-spin ferric ion in each
of these two geometries is a ^6S state so they are indistinguishable.
(iv) The technique is not particularly good for determining the
number of ligands attached to the metal though of course, the
presence of a ligand might be proven if its hyperfine structure is
observed.

These limitations reduce the applicability of e.p.r. of iron
compounds mostly to high-spin, Fe^{2+}, $(S = 2)$ or Fe^{3+}, $(S = \frac{5}{2})$ and
low-spin Fe^{3+}, $(S = \frac{1}{2})$ but happily, large variations in the e.p.r.
signals do permit recognition of the strength and symmetry of the
ligand field in many cases. Sulphur ligands frequently induce an
intermediate spin $S = \frac{3}{2}$ which we shall also consider.

HIGH-SPIN FERROUS; S = 2

The electron configuration of the high-spin ferrous ion in
an octahedral field is $t_{2g}^4 eg^2$. The four electrons in the t_{2g}
set endow orbital magnetism with L = 1. Since S = 2, spin-orbit
coupling (ζ) gives three states whose energies increase in the
order of J = 1, 2, 3. A lower symmetry component in the ligand
field may be assumed to separate the t_{2g} orbitals into a single
and double degeneracy separated by energy δ. Griffith has
combined these two effects and, assuming a fictitious spin $S' = \frac{1}{2}$,
calculated the $g_{//}$ value as a function of distortion $\omega = \delta/\zeta$
$(g_\perp = 0)$. The resultant graph appears in his book (1) and shows
that the $g_{//}$ value should lie between 4 and 7 for positive ω
(singlet lowest) or between 7 and 9.75 for negative ω (doublet
lowest). There are very few instances of this behaviour but for
example, Fe^{2+} in $CaCO_3$ has $g_{//}$ = 9.85, $g_\perp$ <0.2 for which
δ = -550 cm^{-1} (2).

The behaviour in a tetrahedral field with configuration $e^3 t_2^3$ is quite different. The ^5E is expected to show a strong Jahn-Teller distortion and leave ^5A or ^5B lowest in a tetragonal distortion. This singlet state, if not further split, would be expected to behave like octahedral Cu^{2+} which has the same ground state having anisotropic g values with at least one value well above 2.0. For example, ^5A would give $g_{//} = 2.0$; $g_\perp = 2-6k\zeta/\Delta E$ with ζ negative. Of course complications arise when we consider spin-orbit coupling to second order because this usually introduces a large zero-field splitting, removing the spin degeneracy to such a degree that no resonance may be seen. However, when Fe^{2+} was doped into a single crystal of $CdSiP_2$ (3) the zero-field splitting was such that three of the possible ten transitions were seen permitting the deduction of $g_{//} = 2.003$; $g_\perp = 2.12$; $D = 9.9$ cm^{-1}. Incidentally, the e.p.r. spectrum of octahedral Fe^{4+}, whose electron configuration is $t_2^3 e^1$, was also seen; this ion too has $S = 2$ and the same ligand field analysis is applicable. Remarkably both of these ions gave quite sharp e.p.r. spectra at room temperature.

No analysis for powder or frozen solution spectra has been given for these ferrous ions but it is surely only a matter of time before assignments are given to $S = 2$ ions in biological systems.

LOW-SPIN FERRIC; $S = \frac{1}{2}$

The theory of e.p.r in low-spin d^5 systems has been nicely summarized by Abragam and Bleaney (4) and by Griffith (1). The latter's method is briefly summarized here. A rhombic distortion along the Cartesian axes of the octahedrally bound ferric ion is assumed to leave the dxz, dyz and dxy orbitals with energies of $-V/2$, $+V/2$ and Δ respectively; these energies correspond to a hole in t_{2g}^6 rather than to single electron functions. A spin-orbit coupling energy of $\zeta L.S$ is then assumed. This is comparable in magnitude to the distortion, so an exact solution for the energy of the ion in zero field is obtained by setting up the combined secular determinant.

	$\lvert 1\alpha\rangle$	$\lvert -\xi\beta\rangle$	$\lvert -1\alpha\rangle$	$\lvert -1\beta\rangle$	$\lvert \xi\alpha\rangle$	$\lvert 1\beta\rangle$
$\langle 1\alpha \rvert$	$-\zeta/2$	$\sqrt{2}\zeta/2$	$V/2$	0	0	0
$\langle -\xi\beta \rvert$	$\sqrt{2}\zeta/2$	Δ	0	0	0	0
$\langle -1\alpha \rvert$	$V/2$	0	$\zeta/2$	0	0	0
$\langle -1\beta \rvert$	0	0	0	$-\zeta/2$	$\sqrt{2}\zeta/2$	$V/2$
$\langle \xi\alpha \rvert$	0	0	0	$\sqrt{2}\zeta/2$	Δ	0
$\langle 1\beta \rvert$	0	0	0	$V/2$	0	$\zeta/2$

$$= 0 \quad (1)$$

The six states are three Kramers doublets; each is assumed to be of the form

$$\psi^+ = A\,|1\alpha\rangle \quad + \quad B\,|\xi\,\beta\rangle \quad + \quad C\,|-1\alpha\rangle$$

$$\psi^- = A\,|-1\beta\rangle \quad - \quad B\,|\xi\,\alpha\rangle \quad + \quad C\,|1\beta\rangle \qquad\qquad (2)$$

with g values

$$g_x = 2\,\langle\psi + |\,kL_x \;+\; 2S_x\,|\psi-\rangle$$

$$g_y = 2i\langle\psi + |\,kL_y \;+\; 2S_y\,|\psi-\rangle \qquad\qquad (3)$$

$$g_z = 2\,\langle\psi + |\,kL_z \;+\; 2S_z\,|\psi+\rangle$$

which are evaluated as

$$g_x = 2[\,2AC - B^2 \;+\; kB(C-A)\,\sqrt{2}\,]$$

$$g_y = 2[\,2AC + B^2 \;+\; kB(C+A)\,\sqrt{2}\,] \qquad\qquad (4)$$

$$g_z = 2[A^2 - B^2 \;+\; C^2 \;+\; k(A^2-C^2)\,]$$

The procedure is to evaluate A, B, C and k using the experimental g values and the extra equation $A^2 + B^2 + C^2 = 1$; this must define the lowest Kramers doublet. These coefficients naturally satisfy the secular equations which are used to define the secular determinant so these may be written as a set of three simultaneous equations

$$(-\zeta/2-E)A \;+\; \zeta B/\sqrt{2} \;+\; VC/2 \;=\; 0$$

$$\zeta A/\sqrt{2} \;+\; (\Delta-E)B \;+\; 0 \;=\; 0 \qquad\qquad (5)$$

$$VA/2 \;+\; 0 \;+\; (\zeta/2-E)C = 0$$

and used to solve for V/ζ and Δ/ζ

Substitution of V and Δ into the secular determinant then gives the energies of the three Kramers doublets together with the eigen functions (A, B, C) for each. As a check, it is necessary to show that the coefficients A, B, C of the lowest Kramers doublet do indeed correspond to those of the experimental g values. This procedure has been programmed by Hudson and Kennedy (5) and modified by the author (6). For systems with three distinct g values variations in the sign of g and of the assignment of each component to the axes, x, y, and z leads to 48 solutions in 8 groups of six physically identical solutions. Only those solutions with values of k, the orbital reduction factor, close to one, are realistic. Generally the other values of k are either negative, very large ($k > 2$) or very small (< 0.1) and may reasonably be discarded. Simple theory predicts that k should lie in the range $0.75 < k < 1.0$. However, k values greater than unity are sometimes found. This can arise if the observed g values have been inflated by admixture of $t_{2g}^4 e_g^1$ into the configuration t_{2g}^5 or by the presence of charge transfer of an electron from the ligands on to the iron (7). Some examples of low spin haemoglobins analysed by this method are shown in Table 1, and a typical spectrum is shown in Figure 1.

	g_x	g_y	g_z	k	Δ	V	E_1	E_2	E_3
Cowpea leghaemoglobin azide (8)	-2.74	2.19	-1.72	0.97	4.3	-3.6	4.4	1.7	-1.9
Soybean a leghaemoglobin azide (8)	-2.74	2.19	-1.71	0.96	4.2	-3.5	4.3	1.7	-1.9
Sperm whale haemoglobin azide (9)	-2.80	2.22	-1.72	1.06	4.3	-3.7	4.5	1.8	-1.9
Soybean a leghaemoglobin cyanide (8)	-3.25	2.90	-0.30	1.29	1.3	-0.3	1.7	0.1	-0.5
Soybean c leghaemoglobin cyanide (8)	-3.24	2.20	-0.32	1.04	1.3	-1.0	1.7	0.3	-0.7
Cowpea leghaemoglobin cyanide (8)	-3.24	1.99	-0.30	0.97	1.3	-1.2	1.8	0.4	-0.8

Table 1. Analysis using LOSPIN (6). Of the six solutions in each case, that corresponding to large positive Δ was chosen on the basis that this puts the unpaired electron in dxy. This is justified on the basis that the other two orbitals are likely to be lowered by π bonding with high-lying empty orbitals on the ligands while dxy is unlikely to π bond at all, as has been shown for copper phthalocyanine (10). In each case another k value was rejected partly because it had a higher value and partly because it corresponded to lower Δ and V values. E_1, E_2, and E_3 are the energies of dxy, dyz and dxz after the combined effects of Δ, V and the spin-orbit coupling.

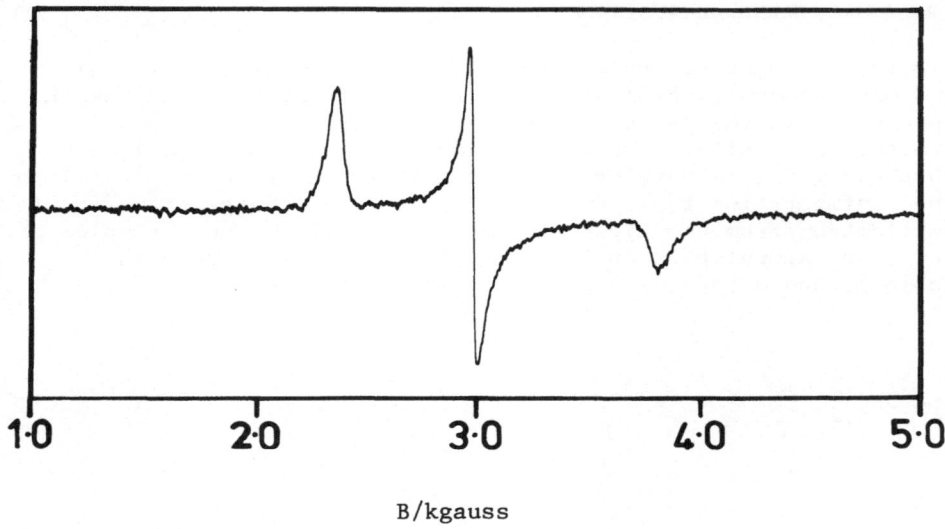

1·0   2·0   3·0   4·0   5·0

B/kgauss

Figure 1.   X-band e.p.r. spectrum of the azide derivative of
cowpea leghaemoglobin (8).

     The low-spin e.p.r. g values have been exploited by Peisach,
Blumberg and Adler to predict what ligand might be attached in
the sixth position of low-spin ferri-haemoproteins (11).   They
defined rhombicity as V/Δ and plotted this against Δ for a number
of low-spin derivatives and found that different ligands plotted
as fairly well-defined groups on this graph.   Thus the nitrogen
donors azide and imidazole fell into region H, protonated imidazole and cytochrome b_5 into region B,hydroxides into region O,
mercaptides and cytochrome P_{450} into region P and so on.   They
used the plot in the study of the green sulphmyoglobin to show
that various low-spin derivatives of this protein were like the
corresponding chlorins and that therefore one pyrrole ring of
the porphyrin ring had indeed been reduced to form the chlorin
in this case.   These ideas have now been widely used.   Another
example is the haemoglobin which may be extracted from the root
nodules of leguminous plants.   Comparison of the OH^-, N_3^- and CN^-
derivatives of leghaemoglobin with those of human haemoglobin
indicate that the environment around the iron is the same, i.e.,
the proximal ligand is histidine (8,12).

HIGH-SPIN FERRIC; $S = \frac{5}{2}$

     Whether octahedral, $t_{2g}^{3}e_{g}^{2}$ or tetrahedral, $e^{2}t_{2}^{3}$ the high-
spin configuration of the free ion is characterized by the ^6S
state.   This state retains its six-fold spin degeneracy in an
octahedral field, or field of lower symmetry even when spin-
orbit coupling is considered to first order but in second order,
the degeneracy is lifted to give three closely spaced Kramers
doublets, see Figure 2.

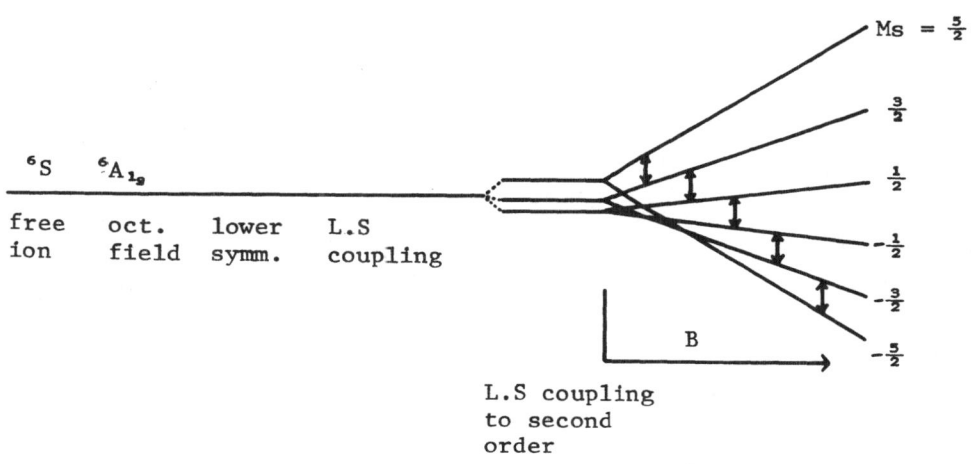

Figure 2.   The five " normally allowed " transitions ($\Delta M_S = \pm 1$)
seen when the zero field splitting is small.

The separation between these doublets is measured by the
zero-field splitting parameters D and E in the spin Hamiltonian.

$$\hat{H} = g\beta B.\hat{S}. + D[\hat{S}_z^2 - \tfrac{1}{3}S(S + 1)] + E(\hat{S}_x^2 - \hat{S}_y^2) \qquad (6)$$

In this section we consider the effect on the e.p.r. spectrum
of a variation in the magnitude of these zero-field splitting
parameters.   Unless otherwise stated the g value is assumed equal
to 2.0, a good approximation for an S state because there can be
no anisotropy in g if there is no orbital contribution to the
magnetic moment.

Small Zero-field Splitting.

When a magnetic field is applied to a system with $S = \tfrac{5}{2}$
and only small D and E, the two-fold spin degeneracy of the
Kramers doublets is further lifted as shown in Figure 2.
The selection rule $\Delta M_S = \pm 1$ permits the five transitions which
are shown and it should be appreciated that a very small zero-
field splitting (about 0.1 cm^{-1} in this case) can have the very
profound effect of spreading the spectrum out over thousands of
gauss.   The transitions are anisotropic and in a single crystal
this anisotropy may be used to estimate D and E.   Biological
samples are usually in the form of solutions which would need to
be frozen to see this spectrum, so all one expects to see is a
very broad absorption (halfwidth anything from a few hundred to a

few thousand gauss) centred on the g = 2.0 region. However,
in some cases sufficient structure may be seen to permit
an estimate of D using the D/B graphs explained in a later section.
An inorganic example, of near octahedral oxygen donors of type
$FeL_6(C\ell O_4)_3$ (L = Me_2SO, urea, DMF) and similar salts, may be
cited (13).

Large Zero-field Splitting

There are two important consequences of a large zero-field
splitting parameter D: first the quantum of microwave energy
may be too small to bridge the gap between the Kramers doublets so
we may expect fewer than five transitions; second, the states
are no longer true spin states because they are mixed by the
zero-field splitting, hence the $\Delta M_S = \pm 1$ selection rule no longer
holds. The resultant effect is that the three transitions from
within the three Kramers doublets become observable, in principal
at least, and these are the only ones seen.

The symmetry of the ligand field around the ion now becomes
an important feature and two types of symmetry, axial and rhombic
become easily recognizable by their e.p.r. signals. Before
discussing these, it is useful to define the distortion parameter
E/D = λ (not to be confused with spin-orbit coupling), and note
its effect on the spacing of the Kramers doublets. In axial
symmetry (E = λ = 0) these are separated by 4D and 2D respectively
as shown to the left of Figure 3.

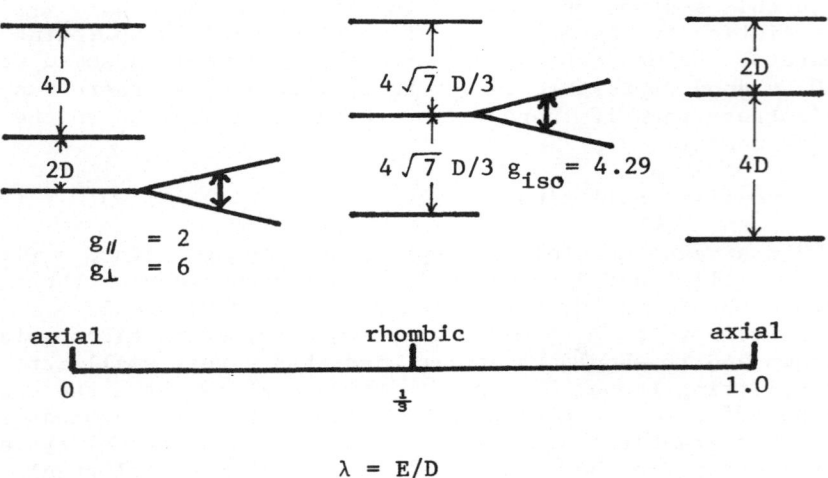

Figure 3. Effect of increasing rhombic distortion.

A small distortion from axial is represented in the spin
Hamiltonian (6) by the appearance of a small value of E. As the
departure from axial symmetry increases (increasing E) so the
spacing between the Kramers doublets changes, approaching that
situation where they are equally spaced and separated by
approximately 2.5D as in the centre of Figure 3. This represents
axial symmetry again though the spacing has been reversed; it
corresponds to an axis of symmetry along a different direction.
The situation at the centre of Figure 3 represents some
environment as far removed as possible from axial and is termed
rhombic. The e.p.r. from this is unique though the symmetry of
the ion is not; this will be dealt with later. It is pertinent
to mention, however, that all degrees of distortion space are
contained within the region $0 < \lambda < \frac{1}{3}$ and it is conventional to
report zero-field splitting parameters within this range (14).

Axial Symmetry. This environment is typified by ferri-
myoglobin whose porphyrin ring defines the xy plane of symmetry
for e.p.r. purposes. A typical frozen solution has an asymmetric
e.p.r. spectrum apparently with $g_\perp = 6.0$ and $g_{//} = 2.0$. Since
we have already said that g = 2.0 this needs some explanation.

The secular determinant associated with equation (6) is

$$
\begin{vmatrix}
 & |\tfrac{5}{2}\rangle & |\tfrac{3}{2}\rangle & |\tfrac{1}{2}\rangle & |-\tfrac{1}{2}\rangle & |-\tfrac{3}{2}\rangle & |-\tfrac{5}{2}\rangle \\
\langle\tfrac{5}{2}| & \frac{10D}{3} & & \sqrt{10}E & & & \\
\langle\tfrac{3}{2}| & & -\frac{2D}{3} & & 3\sqrt{2}E & & \\
\langle\tfrac{1}{2}| & \sqrt{10}E & & -\frac{8D}{3} + \frac{g\beta B_{//}}{2} & \frac{3g\beta B_\perp}{2} & 3\sqrt{2}E & \\
\langle-\tfrac{1}{2}| & & 3\sqrt{2}E & \frac{3g\beta B_\perp}{2} & -\frac{8D}{3} - \frac{g\beta B_{//}}{2} & & \sqrt{10}E \\
\langle-\tfrac{3}{2}| & & & 3\sqrt{2}E & & -\frac{2D}{3} & \\
\langle-\tfrac{5}{2}| & & & & \sqrt{10}E & & \frac{10D}{3}
\end{vmatrix} = 0 \quad (7)
$$

Although strictly speaking the determinant does not factorise, we
may nevertheless see the approximate behaviour of the lowest
Kramers doublet (if positive D) if we pick out the small matrix
associated with $|\pm\tfrac{1}{2}\rangle$. This approximation is valid since other
terms in the same columns or rows are small relative to D which is
about $7 - 10$ cm^{-1} for haemoproteins. Some terms in $B_{//}$ and $B_\perp$
have been left out of the determinant.

For B parallel to z

$$
\begin{vmatrix}
\dfrac{-8D}{3} + g\beta B_{\parallel} & \\
\qquad -\varepsilon & \\
& \dfrac{-8D}{3} - g\beta B_{\parallel} \\
& \qquad -\varepsilon
\end{vmatrix} = 0
$$

from which $\varepsilon = \dfrac{-8D}{3} \pm \dfrac{g\beta B_{\parallel}}{2}$

and

$$\Delta\varepsilon = g\beta B_{\parallel}$$

$$g_{eff} = 2.0$$

For B parallel to x or y

$$
\begin{vmatrix}
\dfrac{-8D}{3} - \varepsilon & \dfrac{3g\beta B_{\perp}}{2} \\
\dfrac{3g\beta B_{\perp}}{2} & \dfrac{-8D}{3} - \varepsilon
\end{vmatrix} = 0
$$

from which $\varepsilon = \dfrac{-8D}{3} \pm \dfrac{3g\beta B_{\perp}}{2}$

$$\Delta\varepsilon = g\beta B_{\perp}$$

and

$$g_{eff} = 6.0$$

Thus we see that the effective g values (normally written g_{eff}) range from $g_{\parallel}$ = 2.0 to $g_{\perp}$ = 6.0 even though the real g = 2.0; the effect arises from the presence of the D terms which separate out one doublet for which the effective spin $S' = \frac{1}{2}$ even though the real spin is $S = \frac{5}{2}$. Two examples are chosen in Figure 4. The first is a leghaemoglobin (8) extracted from soybean root nodules, after oxidation to ferric and reaction with sodium fluoride. Leghaemoglobin is very similar to mammalian myoglobin having one haem only, in which the iron, when ferric, is bound to the four nitrogens of the porphyrin ring, one nitrogen from the proximal histidine and in the absence of fluoride, a water molecule. That fluoride displaces water and binds to the iron is confirmed by the doublet hyperfine structure on the $g_{\parallel}$ peak of figure 4(a); this comes from ^{19}F which has $I = \frac{1}{2}$. Thus, in the presence of fluoride the iron remains high-spin but when stronger ligands such as OH^-, N_3^-, CN^-, displace the water, the ferric ion goes low-spin as already discussed (see Figure 1).

The second part of Figure 4 shows the spectrum of an axially symmetric inorganic molecule $Fe[N(SiMe_3)_2]_3$. For the moment, note the features at g_{eff} = 6.0 and 2.0; the absorptions above 5kG will be discussed later. The important point to note here is

that the immediate environment of the ferric ion is now a plane
of only three nitrogen atoms. Thus the typical e.p.r. spectrum
we have discussed does not distinguish between threefold and
fourfold axis. What it does indicate is the presence of a
high-spin axially symmetric ferric ion with a large zero-field
splitting parameter D ($\gtrsim$ 0.4 cm^{-1} at X band)

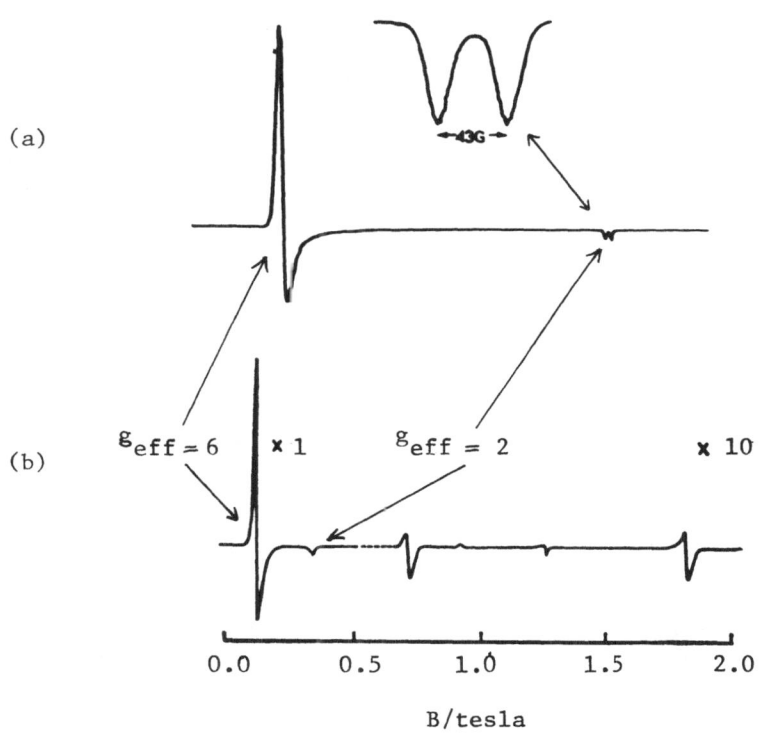

Figure 4. E.p.r. spectra (9.2 GHz) of (a) the fluoride derivative
of leghaemoglobin from cowpea (8) and (b) Fe[N(SiMe$_3$)$_2$]$_3$ (15).

    Rhombic Symmetry. We now return to the situation characterised
by the equally spaced Kramers doublets at the centre of Figure 3.
These reflect the spacing of an excited quartet state which itself
has been split by a low symmetry element of the ligand field.
Griffith has considered (16) what environments might be responsible
for this type of behaviour and has shown that it is necessary for
the environment of the d^5 ion to be more symmetric than the overall symmetry group of the site in a special way. He pointed out
that there may be many of these special environments but three
which seem quite likely to occur frequently are tetrahedral
MA$_2$B$_2$(C$_{2v}$), distorted octahedral MA$_6$(D$_{2h}$) and octahedral MA$_3$B$_3$(C$_{2v}$)

as shown in Figure 5.

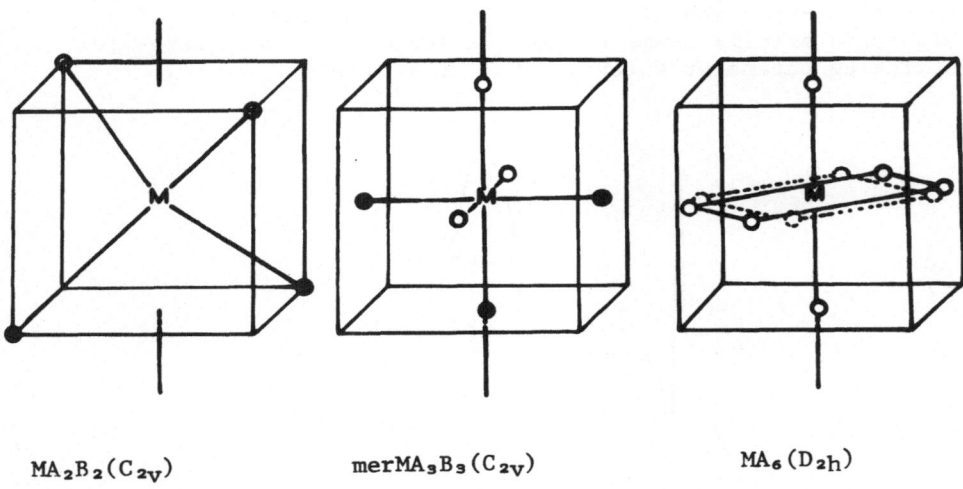

$MA_2B_2(C_{2v})$ merMA$_3$B$_3$(C_{2v}) $MA_6(D_{2h})$

Figure 5. The three special rhombic symmetries proposed by
Griffith (16).

These geometries are all linked then by a common behaviour
among the electronic energy levels and thus all give the same
type of e.p.r. signal which is an isotropic line at $g_{eff} = \frac{30}{7} =$
4.29 arising from the middle Kramers doublet Ms = $\pm \frac{3}{2}$. The
theory of this has been worked out in detail (16) and there is
unhappily no simple description we can make as in the case of
axial symmetry. E.p.r. signals in the g_{eff} = 4.3 region are
very common in biological materials and could be associated with
a variety of environments provided these are very far removed from
axial symmetry. However, at least two of Griffith's symmetries
(the C_{2v} ones) are known to give this resonance in the isoelectric
manganese(II) compound MnCℓ_2(OPPh$_3$)$_2$ (17) and in
Fe(3-methylpyrazole)$_3$Cℓ_3 (18).

A biological example of the latter is to be found in enterochelin the iron-scavenging cyclic triester of 2,3-dihydroxybenxoyl-
serine

HOH$_2$C—CH—CO$_2$H
|
NH
|
C=O
OH
OH

    The acid and alcohol groups of three of these molecules
esterify to form a 12 membered ring, and when the three aromatic
amide residues are all to one side of this ring, six catecholic
hydroxyl groups may bind to a ferric ion. The three ortho
oxygens possess a slightly different electronic structure to those
in the meta position and in this way the six oxygens are presented
in two groups of three and the iron complex may be represented
by MA_3B_3 of which there are two isomers. Llinas, et al., (19)
have shown that the gallium(III) complex has the facial isomer
with C_3 symmetry. The iron(III) complex would likely have $g_\perp = 6$,
$g_{||} = 2$ if it too adopted this axial symmetry but it does not;
it has $g = 4.3$ as shown in Figure 6. Thus the conformation of
ferric enterochelin may be assumed meridional MA_3B_3 (Figure 5),
a symmetry predicted by Griffith(16) to give this unique $g_{eff} = 4.3$
e.p.r. signal.

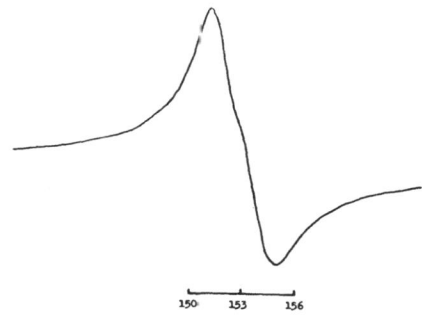

150   153   156

Field/mT

Figure 6. X-band e.p.r. spectrum of a 10mM solution of ferric
enterochelin at pH 7.0 in 20% methanol (20).

    In fact there is a very slight departure from pure rhombic as
the inflection in the figure shows. Butterworth has simulated
this spectrum and finds $D = 0.6$ cm^{-1}, $\lambda=0.32$, $g_{eff} = 4.26$ (20).
He also discusses the interesting proposal by Spartalian, et al.,
(21) that any complex with spin-Hamiltonian parameters which fit
the equation

$$4a = 3D - 9E, \tag{8}$$

where a is a quartic term to be added to equation (6), will give
an isotropic line at $g_{eff} = 4.3$, even though λ is not equal to $\frac{1}{3}$.
He found (20) that for the $g_{eff} = 4.3$ isotropic line to result
when this equation is satisfied a constraint must be applied to
the equation such that D must be greater than a certain limiting
value depending on λ. Thus the equation is unimportant for high

D values because this would imply that the a values were unreason-
ably high, at least if λ is well away from one third.

Intermediate Symmetry. The effect of a small distortion
applied to an axial molecule having high D is to split the $g_\perp$
peak into two since the x and y directions are then inequivalent
(see Figure 8). There are several examples of this among
haemoglobins for which the rhombicity may be redefined in terms
of the splitting between g_x and g_y as $E/D = \Delta g/48$. Since the
maximum value of E/D is one third, the percentage rhombicity may
be written (22) as

$$R = \frac{\Delta g}{16} \times 100\% \qquad\qquad (9)$$

Using this definition, many data have been collected and shown
to fall in distinct groups when plotted on sucha an R scale. In
order to make proper comparisons these measurements should be done
at very low temperatures, e.g., 1.8K. At higher temperatures,
e.g., 77K,line broading may blur the separation between g_x and
g_y which could appear as a single $g_\perp$ line. A plot of e.p.r.
intensity at very low temperatures is also a very good way of
measuring the magnitude and the sign of D (22).

The maximum rhombicity noted by Peisach, et al., (22) for
the minority constituent of base-treated myoglobin at pH 10.1 was
17%; the salicylated derivative of ferrihaemoglobin A was nearby.
In these, the departure from tetragonality probably arises from
a change in the ferric environment resulting in turn from a
disruption of the tertiary structure of the protein, rather than
from ligation of the denaturing agent to the haem. It is likely
that the haem environment is the same in these two cases. For
some of the other groupings on the R scale one cannot say the
same, but at least it seems that the environment of the high-spin
haem is determined by the structure of the protein at primary,
secondary and tertiary levels. In contrast, the environment of a
low-spin ferrihaem is largely determined by the immediate ligands
of the iron.

Mutant haemoglobins of type M which have structural abnormalities
near the iron atom might well be expected to show deviations from
tetragonality in the $g_{eff} = 6.0$ region. This indeed has been
shown to be the case and the several variants known may be
instantly recognized by the rhombicity and shape of the e.p.r.
spectrum. The ferric ion which these mutants contain in their
natural state either on the α or on the β chains acts as a
natural probe for the protein and has been used as such to measure
the haem-haem cooperative effect (22, 23). Thus two of the mutants,
Milwaukee and Hyde Park, show a change in the e.p.r. spectrum of
the mutated β chains on deoxygenation of the normal α chains.
Measurements on cooperativity in these various mutants, some of
which have proximal histidine replaced by tyrosine, have been

used (23) to support the Perutz theory that the proximal histidine
movement triggers the haem-haem interaction in haemoglobin.

Near-axial e.p.r. signals are not restricted to porphyrins.
Lipoxygenase, an enzyme which catalyses the oxidation of poly-
unsaturated fatty acids contains one iron atom per molecule, but
no porphyrin ring. The split $g_\perp$ e.p.r. signal exhibited when
hydroperoxide is present is unusual but not unprecedented in a
non-haem protein (24).

At the other end of the λ scale (Figure 3), a small departure
from pure rhombic ($\lambda = \frac{1}{3}$) causes the isotropic $g_{eff} = 4.29$ signal
to split into three; Blumberg's graphs show the effect quite nicely
(25) though this and the splitting in the axial case are both
shown in the D/B graphs discussed later (Figure 8). The
departure from fully rhombic is displayed by the enzyme proto-
catechuate 3,4-dioxygenase (PCAase) which normally gives a sharp
line at $g_{eff} = 4.28$. In the presence of 4-nitrocatechol, however,
a known inhibitor which presumably binds the enzyme, this e.p.r.
line splits into 4.77, 4.28 and 3.74 (26). A helium pumping
experiment on this enzyme was used to measure the zero-field
splitting parameter D which, being large (5.4 cm⁻¹) was thought to
arise from a tetrahedral arrangement of sulphurs as in the case
of rubredoxin (27). The 4-nitrocatechol presumably binds the
enzyme causing a small distortion of the environment of the iron
but not a ligand substitution. Addition of a different substrate
however, homocaffeic acid, partially converted the 4.28 signal
into a near axial signal with $g_{eff} = 6.6$ and 5.5; this was thought
to be caused by a severe flattening of the four sulphur atoms or
possibly by ligand exhange (27). Another enzyme, pyrocatechase
also shows the $g_{eff} = 4.28$ e.p.r. signal which this time simply
diminished in amplitude on binding 4-nitrocatechol. Thus the
e.p.r. behaviours, and incidently the optical behaviours, differ
between these enzymes yet they both cleave the same substrates at
the same carbon-carbon bond. Tyson is surprized that this should
be so (26). Perhaps the answer is that in fact the ferric
environment differs between the two proteins; we have already
pointed out that many different symmetries can give the same
$g_{eff} \approx 4.3$ e.p.r. signal.

Before leaving the subject of the $g \approx 4.3$ signal, it will be
instructive to consider the interesting thought that this signal
might, contrary to all the preceding, arise from an axially
symmetric molecule!

The paper by Peterson, et al., (28) translated so as to be
more digestible to the biochemist goes something like this. In
an assembly of molecules there is a large variation in the ligand
field at the iron atom. So large is this variation that the g
value measured in say the parallel direction of an axially symmetric
molecule is best described by a probability density function rather

than a fixed value; furthermore this function overlaps a similar
function for $g_\perp$ at some field in between the two. It is
necessary to note how $g_{//}$ and $g_\perp$ vary when the ligand field is
varied. If the two are related by a negative correlation
coefficient (ρ) then a sharp spike might occur in the e.p.r.
spectrum where the two probability density functions overlap.
In other words, perhaps it could happen that with ferrihaemoproteins the $g_\perp$ = 6 line goes upfield to meet the $g_{//}$ = 2 line
coming down as some ligand field parameter varies as a consequence
of perturbation within the protein; they would meet at g = 4.
The question is does it happen? The D/B graphs (2 9) described
later show that there is no obvious way in which increasing D
could cause a negative ρ for S = $\frac{5}{2}$. The same is true for B//z if λ is
varied rather than D ; an increase (from zero) with applied field
parallel to z shows no obvious negative correlation. However,
if the field is normal to the symmetry axis of the molecule the
D/B graphs show that there is negative correlation between the
$\left|\frac{1}{2}\right\rangle$ → $\left|-\frac{5}{2}\right\rangle$ and the $\left|\frac{5}{2}\right\rangle$ → $\left|\frac{3}{2}\right\rangle$ transitions for field parallel
to the x direction. They would coincide at about λ = 0.13 for
high D values giving g $\sim$ 3. At the same time the $\left|\frac{1}{2}\right\rangle$ → $\left|-\frac{1}{2}\right\rangle$
transition (which is now primarily $\left|\frac{5}{2}\right\rangle$ → $\left|-\frac{5}{2}\right\rangle$) would give
g = 8.25 for the y direction. Its intensity is relatively weak but
its field position changes very little with λ and not at all with
D so it is likely to be fairly intense. The answer to our
question therefore is no, it is not likely that the axially
symmetric $g_\perp$ = 6, $g_{//}$ = 2 type of molecule could give $\tilde{g} \sim 4$
through large variance and negative correlation. However, the
effect can easily explain the g_{eff} = 8.3, 3.3 signals found in
cytochrome P_{450} (30, 31). This signal has been used to monitor the
relative amounts of high-spin and low-spin iron in cytochrome P_{450}
of bovine adrenal mitochondria which is involved in cholesterol
side-chain cleavage. The P_{450} is 80% high-spin as isolated and
this particular form of the iron is believed to derive from
endogenous cholesterol. The e.p.r. signal may be described as
having 31% rhombicity on the basis of the quoted g values. This
may be true but the statistical approach (28) shows also that it
might represent a flexible molecule distorted into a myriad of
environments around the iron resulting in large variations in
rhombicity and possibly also D.

Figure 7 shows the e.p.r. spectrum of rubredoxin. Two
interpretations could be put on this spectrum: it could be two
signals, a sharp one on top of a broad; it could be one signal
which is only near-rhombic and therefore has three g-values.
Tha latter interpretation seems to be right (27). But could the
statistical approach (28) account for the sharp spike too? The
answer is yes but it doesn't make a lot of difference. Reference
to the D/B graph of Figure 8 shows (bottom right diagram) that
g_z and g_y approach g_x (which is almost stationary) as λ increases
(i.e., we have negative correlation). However, we do not need to

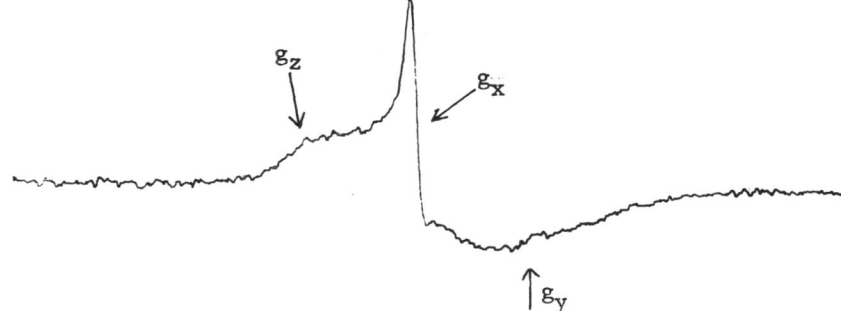

Figure 7. E.p.r. of rubredoxin in the g = 4 region (X-band).

invoke it because if we simply imagine that g_x, g_y, g_z are each
represented by a probability density function in the variable
λ, then of necessity (see graph) the variance of g_x would be
very small giving a sharp line while the variances of g_y and
g_z would be larger giving broader lines. This seems to explain
the spectrum without requiring that the variances be so large as
to overlap. In conclusion perhaps we should more frequently
consider contributions from varying environments in our e.p.r.
interpretations; after all proteins can be very flexible molecules.

Intermediate Zero-field Splitting.

     The theory which has been used earlier to derive g_{eff} = 6.0
for the direction perpendicular to the axis of symmetry, likewise
the theory for the g_{eff} = 4.29 rhombic signal, depended on magnetic
field B being a small perturbation relative to the zero-field
splitting. But when this is no longer true, i.e. $D \lesssim 0.3$ cm^{-1} for
X band, the secular determinant (7) has to be solved exactly,
which is a very simple matter with current computer methods. An
early approach to the problem (29) was to construct, within the
computer, a plot of the energy of each of the six spin states as
a function of field for given values of D and E. A check was
then made disregarding the selection rules to find the resonance
fields at which the quantum of microwave energy spanned the gap
between all the 15 pairs of energy levels possible within 6 energy
states. Having found the transitions their transition probabilities
were then computed before going on to the next D,E values. In
this way curves representing the resonance positions in field
were built up as a function of D. For each λ value computations
with field parallel to the x, y and z directions were given (29).
Typical curves are shown in Figure 8 which also shows similar
curves for S = $\frac{3}{2}$.

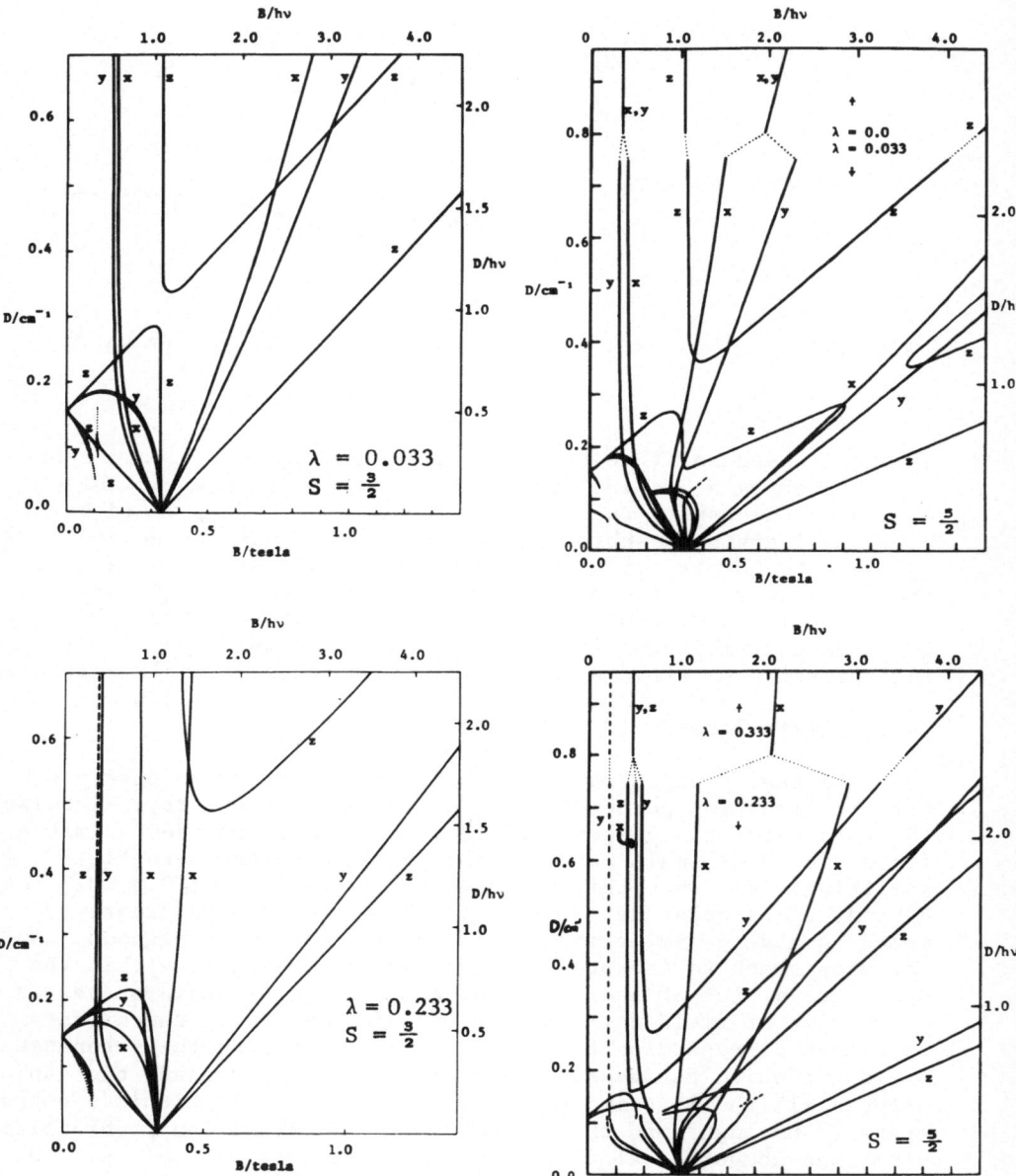

Figure 8. Predicted e.p.r. transitions as a function of the zero-field parameter D for $S = \frac{3}{2}$ on the left and for $S = \frac{5}{2}$ on the right.
/continued.

The method of using these D/B curves (29) is to search for that horizontal line for which a feature in the experimental spectrum occurs at every point where the horizontal line cuts the D/B curves. If a good fit is not found a search is made on the D/B curves corresponding to a different λ value.

There are two noteworthy disadvantages of these curves which may militate against finding a good fit for powders or rigid solution spectra. First, a feature may not appear in the experimental spectrum even though it is predicted. This could happen for example if a large linewidth is coupled with a weak transition probability or if the line is highly anisotropic. After all the intensity in a powder spectrum is high only when the function $\partial B/\partial\theta$ is low. Second, features may appear even though they are not predicted. This is because the published D/B curves (29) represent the x, y and z axes only, and for low D values, off-axis turning points can also occur. Blumberg discusses this point and shows that the g ellipsoid characteristic of an $S = \frac{1}{2}$ system can look more like a symmetrical but knobbly potato for $S = \frac{5}{2}$ (25). Nevertheless, the D/B curves may be quite useful for a tentative assignment subject to confirmation by further computing.

The D/B curves of Figure 8 do show quite nicely the effect of departure from axial symmetry ($\lambda = 0.033$, top right) where the $g_\perp = 6.0$ line at about 1.2kG is clearly split for the x and y directions at high D values. (The unsplit portion at the top of the graph corresponds to axial symmetry with $\lambda = 0$). Note too how these two lines curve down towards g = 2.0 (3.3kG) for D values less than about 0.2 cm^{-1} as D goes to zero. This part of the curve is the region where perturbation theory is quite unsuitable. Likewise for near-rhombic symmetry ($\lambda = 0.233$, bottom right) the $g_{eff} = 4.3$ region (1.5kG) is split into three lines (x, y, z) but that these may shift or disappear altogether as the D value diminishes. At the top of the graph these three lines are seen to be coincident at $\lambda = \frac{1}{3}$ corresponding to fully rhombic symmetry.

The interpretation of the transferrin e.p.r. spectrum has interested many people since it provides some clues about the environment of the iron which is as yet unknown; the spectrum is shown in Figure 9.

Caption to Figure 8 continued.

Rhombicity parameter $\lambda(= E/D)$ is indicated on each graph. Dashed lines represent weak transitions. The letter x,y or z close to a line represents the axis to which the magnetic field was assumed parallel in the calculation for that line. On the right, note how separated transitions coalesce as the rhombicity approaches zero (top) or maximum value of $\frac{1}{3}$ (below).

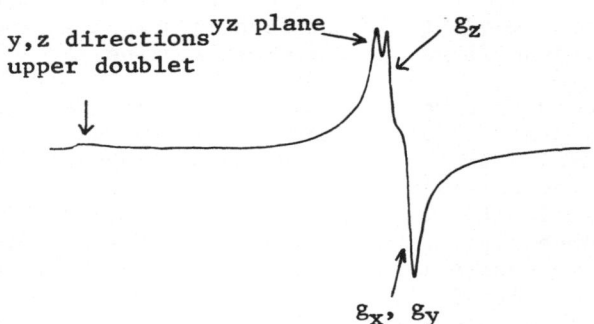

Figure 9. E.p.r. spectrum (9.2 GHz) of human serum transferrin.

The most obvious suggestion is that this spectrum might
represent g_x, g_y, g_z split apart from g_{eff} = 4.29 by a small
departure form fully rhombic. However, various experiments including
Mössbauer and e.p.r. at a higher frequency indicate that this is not
so. In fact, it seems that this is an example of rather low D
where Blumberg's graphs (25) derived from perturbation treatment
are not applicable; indeed, detailed computer simulation has shown
that the left hand bump arises from a turning point in the yz plane
rather than along one of the axes (32,33). This is one of those
examples where even the D/B graphs, which are not dependent on a
perturbation approximation, have failed to explain all the features
because they were computed for fields along the axes only. The
best fit computed parameters were D = 0.23 cm⁻¹, λ = 0.325, g_x =
2.038, g_y = 2.019, g_x = 2.044 (33).

With an understanding of the D/B graphs we are now in a
position to return to the spectrum of Fe[N(SiMe₃)₂]₃ which displays
features at high field (Figure 4); these have been very useful
in estimating D. The two larger features occur at 7125 gauss and
18,200 gauss and are predicted from Figure 8 (top right, upper
part with λ = 0) for a D value of 1 cm⁻¹, but the D/B graph does
not predict the intermediate weaker features at 9170 gauss and
12,650 gauss. Exact calculations for g = 2.0, D = 1 cm⁻¹ and
λ = 0 do predict these lines but with zero intensity (15). Presumably
they would have been found to be weakly allowed if quartic terms
(terms in a) had been included in the spin Hamiltonian (6), or it
calculations close to the principal directions had been performed.

SPIN EQUILIBRIA; $S = \frac{5}{2}$, $S = \frac{1}{2}$

Purified ferrihaemoproteins frequently show both high- and

low-spin behaviour though one of these is usually very much the
major component; the signals probably reflect differences in the
conformation around the iron. Such conformation changes may be
induced reversibly by temperature: for example, leghaemoglobin
is almost entirely high-spin at room temperature whereas at 1.6K
a major part of it is low-spin (2.69, 2.24, 1.72). By comparison
of these g values with those of known proteins, both proximal and
distal ligands are identified as imidazole. The molecule is
imagined as flexible and hinged such that at low temperature it
closes to enable the distal histidine 61 to bind to the $_o$iron while
at higher temperatures it opens. A movement of about 3Å would
be required to $_{draw}$ the distal histidine away and favours the
high-spin configuration (12). The spectrum of the related soya-
bean leghaemoglobin ¢ showing both high-spin and low-spin forms is
shown in Figure 10. The low-spin form here is also thought to be
the bis-imidazole haemichrome (8).

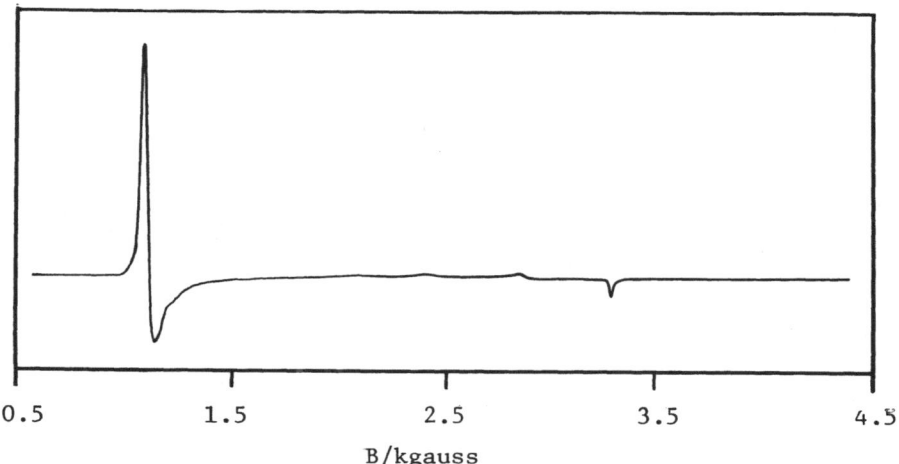

0.5   1.5   2.5   3.5   4.5

B/kgauss

Figure 10. E.p.r. (9.2 GHz) of leghaemoglobin c taken at 11K.
Both high-spin and low-spin signals may be seen.

   The reverse type of behaviour is found with cytochrome P_{450}
from Pseudomonas Putida. As isolated, this enzyme is typically
low-spin with sulphur ligated to the iron. Below 20K about 7%
of the haem is high-spin, while binding a substrate (+ camphor)
causes the high-spin form to increase to 60%. It is believed that
the high-spin form is favoured by displacement of the sulphur to
give a weaker ligand field as the camphor binds (34).

   Reversible changes may also be brought about by other means.
For example, Maltempo describes several new low-spin signals which
are formed in chromatium cytochrome c' above pH 11.6 (33). These
reconvert to high-spin below pH 11.0 though there is a slight pH

hysteresis effect. Cytochrochrome c oxidase has been the subject
of similar very detailed e.p.r. studies. The temperature was
varied between 6K and 100K at various states of oxidation and at
different conditions of pH and presence of solute. The weaker
of two low-spin components was noticed to be in a pH-dependent
equilibrium with a high-spin form; the former was favoured at
high pH.The equilibrium may represent a single component whose
spin state changes as a function of some proton ionization of
the protein. However, careful intensity measurements were made
and the appearance of the one signal did not completely match
the disappearance of the other so perhaps additional (undetectable?)
components also participate in the equilibrium (36).

INTERMEDIATE SPIN; $S = \frac{3}{2}$

As mentioned in the introduction, and as has been exemplified
in the text, sulphur is a ligand which might be strong enough to
cause a ferric ion to adopt the low-spin state. Reference to the
appropriate Tanabe Sugano energy level diagram shows that at the
ligand field required to cause $^6A_1$ to go over to $^2T_2$ there is a
fairly low lying $^4T_1$ state representative of an intermediate spin
of $\frac{3}{2}$. In octahedral symmetry this state never becomes the ground
state but in fields of lower symmetry it can and Harris has
discussed the strength and symmetry of the ligand field required
(37). The e.p.r. signal from such a state is characterized by
the zero-field splitting parameters D and E of equation (6) and
as with $S = \frac{5}{2}$, rhombicity, defined as E/D covers the range 0 to $\frac{1}{3}$.
As with high-spin ferric D/B graphs may be prepared and utilized
to predict a spectrum for given values of D and E. The left
hand side of Figure 8 shows two such graphs corresponding to
rhombicities close to either end of the λ scale. The iron(III)
bis-dithiocarbamates of formula $[Fe(dtc)_2X]$ (X = Cℓ, Br, I, SCN)
are well known members of this class and having approximately
square pyramidal symmetry are expected to have low λ values.
The top left diagram of Figure 8 indicates that at very high D
values their spectra should have a g $\approx$ 4 feature split into two
and a $g_{//}$ feature near g = 2. The thiocyanate e.p.r. spectrum
was sharp enough to show g_x = 3.27, g_y = 4.65 but the $g_z \approx 2$
feature was too broad to see as a result of exchange interactions
(38); this corresponds to λ = 0.12. A well resolved spectrum
showing all three features was found for thiooxalate
$[Fe(S_2C_2O_2)_2Br]^{2-}$ of similar structure though the corresponding
iodide was axial (39). Graphs of g_{eff} against λ for the two
doublets, similar to those for the three doublets of $S = \frac{5}{2}$ (25)
are also to be found in this paper.

Exchange broadening between the paramagnetic centres in proteins
is not usually a problem because the centres are naturally
magnetically diluted. Thus nitrogenase gives various sharp e.p.r.

spectra and one of these (g_{eff} = 4.28, 3.78, 2.0) is assigned to this unique S = $\frac{3}{2}$ state. Indeed, it is associated with the MoFe protein and also with the MoFe cofactor which can be extracted from the MoFe protein (40). The signal does not prove that there is an isolated ferric ion with a rather special symmetry however. As is well known among the non-haem iron sulphur proteins (see next section), ferrous and ferric ions may interact to give S = $\frac{1}{2}$. It would be an interesting exercise to work out how many ways about four to six high- or low-spin ferric or ferrous ions might interact ferromagnetically or antiferromagnetically in pairs, or otherwise to give a ground state with S = $\frac{3}{2}$. To make the problem more interesting replace one of the iron atoms by a molybdenum atom in the oxidation state of your choice. A non-haem iron sulphur cluster or iron/molybdenum sulphur cluster of this kind might well be responsible for the S = $\frac{3}{2}$ state in the MoFe cofactor. An inorganic anion $\{[(PhSFe)_3MoS_4]_2(SPh)_3\}^{3-}$ which contains two Fe_3MoS_4 cubes linked by their molybdenum atoms through three bridging -SPh groups (41) is known to have g values of 5.1, 4.3 and 2.0 (42) after two-electron reduction by acenaphthylenide. It is likely that this too corresponds to S = $\frac{3}{2}$ but note that the nitrogenase MoFe cofactor cannot contain a dimeric unit like this one since it contains only one molybdenum atom.

Another class of biological metallocomplex has been discovered by its e.p.r. signal characteristic of the S = $\frac{3}{2}$ state. This is probably an iron transport or iron-storage system of mitochrondria from N. Crassa, normally unseen but made detectable by reaction with NO; the g values are 4.11, 3.95 and 2.00. It is not known if the source of the signal is a single iron or a cluster but the ferrous EDTA complex when similarly treated with NO gives an almost identical signal (43), so presumably the former suggestion is favoured.

A rather special case of a quartet ground state has been proposed to account for the $g_\perp$ = 4.75, $g_\parallel$ = 1.99 parameters reported for the major paramagnetic component observable in chromatium ferricytochrome c' at pH 7.0 (44). It is proposed that the $^4A_2$ ground state is quantum mechanically admixed with the low-lying $^6A_1$ state to give $g_\perp$ values in between 4.0 (characteristic of the former) and 6.0 (characteristic of the latter). The protein is regarded magnetically (and structurally too) as intermediate between high-spin (S = $\frac{5}{2}$, iron out of plane) and low-spin (S = $\frac{1}{2}$, iron in plane), a view which is supported by its ability to undergo transformation either way.

Methods of calculation for both S = $\frac{3}{2}$ and S = $\frac{5}{2}$ have been reviewed by Pilbrow (45) who also derives a new perturbation formula applicable to S = $\frac{3}{2}$ provided D is large. He makes the point that the best determinations of effective g values are by computer simulations of the whole spectrum, not forgetting the 1/g factor (46), rather than by measurement of peak positions.

NON-HAEM IRON SULPHUR CLUSTERS

The e.p.r. signal centred on g = 1.94 was very important
in the characterization of the non-haem iron proteins containing
acid-labile sulphur. We now know these to contain either [4Fe4S]
or [2Fe2S] clusters each of which gives an e.p.r. signal
characteristically giving an average g value less than two. This
signal is unusual for monomeric iron as may be seen by reference
to the earlier sections of this chapter.

An explanation of this unusual signal was found in the
hypothesis of antiferromagnetic coupling between neighbouring
iron atoms of the 2Fe2S protein spinach ferredoxin (47). The
hypothesis has been amply borne out by subsequent experiments of
many kinds so it is worth looking at the theory used to explain
this signal. In the oxidized form it is assumed that two ferric
ions are situated close enough to each other for their spins
($S = \frac{5}{2}$) to interact to give a ladder of levels from S = 0 to S = 5.
The assumption of antiferromagnetic coupling requires the ground
state to be S = 0 consistent with the absence of an e.p.r.
signal. The protein accepts only one electron per molecule on
reduction by dithionite. If this is assumed to reduce one ion
to ferrous, the same antiferromagnetic coupling leaves a paramagnetic ground state with $S = \frac{1}{2}$ with excited states up to
$S = \frac{9}{2}$.

In order to discuss what e.p.r. signal is expected from such
a ground state, we make the assumption that the resultant g
value is related to those of the two interacting ions S_1 (with
g_1) and S_2 (with g_2) and write $g\hat{S} = g_1\hat{S}_1 + g_2\hat{S}_2$. Multiplying
each side by $\hat{S} = \hat{S}_1 + \hat{S}_2$ we get

$$g\hat{S}^2 = g_1\hat{S}_1^2 + g_2\hat{S}_2^2 + \hat{S}_1\hat{S}_2(g_1 + g_2) \tag{10}$$

The factor $\hat{S}_1\hat{S}_2$ may be replaced by $\frac{1}{2}(\hat{S}^2 - \hat{S}_1^2 - \hat{S}_2^2)$ obtained
by squaring the equation for $\hat{S}$, after which, substitution
and rearrangement gives

$$g = \frac{g_1 + g_2}{2} + \frac{(g_1 - g_2)(\hat{S}_1^2 - \hat{S}_2^2)}{2\hat{S}^2} \tag{11}$$

This is a general formula into which in this case the following
expectation values should be inserted:

$\hat{S}_1^2 = S_1(S_1 + 1) = \frac{35}{4}$ for the ferric ion with $S_1 = \frac{5}{2}$

$\hat{S}_2^2 = S_2(S_2 + 1) = \frac{24}{4}$ for the ferrous ion with $S_2 = 2$, and

$\hat{S}^2 = S(S + 1) = \frac{3}{4}$ for the combined ground state with $S = \frac{1}{2}$.

The resultant equation is $g = (7g_1 - 4g_2)/3$ from which we can
see that an effective g value less than 2.0 may certainly be
derived in principle. In practice a reasonable value for
g_1 is about 2.02 the value found for tetrahedral Fe^{3+} in ZnS,
and g_2 may be considerably greater than 2.00 as explained in
the section on the tetrahedral ferrous ion, at least for the
x and y directions. Thus two g values below the free spin
value are predicted and this has become the hall-mark of this
type of protein.

The exchange integral J corresponding to the assumed antiferromagnetic exchange in [2Fe2S] clusters has been measured
by e.p.r. linewidth measurement assuming an Orbach spin-lattice
relaxation mechanism (48) and by intensity measurements of
depopulation of the ground state with increasing temperature (49).
Similar ideas of exhange interaction have been applied to [4Fe4S]
clusters too (50) but the molecular orbital treatment of this
cluster is also very attractive (51). It is assumed in this
theory that the C_{3v} symmetry offered by the four sulphur atoms
partially lifts the d orbital degeneracy of each iron atom in
the cube. Metal-metal bonding may then be introduced by forming
molecular orbitals from the three separated sets of d orbitals
z^2, (xz, yz) and (xy, $x^2 - y^2$) their relative energies being
calculated as a function of iron-iron distance according to a
Hückel method. In this way a qualitative ordering of energy levels
of type a, e or t is built up in keeping with the experimental
requirement that the C state of the [4Fe4S] cluster, possessing
22 electrons, should be diamagnetic. A scheme involving heavy
metal-metal bonding for this state has the configuration
$t^6 < t^6 < e^4 < a^2 < e^4 < t^0$. On reduction to C^- one electron
in t will give g values below 2.0 as does the Ti^{3+} ion. But on
oxidation the ground state C^+ would be e^3 from which positive g
shifts are expected because the shell is more than half filled,
as with Cu^{2+}. Thus these ideas, although rather qualitative
explain $g_{avg} < 2.0$ in reduced ferredoxins and $g_{avg} > 2.0$ in
oxidized.   HiPIP.\

Some proteins contain two [4Fe4S] clusters and their e.p.r.
behaviour is more complicated. When partially reduced a simple
rhombic g tensor is seen($g_{avg} = 1.96$); it is the same at both X
and Q band frequencies. However, when fully reduced a new,more
complex spectrum appears which is different again at the higher
Q band frequency (35 GHz). The behaviour is explained assuming
only one reduced cluster per molecule for the simple spectrum
but that when both sites are made paramagnetic in the fully
reduced protein the clusters interact either as two magnetic
dipoles or through ferromagnetic electrostatic exchange with
S = 1. Either way, there should be a half-field line at about
3.92 (= 2 x 1.96) and experimentally this is observed (52).

REFERENCES

1. J.S. Griffith, " The Theory of Transition-Metal Ions",
    Cambridge Press, 1961.

2. V.F. Meshacheryakov, B.N. Grechushnikov and J.N. Kalinkina,
    Zhur. eksp. tear. Fiz., 1974, 66, pp1870 (Soviet Physics,
    J.E.T.P., 1974, 39, pp920).

3. U. Kaufmann, Phys. Rev. B., 1976, 14, pp1848.

4. A. Abragam and B. Bleaney, " Electron Paramagnetic Resonance
    of Transition Ions", Clarendon Press, Oxford, 1970.

5. A. Hudson and M.J. Kennedy, J. Chem. Soc. (A), 1969, pp1116.

6. The modified version of LOSPIN has a tidy printout, contains
    various checks, is more rapid for axial symmetry and works
    also for d^1. A listing is available on request.

7. J.H.M. Thornley, J. Phys. (C), 1968, 1, pp1024.

8. C.S. Maskall, J.F. Gibson and P.J. Dart, Biochem. J., 1977,
    167, pp435-445.

9. J.F. Gibson and D.J.E. Ingram, Nature, 1957, 180, pp29-31.

10. C.M. Guzy, J.B. Raynor and M.C.R. Symons, J. Chem. Soc. (A),
    1969, pp2299.

11. J. Peisach, W.E. Blumberg and A. Adler, Ann. New York Acad.
    Sci., 1973, 206, pp310-327.

12. C.A. Appleby, W.E. Blumberg,J. Peisach, B.A. Wittenberg and
    J.B. Wittenberg, J. Biol. Chem., 1976, 251, pp6090-6096.

13. S.A. Cotton and J.F. Gibson, J. Chem. Soc. (A), 1971,
    pp 1690-1693.

14. P.L. Hall, B.R. Angel and J.P.E. Jones, J. Magn. Resonance,
    1974, 5, pp64-68.

15. D.C. Bradley, R.G. Copperthwaite, S.A. Cotton, K.D. Sales,
    and J.F. Gibson, J.C.S. Dalton, 1973, pp191-194.

16. J.S. Griffith, Mol. Phys., 1964, 8, pp213-224.

17. D. Vivien and J.F. Gibson, J.C.S. Faraday II, 1975, 71,
    pp1640-1653.

18.   S.A. Cotton and J.F. Gibson, J. Chem. Soc. (A), 1971,
      pp1696-1699.

19.   M. Llinas, D.M. Wilson and J.B. Neilands, Biochem., 1973,
      12, pp3836-3843.

20.   R.M. Butterworth, Ph.D. thesis, University of London, 1976.

21.   K. Spartalian, W.T. Oosterhuis and J.B. Neilands, J. Chem.
      Phys., 1975, 62, pp3538-3543.

22.   J. Peisach, W.E. Blumberg,S. Ogawa, E.A. Rachmilewitz and
      R. Oltzik, J. Biol. Chem., 1971, 246, pp3342-3355.

23.   K. Nishikura, Y. Sugita, M. Nagai and Y. Yoneyama,
      J. Biol. Chem., 1975, 250, pp6679-6685. See also K. Nagai
      and H. Hori, FEBS Letters, 1978, 93, pp275-277, for similar
      experiments with valency hybrids.

24.   J.J.M.C. de Groot, G.A. Veldink, J.F.G. Vliegenthart,
      J. Boldingh, R. Wever and B.F. van Gelder, Biochem.
      Biophys. Acta, 1975, 377, pp71-79.

25.   W.E. Blumberg " Magnetic Resonance in Biological Systems"
      Pergamon Press, 1967, ed., A. Ehrenberg, B.G. Malmström
      and T. Vänngård.

26.   C.A. Tyson, J. Biol. Chem., 1975, 250, pp1765.

27.   W.E. Blumberg and J. Peisach, Ann.of New York Acad. Sci.,
      1973, 222, pp539-560.

28.   G.E. Peterson, C.R. Kurkjian and A. Carnevale, Phys. and
      Chem. of Glasses, 1974, 15, pp52-58.

29.   R.D. Dowsing and J.F. Gibson, J. Chem. Phys., 1969, 50,
      pp294-303. Copies of the D/B graphs published in this paper
      and also those for S = $\frac{3}{2}$ are available on request.

30.   P.R. Rich, R. Cammack and D.S. Bendall, Eur. J. Biochem.,
      1975, 59, pp281-286.

31.   C.R. Jefcoate, W.H. Crme-Johnson and H. Beinert, J. Biol.
      Chem., 1976, 251, pp3706-3715.

32.   R. Aasa, J. Chem. Phys., 1970, 52, pp3919-3930.

33.   J.F. Gibson, R.M. Butterworth and E.M. Price, to be published.

34. R. Tsai, C.A. Yu, I.C. Gunsalus, J. Peisach, W.E. Blumberg,
     W.H. Orme-Johnson and H. Beinert, Proc. Nat. Acad. Sci.,
     (U.S.A.), 1970, 66, pp1157-1163.

35. M.M. Maltempo, Biochim. Biophys. Acta, 1975, 379,pp95-102.

36. C.R. Hartzell and H. Beinert, Biochim. Biophys. Acta,
     1974, 368, pp318-

37. G. Harris, Theoret. Chim. Acta, 1968, 10 pp119-

38. G.E. Chapps, S.W. McCann, H.H. Wickman and R.C. Sherwood,
     J. Chem. Phys., 1974, 60, pp990-

39. D. Niarchos, A. Kostikas, A. Simopoulos, D. Coucouvanis,
     D. Piltingsrud and R.E. Coffman, J. Chem. Phys., 1978, 69,
     pp4411-4418.

40. J. Rawlings, V.K. Shah, J.R. Chisnell, W.J. Brill,
     R. Zimmerman, E. Münck and W.H. Orme-Johnson, J.Biol. Chem.,
     1978, 253, pp1001-1004.

41. G. Christou, C.D. Garner, F.E. Mabbs and T.J. King,
     J.C.S. Chem. Comm., 1978, pp740.

42. G. Christou, personal communication.

43. P.R. Rich, J.C. Salerno, J.S. Leigh and W.D. Bonner,
     FEBS Letters, 1978, 93, pp323-326.

44. M.M. Maltempo, J. Chem. Phys., 1974, 61, pp2540-2547.

45. J.R. Pilbrow, J. Magn. Resonance, 1978, 31, pp479-490.

46. R. Aasa and T. Vänngård, J. Magn. Resonance, 1975, 19,
     pp308-315.

47. J.F. Gibson, D.O. Hall, J.H.M. Thornley and F.R. Whatley,
     Proc. Nat. Acad. Sci. (U.S.A.), 1966, 56, pp987-990.

48. J-P. Gayda, J.F. Gibson, R. Cammack, D.O. Hall and
     R. Mullinger, Biochim. Biophys. Acta, 1976, 434, pp154-163.

49. J.C. Salerno, T. Ohnishi, H. Blum and J.S. Leigh,
     Biochim. Biophys. Acta, 1977, 494, pp191.

50. H. Blum, J.C. Salerno, R.C. Prince, J.S. Leigh and
     T. Ohnishi, Biophys. J., 1977, 20, pp23.

51. A.J. Thomson, Biochem. Soc. Trans., 1975, 3, pp468.

52. G. Palmer " Iron Sulphur Proteins", pp1-56 in " The Enzymes" 1975, ed., P.D. Boyer XII Oxidation Reduction Part B.

ESR AND NMR SPECTRA OF EXCHANGE COUPLED METAL IONS.

Dante Gatteschi

Istituto di Chimica Generale e Inorganica,
University of Florence, Italy.

1. THE NATURE OF THE EXCHANGE INTERACTION

1.1. The Valence Bond Approach

It is well known from experiments that when two metal ions
are sufficiently close to each other in a molecule, the magnetic
properties of the couple can differ appreciably from the sum of
the properties of the two individual ions. This can be observed
by magnetic susceptibility measurements, ESR and NMR spectroscopy,
heat capacity measurements, etc. The interactions which occur in
the couple are called exchange interactions. It will be the purpose of this short review to summarize the foundations of exchange
interactions and to show how they can be relevant in the interpretation of the ESR and NMR spectra of paramagnetic complexes.
The term exchange has its origin in the Heisenberg-Dirac-van
Vleck Hamiltonian (1), which was developed within a Heitler-London
(Valence Bond) formalism. Essentially the exchange interaction is
an extreme case of weak bonding interaction between the two atoms.
As such it is electrostatic in nature. The simplest way to understand this point is to consider the case of two atoms, A and B,
separated by some distance r_{AB}, each with one unpaired electron

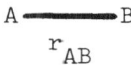

with spin-orbit coupling effects which are neglectable, so that
the wavefunctions can be factored into a product of orbital and
spin functions. The Hamiltonian appropriate to this system is:

*I. Bertini and R.S. Drago (eds.), ESR and NMR of Paramagnetic Species in Biological and
Related Systems, 255–273.*
Copyright © 1979 by D. Reidel Publishing Company

$$\hat{H} = - \frac{\hbar^2}{2m}(\nabla_1^2 + \nabla_2^2) - Ze^2(\frac{1}{r_{A1}} + \frac{1}{r_{B1}} + \frac{1}{r_{A2}} + \frac{1}{r_{B2}}) + \frac{e^2}{r_{12}} + \frac{Z^2e^2}{r_{AB}} \quad (1)$$

where 1 and 2 refer to the two electrons and A and B to the two atoms. In the Valence Bond formalism the eigenfunctions of the Hamiltonian (1) properly antisymmetrized are:

$$\Psi_1 = (2 - 2S^2)^{-\frac{1}{2}}\left[\phi_A(1)\phi_B(2) - \phi_B(1)\phi_A(2)\right]\alpha(1)\alpha(2)$$

$$\Psi_2 = (2 - 2S^2)^{-\frac{1}{2}}\left[\phi_A(1)\phi_B(2) - \phi_B(1)\phi_A(2)\right]\beta(1)\beta(2) \quad (2)$$

$$\Psi_3 = 2(1 - S^2)^{-\frac{1}{2}}\left[\phi_A(1)\phi_B(2) - \phi_B(1)\phi_A(2)\right]\left[\alpha(1)\beta(2) + \beta(1)\alpha(2)\right]$$

$$\Psi_4 = 2(1 + S^2)^{-\frac{1}{2}}\left[\phi_A(1)\phi_B(2) + \phi_B(1)\phi_A(2)\right]\left[\alpha(1)\beta(2) - \beta(1)\alpha(2)\right]$$

where S is the overlap integral, and α and β are the spin functions. Ψ_1, Ψ_2, and Ψ_3 correspond to a triplet state, while Ψ_4 corresponds to a singlet state. Resolving the appropriate secular determinant gives the energy difference between these two states:

$$E_{triplet} - E_{singlet} = \frac{2(K_{AB} - J_{AB}S^2)}{1 - S^4} \quad (3)$$

where J_{AB} and K_{AB} are the coulomb and exchange integrals respectively:

$$J_{AB} = \langle\phi_A(1)\phi_B(2)|\frac{Ze^2}{r_{A2}} - \frac{Ze^2}{r_{B2}} + \frac{e^2}{r_{12}} + \frac{Z^2e^2}{r_{AB}}|\phi_A(1)\phi_B(2)\rangle$$

$$\quad (4)$$

$$K_{AB} = \langle\phi_A(1)\phi_B(2)|\frac{Ze^2}{r_{A2}} - \frac{Ze^2}{r_{B2}} + \frac{e^2}{r_{12}} + \frac{Z^2e^2}{r_{AB}}|\phi_B(1)\phi_A(2)\rangle$$

It is customary to express the energy separation (3) within a spin Hamiltonian framework, using the operator:

$$\hat{H} = - J\,\hat{S}_1 \cdot \hat{S}_2 \quad (5)$$

where $\hat{S}_1$ and $\hat{S}_2$ are the spin operators for electron 1 and 2 respectively and J is a constant. It is easy to show that the Hamiltonians (1) and (5) yield the same energy separation between the singlet and triplet states provided that

$$J = 2\frac{(K_{AB} - J_{AB}S^2)}{1 - S^4} \quad (6)$$

A positive $\underline{J}$ means that the spin triplet has lower energy, while the reverse is true for negative $\underline{J}$. The former case is referred to as ferromagnetic and the latter as antiferromagnetic coupling. It must be noted that several different forms of the Hamiltonian (5) are used in the literature. They are all equivalent to each other, but one must be much careful in comparing data from different papers, since different conventions may have been used. In particular both the + and − signes have been used, as well as $\underline{J}$ and $2\underline{J}$ as constants.

In the above treatment it was assumed that direct exchange interaction is operative between the two metal atoms A and B. However it is well known that magnetic coupling can occur even if the two paramagnetic atoms are separated by intervening diamagnetic atoms. It was suggested that the model of direct exchange could be extended also to this case (2), and the interaction was then called superexchange interaction.

The essential physical basis of the superexchange coupling is that the metal d orbitals, in which the unpaired spins originate, overlap with filled s and p orbitals of intermediary atoms. As a consequence the orbitals containing the unpaired spin encompass both the metals and the intervening ligand orbitals. Considering a simple complex as the one shown below

$$A \text{\textemdash} L \text{\textemdash} B$$

one must consider that the orbital containing the unpaired electron on atom A will overlap with some orbital of L of appropriate symmetry. If this orbital is overlapping the orbital which contains the unpaired electron on B an antiferromagnetic interaction will originate, while if it is orthogonal to that a ferromagnetic coupling will be obtained.

It must be stressed that from these considerations only qualitative guesses of the sign of $\underline{J}$ can be made, and that calculations are actually impossible. However they are much useful, and it is possible to make predictions on the expected magnetic coupling in dinuclear and polynuclear complexes. Several text books and review articles (3,4) show how this can be done. In particular a paper by Ginsberg (5) gives a very detailed discussion of the expected magnetic coupling in several octahedral complexes.

1.2. The Molecular Orbital Approach

The approach outlined above is essentially a Valence Bond approach. There are some theoretical considerations which suggest that the Valence Bond approach is superior to the Molecular Orbital approach for predicting the ground state of weakly interacting centers (6). The main difference between the two models, at their lowest degree of approximation, which is the one commonly used, is that Valence Bond considers strong electron correlation, while low electron correlation is inherent to the Molecular Orbital mod-

el. In other words ionic terms, which put both the electrons on
the same center, have the same weight as covalent terms in the
Molecular Orbital scheme, while only covalent terms are introduced in the zero order Valence Bond functions. It is apparent that
in the limit of two weakly interacting metal centers the probability of finding both the electrons on the same atom must be low,
making the Molecular Orbital approximation inadequate.

Several authors have suggested Molecular Orbital approaches
to the evaluation of the magnetic coupling constant, trying to
cope with the intrinsic inconsistencies of the model outlined a-
bove. The most thorough attempts to correlate J with Molecular
Orbital integrals have been made by Dance (7), Kahn and Briat (8)
and Hoffmann (9). The approaches of the first three authors are
essentially the same, even if they start from different points of
view, while the results of Hoffmann's approach are somewhat different. However all the methods express the observed J constant
as a sum of ferromagnetic and antiferromagnetic contributions,
and at least qualitatively they are all equivalent. In the following we will use the approach by Hoffmann.

Let us consider the electronic structure of two weakly interacting metal ions in a dinuclear complex. Let the unpaired electron in each monomeric unit be in an orbitally non degenerate molecular orbital. Due to the interaction through the bridging ligands the two metal orbitals will be split, giving a bonding and
an antibonding molecular orbital

$$\phi_2$$

A _____   _____ B

$$\phi_1$$

The two electrons can be fed into the two molecular orbitals,
yielding one triplet (from the configuration $\phi_1 \phi_2$) and three
singlet spin states (from the configurations $\phi_1 \phi_2$, ϕ_1^2, ϕ_2^2).
The energy difference between the triplet and the lowest singlet
spin state is given by

$$E_t - E_s = J_{12} - K_{12} - \tfrac{1}{2}(J_{11} + J_{22})+ \tag{7}$$
$$+ \tfrac{1}{2}\left[(2h_1 + J_{11} - 2h_2 - J_{22})^2 + 4K_{12}^2\right]^{\tfrac{1}{2}}$$

where J and K stand for coulomb and exchange integrals respectively relative to the indicated ϕ molecular orbitals, and h is the
energy relative to the core Hamiltonian, i. e. the part of the
Hamiltonian (1) which was not included in (4). In the above derivation it was assumed to have obtained the molecular orbitals

from a SCF calculation on the triplet state.

In order to express $\underline{J}$ it is more convenient to use localized molecular orbitals which contain both metal and ligand character but will be essentially d orbitals on metal A and B respectively:

$$\phi_a = \tfrac{1}{2}(\phi_1 + \phi_2) \qquad\qquad \phi_b = \tfrac{1}{2}(\phi_1 - \phi_2) \tag{8}$$

Using these orbitals and the orbital energies ε_1 and ε_2 one ob-

$$E_s - E_t = 2K_{ab} - \frac{(\varepsilon_1 - \varepsilon_2)^2}{J_{aa} - J_{ab}} = \underline{J} \tag{9}$$

K_{ab} is positive giving a ferromagnetic contribution to the coupling constant, $J_{aa} - J_{ab}$ is also positive, since J_{aa} is a one center coulomb integral, while J_{ab} is a two center integral. Therefore the second term in (9) gives the antiferromagnetic contribution.

Equation (9) can be made more general for the case of identical atoms containing more than one unpaired electron according to the relation:

$$\underline{J} = \frac{1}{m^2}\, \Sigma_{i\in A}\ \Sigma_{j\in B}\ K_{ij} - \frac{1}{m^2}\, \Sigma_i\ \frac{\tfrac{1}{2}(\varepsilon_{2i} - \varepsilon_{2i-1})^2}{J_{a_i,a_i} - J_{a_i,b_i}} \tag{10}$$

Hoffmann suggests that in trying to use these formulae for the interpretation of the magnetic properties of series of dimers the most relevant quantity is $(\varepsilon_1 - \varepsilon_2)$, since the other values should be a fairly slowly varying quantity as a function of distortions and substituent effects for closely related compounds. The antiferromagnetic coupling is expected to increase when the orbital energy difference is increased. This is easily obtained from Molecular Orbital calculations, and Hoffmann used an Extended Huckel formalism to intrpret the variation of the coupling constant in series of copper(II) complexes.

1.3. The Ligand Field Approach

It is well known that Ligand Field models are the simplest approach to the orbital energies of transition metal complexes. If one takes into consideration a dinuclear complex of the type

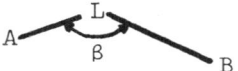

and assumes that no direct interaction is operative between the two metal atoms A and B, the assumptions which support the Ligand Field treatment on one metal center keep their validity.

Glerup (10) and Schmidtke (11) had given some particular solutions of the Ligand Field problem in dinuclear complexes . Bencini and Gatteschi (12) have recently provided a general formulation for the calculation of the orbital energies according to the Angular Overlap Model, which is a particular case of Ligand Field parameterization (13).

In this scheme the antibonding effect on the metal d orbitals determined by the ligands is considered to be proportional to the squared metal ligand diatomic overlap (13):

$$e_\lambda = K \, S_\lambda^2 \tag{11}$$

where $\lambda = \sigma$, π, δ... and S is the overlap integral. The d orbital energies can be obtained by a first order perturbation treatment on the metal d orbitals, by calculating matrix elements of the type:

$$<du|\hat{A}|dv> \tag{12}$$

where $\hat{A}$ is a suitable one electron operator.

For dinuclear complexes the Angular Overlap operator can be conveniently written (12) according to the relation:

$$\hat{A} = |p_z><p_z| + |p_x><p_x| + |p_y><p_y| \tag{13}$$

if one wants to consider only the interaction with the p ligand orbitals for the sake of simplicity. The orbital energies of the dimer can be evaluated by a first order perturbation treatment on the basis of ten d orbitals (five for each metal center) according to the relations for the matrix elements:

$$<Adv|\hat{A}|Adu> = \Sigma_i \, F_{vi}^A \, F_{ui}^A \, e_{iA}$$

$$<Bdv|\hat{A}|Bdu> = \Sigma_i \, F_{vi}^B \, F_{ui}^B \, e_{iB} \tag{14}$$

$$<Adv|\hat{A}|Bdu> = \Sigma_i \, \Sigma_j \, F_{vi}^A \, F_{uj}^B \, F_{p,ij} \, (e_{iA} e_{jB})^{\frac{1}{2}}$$

where A and B refer to the two metal centers, u and v are indices of d orbitals, F^A and F^B are the angular overlap matrices for metal atom A and B respectively, and F_p is such that

$$F_{p,ij} = \left\{ \begin{array}{l} F_{p,ij}^L \quad \text{for i and j} = \sigma, \ \pi s, \ \pi c \\[10pt] 0 \ \text{for i or j} = \delta s, \ \delta c \end{array} \right. \tag{15}$$

where F_p^L is the angular overlap matrix for the ligand p orbitals.

The advantage of this parameterization of the orbital energies of dinuclear complexes is that it is obtained through a Ligand Field calculation, which is simple to handle, since only a
10x10 matrix must be diagonalized, and the values of the parameters are in principle the same which are obtained from the interpretation of the spectral properties of simple mononuclear complexes.

Several sample calculations on model dinuclear copper(II)
complexes have been reported (12). In particular a series of
square planar and square pyramidal di-hydroxo bridged copper(II)
complexes studied by Hatfield and Hodgson (14) have been taken into consideration. The authors related the experimental $\underline{J}$ values
to the variation of the Cu–O–Cu, β, angle throughout the series.
In particular they showed (15) that the experimental $\underline{J}$ is increasingly ferromagnetic for $\beta < 98°$, while it is antiferromagnetic for
$\beta > 98°$ (Figure 1).

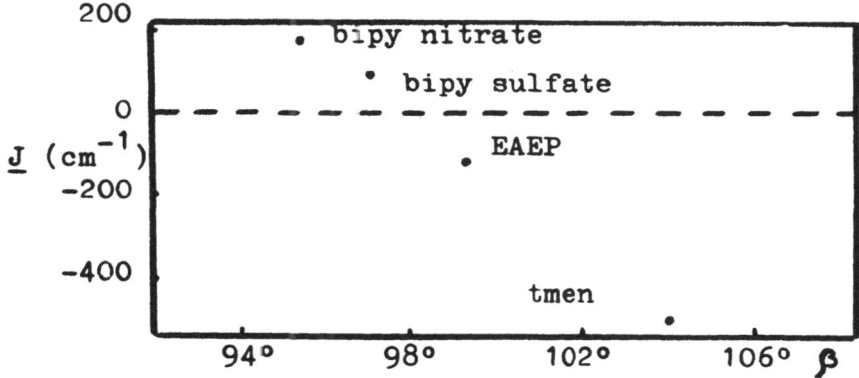

Figure 1. The dependence of $\underline{J}$ on the Cu–O–Cu angle, β, in a series
of di-hydroxo bridged copper(II) complexes. After ref. 15.

Our calculations, shown in Figure 2 confirm essentially this
observation, since the energy difference $E(b_{3u}) - E(b_{2g})$ increases
dramatically as β departs from 90°, and as stated above an increase
of the orbital energy separation increases the tendency to antiferromagnetic coupling. From Figure 2 it is also apparent that no
other parameter has the same dramatic effect on the energies of
the two highest levels, so that it can be understood how a satisfactory correlation could be made between the experimental $\underline{J}$ values
and the β angle, without worrying too much about the nature of the
other ligands. I want to make here the point that a relation as
simple as this can be expected only for copper(II), since the unpaired electrons are in the $d(x^2-y^2)$ orbitals which are well sepa-

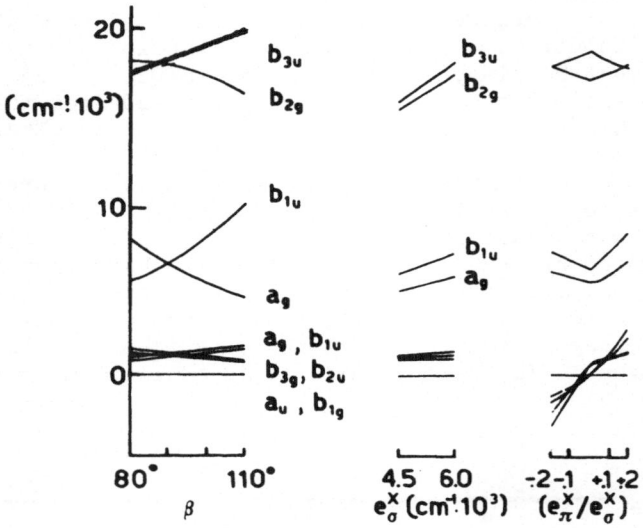

Figure 2. Energy level diagrams for square planar dinuclear complexes showing the effect of varying the M–L–M angle, the e_σ and e_π parameters of the bridge ligand.

rated from all the others, being σ antibonding in character.

Another well documented series of dinuclear complexes is that of chromium (III) (16). In this case however there are three unpaired electrons, which make all the treatment even qualitatively more complicated. Furthermore the unpaired electrons are in π antibonding orbitals, whose relative energies will depend drastically on the π bonding ability of both the bridging and the terminal ligands. It is my feeling that for these complexes it will not be possible to find any simple correlation between one single parameter, e.g. the Cr–L–Cr angle, and the experimental $\underline{J}$ values observed in series of complexes. As a matter of fact, although several groups have been working on the problem no such correlation has been proposed up to now.

2. EFFECT OF EXCHANGE ON THE MAGNETIC RESONANCE SPECTRA

2.1. ESR Spectra of Couples of S = ½ Systems.

When two S = ½ ions are magnetically coupled the ESR spectra of the dinuclear complex can be interpreted using the spin Hamiltonian (17):

$$\hat{H} = \mu_B \, \underset{\sim}{B} \cdot \underset{\sim}{g}_1 \cdot \hat{\underset{\sim}{S}}_1 + \mu_B \, \underset{\sim}{B} \cdot \underset{\sim}{g}_2 \cdot \hat{\underset{\sim}{S}}_2 + \underline{J} \, \hat{\underset{\sim}{S}}_1 \cdot \hat{\underset{\sim}{S}}_2 + \underset{\sim}{d} \cdot \hat{\underset{\sim}{S}}_1 \times \hat{\underset{\sim}{S}}_2 + $$

$$+ \hat{\underset{\sim}{S}}_1 \cdot \underset{\sim}{D} \cdot \hat{\underset{\sim}{S}}_2 \tag{16}$$

The first two terms are the Zeeman operators for the isolated ions, the third term is the isotropic exchange described in the previous section, the fourth is the antisymmetric exchange term, which goes to zero when the two ions are related by a center of symmetry (18) and the fifth term is the anisotropic exchange term, whose main effect is to produce a zero-field splitting for the resulting triplet. The origin of the antisymmetric and of the anisotropic terms is essentially in the spin orbit coupling interaction for the metal ions which had been neglected in the previous section.

Of the three terms which contain the spin coordinated of both the metal ions the isotropic is the leading one, determining the nature of the ground state, either singlet or triplet.

Experimentally the most common case is that of antiferro mag-neticexchange so that the ground state is a singlet. When the se-paration between the singlet and the excited triplet is large com-pared to the thermal energy no ESR signal will be detected. This situation has been found in several copper(II) dimers. For instan-ce the complexes

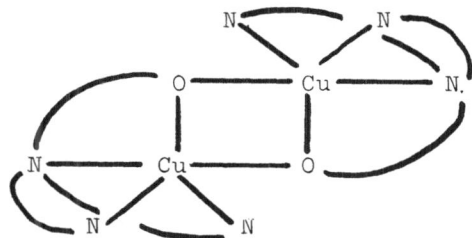

have been reported to be antiferromagnetically coupled, with a $\underline{J}$ value which has been estimated from magnetic susceptibility mea-surements to be $\sim$ –500 cm^{-1} (19). At room temperature less than 10% of the molecules are in the triplet state; since the linewidth of the triplet ESR spectra are in general pretty broad, no ESR signal could be detected for these complexes.

A classic example of well resolved triplet spectra is given by dimeric copper(II) acetate monohydrate. Its dimeric nature has been established through X-ray diffraction experiments (20), how-ever it must be mentione that the first suggestion of such a struc-ture was given on the basis of the ESR spectra (21). The value of $\underline{J}$ was determined accurately through magnetic susceptibility measurements (22) and recently it has been confirmed through an-elastic neutron scattering experiments (23) by which the singlet-triplet transition was detected.

It is apparent that the higher frequency spectra show a larger number of transitions, indicating that the zero-field splitting is of the same order of magnitude of the X-band frequency (0.3 cm^{-1}). At Q-band frequency up to six $\Delta M = 1$ bands are expected, two for each principal direction, if the symmetry of the complex is lower than axial. The extent to which the two transitions corresponding to the same principal direction are split gives information on the value of the zero-field splitting. A quantitative analysis of the bands can be made by using the relations (24):

$$HX1 = (g_e/g_x) \left[(H_o - D' + E')(H_o + 2E')\right]^{\frac{1}{2}}$$

$$HX2 = (g_e/g_x) \left[(H_o + D' - E')(H_o - 2E')\right]^{\frac{1}{2}}$$

$$HY1 = (g_e/g_y) \left[(H_o - D' - E')(H_o - 2E')\right]^{\frac{1}{2}}$$

$$HY2 = (g_e/g_y) \left[(H_o + D' + E')(H_o + 2E')\right]^{\frac{1}{2}} \quad (17)$$

$$HZ1 = (g_e/g_z) \left[(H_o - D')^2 - E'^2\right]^{\frac{1}{2}}$$

$$HZ2 = (g_e/g_z) \left[(H_o + D')^2 - E'^2\right]^{\frac{1}{2}}$$

where the left hand side gives the resonant fields in the indicated principal directions, g_e is the free electron gyromagnetic ratio, H_o is the resonant field for the free electron, $D' = D/(g_e\mu_B)$, $E' = E/(g_e\mu_B)$. D and E are expressed in cm^{-1}, while D' and E' are in gauss.

The spectra of Figure 3 show more features than are predicted by equations (17). The extra bands can be assigned to $\Delta M = 2$ transitions. At X-band one of them is resolved, while at Q-band two of them are present. One is attributed to a transition between two states with M = -1 and +1 respectively which is determined by the absorption of one photon. This transition would be forbidden if the states were pure M = -1 and +1 states. However in low field and low symmetry they are admixed with the M = 0 state so that the transition probability is different from zero. The resonant field for this transition in a polycrystalline spectra is given by (25):

$$H_{min} = (g_e/g_{min}) \left[H_o^2/4 - D'^2/3 - E'^2\right]^{\frac{1}{2}} \quad (18)$$

The highest probability of transition is not for crystallites with principal axes parallel to the static magnetic field, but with axes lying on the surface of a cone defined by an angle

$$\alpha = \cos^{-1} \left[(9 - 4D/h\nu)/(27 - 36D/h\nu)\right]^{\frac{1}{2}} \quad (19)$$

so that

$$g_{min} = \left[(\sin^2\alpha\, g^2 + \cos^2\alpha\, g_z^2)\right]^{\frac{1}{2}}; \qquad g^2 = g_x g_y \quad (20)$$

At Q-band frequency also the resonance for crystallites with the magnetic field in the molecular plane is resolved.

Another $\Delta M = 2$ band can be detected only at high microwave power, and is determined by the absorption of two photons for the orientation of crystallites in the magnetic field such that the separation of the $M = -1$ and 0 and $M = 0$ and $+1$ levels are almost equal (26). The theoretical expressions for the resonance fields of these transitions are:

$$H_{dq} = (g_e/g_{av}) \left[H_0^2 - D'^2/3 - E'^2 \right]^{\frac{1}{2}} \tag{21}$$

with $g_{av} = \left[2/3 \; g^2 + 1/3 \; g_z^2 \right]^{\frac{1}{2}}$.

Using equations (17), (18), (19), (20), and (21) the spin Hamiltonian parameters are found to to be $g_z = 2.39$, $g_x = g_y = 2.07$, $D = 0.34$ cm^{-1}, $E \simeq 0$.

2.2. ESR Spectra of Mixed Valence Complexes.

A topic which is relevant to the interpretation of the ESR spectra of several metallo-enzymes is that of mixed valence dinuclear complexes, i.e. complexes where the same metal is present in two different formal oxidation states. An information which in general can be obtained from the ESR spectra is whether the two metal ions are equivalent or not. The equivalence can be checked easily through the metal hyperfine splitting: if the unpaired electrons are coupled to one metal nucleus the usual $2I + 1$ hyperfine lines will be detected, where I is the nuclear spin; if on the other hand the two ions are equivalent a resulting nuclear spin $F = 2I$ will determine the presence of $2F + 1$ lines of different intensities. The equivalence can be established only to the limit of the ESR time scale, which is of the order of 10^{-7} sec. In other words an ESR experiment can show that the unpaired electron exchange is faster than 10^{7} sec^{-1}.

As an example we may consider the complex trichlorobis(4-methyl-1,8-naphthyridine)dicopper (27), which has the structure shown in Figure 4. The ESR spectra of a glassy solution of this complex yield $g_1 = 2.02$, $g_2 = 2.07$, $g_3 = 2.33$, $A_1 = 16$ G, $A_2 = 13$ G, and $A_3 = 72$ G. Figure 5 shows that the ESR features are split into seven components with relative intensities 1:2:3:4:3:2:1. The g values compare well with those reported for pseudotetrahedral copper(II) complexes, and the hyperfine splitting is almost half that usually observed, as it must be expected (29) since the electron is spending only half its time on a particular nucleus.

Another example of ESR characterization of a mixed valence complex is that of tetra(1,8-naphthyridine)-bis(bromonickel)tetraphenylborate, whose structure is shown in Figure 5 (30). The two nickel ions have on the whole three positive charges, so that they can be regarded as a nickel(I) and a nickel(II). The ESR spectra of this complex are typical of a $S = 3/2$ system split by

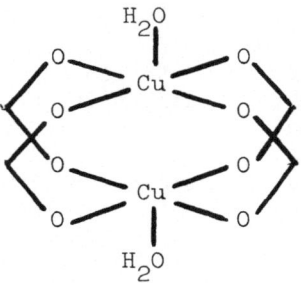

Since $\underline{J}$ is 290 cm^{-1} the ESR spectra can be recorded down to 20 K, but also at room temperature the lines are reasonably narrow to be easily detected. We have recently studied the spectra of a similar complex, $Cu_2(\alpha\text{-chloroacetate})_4(\alpha\text{-picoline})_2$, both at X-band and Q-band frequency. The ESR spectra are shown in Figure 3.

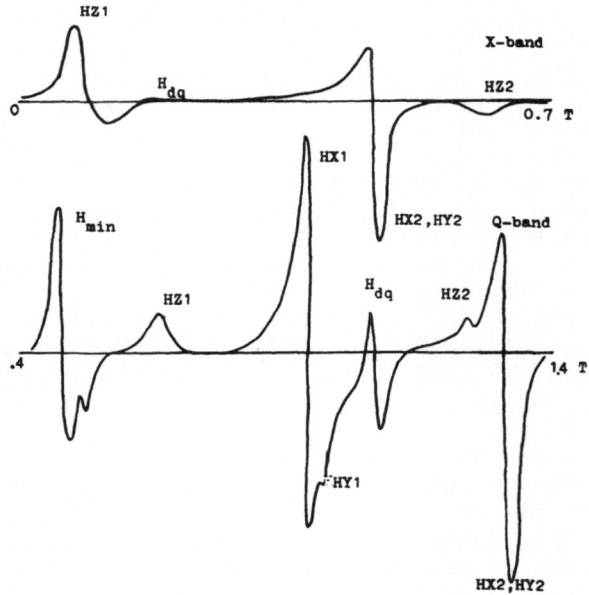

Figure 3. Polycrystalline powder spectra of $Cu_2(\alpha\text{-chloroacetate})_4$ $(\alpha\text{-picoline})_2$ at room temperature.

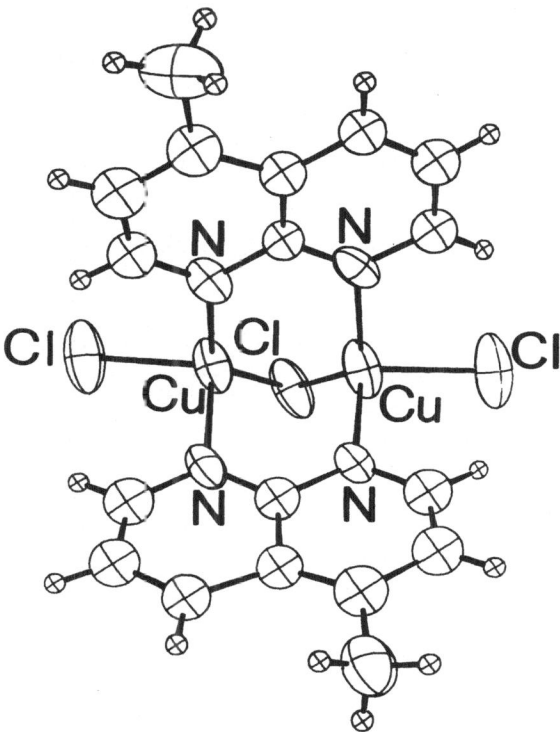

Figure 4. The structure of the complex trichlorobis(4-methyl-
1,8-naphthyridine). After ref. 27

a large zero-field splitting, so that the only transition observed is that within the lowest Kramers doublet (31).

In these cases it is customary to interpret the spectra using a fictitious spin Hamiltonian with $S = \frac{1}{2}$. In this scheme the
observed g values are g = 2.19 and g = 4.30. The spectra remain
practically unchanged in the range 4.2-300 K. These data can be
interpreted considering that nickel(II) is a d^8 ion, with $S = 1$,
and nickel(I) is a d^9 ion, with $S = \frac{1}{2}$. The two nickel ions yield
an $S = 3/2$ ground state, and an excited $S = \frac{1}{2}$ state separated by
$3\underline{J}$. The fact that only one transition belonging to the $S = 3/2$
manifold is detected in all the temperature range suggests a
large separation of the ground and the excited multiplet. In other
words a strong ferromagnetic coupling is operative between the two
ions, with a lower limit to the value of $\underline{J}$ of $\sim$ 300 cm^{-1}. It is
woth mentioning that an extension of the treatment suggested by
Hoffmann to the present case (31) was able to rationalize such a
high coupling.

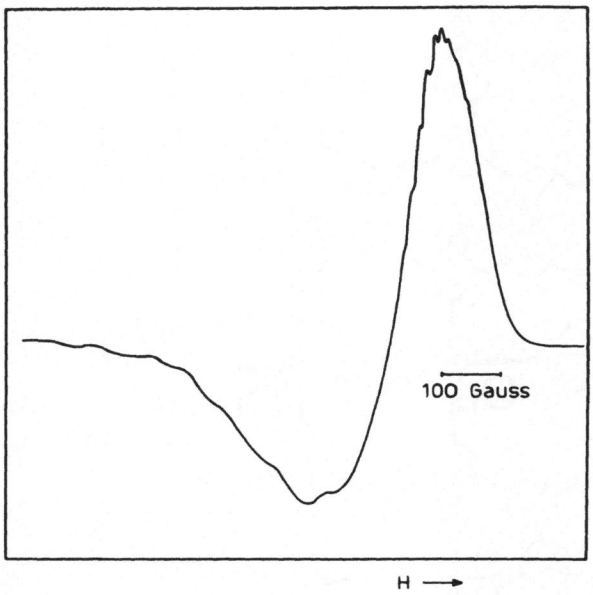

Figure 5. Glassy solution spectra of the complex trichlorobis(4-methyl-1,8-naphthyridine) at 77 K. After ref. 27.

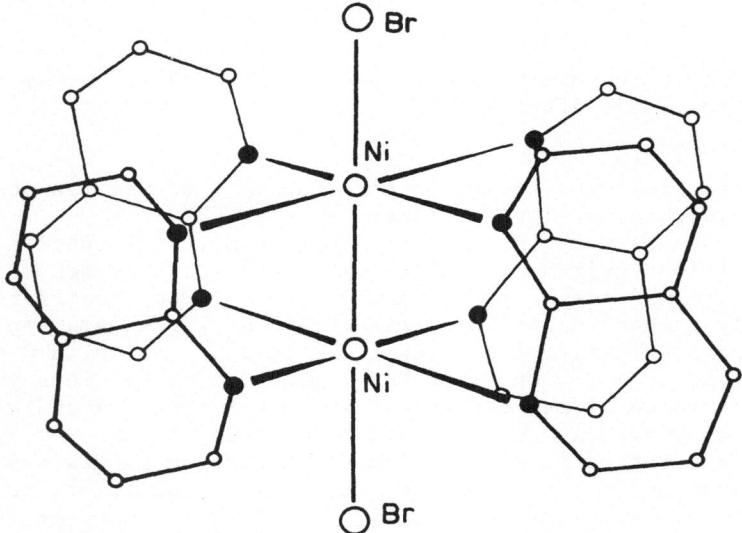

Figure 6. A sketch of the complex tetra(1,8-naphthyridine)-bis(bromonickel)tetraphenylborate.

2.3. ESR Spectra of Weakly Coupled S = $\frac{1}{2}$ Systems.

The copper acetate spectra are typical for large exchange
interactions. However the ESR spectra can be largely modified
from the spectra of the mononuclear complexes also in the case of
far less intense interactions. For instance metalloporhyrins are
known to undergo aggregation processes both in the solid state
and in solution, which have been studied by ESR (32) and NMR(33)
spectroscopy. In particular Boyd et al. (34) studied the ESR spectra of copper(II) and oxovanadium(IV) porphyrins. They observed
broad spectra in the region g = 2, and also some features at half
field (see Figure 7) which were attributed to ΔM = 2 transitions

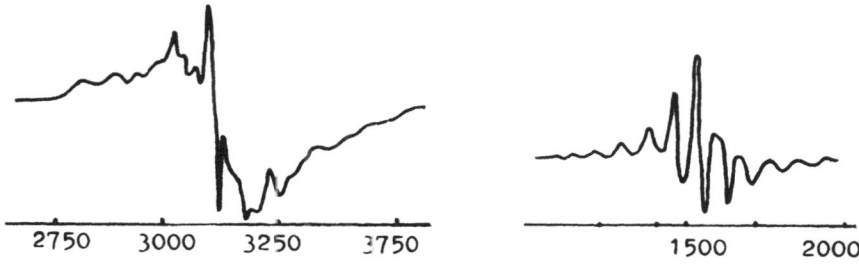

Figure 7. Glassy solution spectra of oxovanadium(IV) deuteropor-
phyrin(IX) at 77 K.

of dinuclear species. This transition is split into fifteen components since the unpaired electrons are shared by the two equivalent ^{51}V nuclei, each with I = 7/2.

For the interpretation of the spectra the authors assumed
that the interaction between the two metal ions is essentially
dipolar, i.e. in this case the exchange interaction can be neglected due to the large distance of the metal centers and the fact
that no intervening atom is supposed to be present between the
two metal ions. The spin Hamiltonian the authors considered was
of the type:

$$\hat{H} = \mu_B \underset{\sim}{B} \cdot \underset{\sim}{g} \cdot \hat{S}_1 + \mu_B \underset{\sim}{B} \cdot \underset{\sim}{g} \cdot \hat{S}_2 + \hat{H}_D \tag{22}$$

where $\hat{H}_D$ is the dipolar Hamiltonian, whic for simple magnetic dipoles takes the form:

$$\hat{H}_D = (\mu_B^2/r^3) \, |g_x^2 \, \hat{S}_{1x}\hat{S}_{2x} + g_y^2 \, \hat{S}_{1y}\hat{S}_{2y} - 2g_z^2 \, \hat{S}_{1z}\hat{S}_{2z}| \tag{23}$$

where r is the metal-metal distance. By solving for this Hamiltonian in the manifold of $|S_1 S_2 m_1 m_2 I_1 I_2 m_{I1} m_{I2}>$ functions the authors calculated the expected transition fields and probabilities as a function of the metal-metal distance. Although the model cannot be too precise, they were able to obtain distance values which compare well with the values obtained from different techniques.

2.4. ESR and NMR Spectra of Weakly Coupled Homo- and Heterodinuclear Complexes.

A different example o- studies of magnetic resonance spectra of dinuclear complexes is proviced by the series corresponding to the formula shown below (35):

M, M' = Co, Ni, Cu, Zn

These complexes are known to show exceedingly low metal-metal interactions, so that the magnetic moments are practically the sum of the moments of the individual ions.

By adding aqueous solutions containing two different metal salts in various molar ratios to suspensions of the ligand in water both the M-M and the M-M' (M' ≠ M) dinuclear complexes were obtained (36). The best evidence for this is given by the ^1H nmr spectra, as shown in Figure 8. The spectra labelled Co-Cu are of a mixture containing large excess of cobalt, so that it can be assumed that very low concentrations of Cu-Cu are present. It is apparent that beyond the signals of the Co-Co species, which are marked by an asterisk, different signals are present, which must be attributed to the Co-Cu species.

The spectra of the M-Zn complexes (M = Co, Ni, Cu) showed a sizeable isotropic shift of the protons of both the pyridine rings closer and farther from the paramagnetic ion. The overall delocalization mechanism of the unpaired spin on the closer pyridine ring is essentially σ (37), while in the farther ring it is essentially π. The origin of this difference is that for the closer ring the delocalization occurs through the lone pair of the heterocyclic nitrogen, which is orthogonal to the π system of the ring, while for the farther ring the unpaired spin is delocalized through the N--N group which is not orthogonal to the π orbitals of the pyridine ring.

The spectra of the M-M' complexes, where both M and M' are

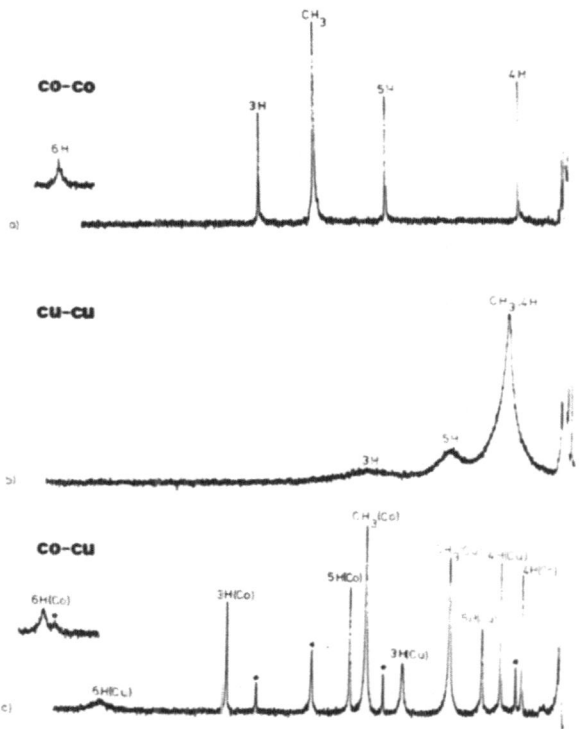

Figure 8. ^{1}H nmr spectra cf the Co-Co (a), Cu-Cu (b), and of
solutions containing both Co-Co and Co-Cu species (c). After
ref. 36.

paramagnetic ions, can be interpreted assuming that the isotropic
shift of a particular proton are the sum of the contributions of
the two metal ions. These in turn can be evaluated from the spectra of the M-Zn complexes. The observed isotropic shifts have
been found to be reproduced within error with the above model.
        Another feature of interest is that the NMR spectra of the
Co-M complexes (M = Ni, Cu) show a neat decrease of the linewidth
of the signals of the proton closer to M as compared to the signals of either M-M and M-Zn complexes. It is well known that octahedral cobalt(II) complexes have very fast relaxation times,
giving narrow NMR lines (see Chapter 10). Since also the protons
farther from the cobalt are under the effect of the cobalt nucleus,
as shown by the isotropic shifts, it is apparent that although at-
tenuated by the large distance the short electronic relaxation
time of cobalt determines the linewidth in these complexes.
        A final remark concerns the ESR spectra. We recorded the ESR

spectra of the Co-Cu complex doped into the Co-Co lattice at room
temperature and were able to obtain well resolved copper hyperfine splitting. This is somewhat surprising since also in the
zinc lattice the room temperature spectra were rather broad. A
behavior somewhat similar to this was previously observed in the
ESR spectra of vanadyl doped $K_2Co(SO_4)_2 \cdot 6H_2O$ (38). These spectra
were explained as follows. The copper (or vanadyl) ion is under
the effect of a dipolar and exchange perturbation from the cobalt
ions. The perturbation interaction must be of the order of 10^{-1}-
10^{-2} cm^{-1} (10^9- 10^8 Hz in frequency units) so that it should result
in a broadening of the copper lines. Since however this interaction is modulated at a frequency of 10^{11}- 10^{13} Hz, corresponding
to the spin lattice relaxation time of cobalt(II), the copper
lines are in the fast exchange region, so that a narrowing occurs.
A proof of this mechanism is obtained by recording the spectra at
lower temperatures. Since the spin lattice relaxation time of co-
balt(II) is expected to become shorter on cooling, the copper
lines are predicted to become broader at lower temperatures, as
observed.

REFERENCES

1) Dirac, P. A. M.: 1926, Proc. Roy. Soc., A112, p. 661; 1929,
   A123, p. 714; Heisenberg, W.: 1926, Z. Physik, 38, p. 441;
   van Vleck, J. H.: 1934, Phys. Rev., 45, p. 405.
2) Kramers, H. A.: 1934, Physica, 1, p. 82; Anderson, P. W.:
   1950, Phys. Rev., 45, p. 405.
3) Goodenough, J. B., "Magnetism and the Chemical Bond" Interscience Publishers Inc., New York, 1963.
4) Kanamori, J.: 1959, Phys. Chem. Solids, 10, p. 87.
5) Ginsberg, A. P.: 1971, Inorg. Chim. Acta Review, 5, p. 45.
6) Coulson, C. A., and Fischer, I.: 1949, Phil. Mag., 40, p. 386.
7) Dance, I. G.: 1974, Inorg. Chim. Acta, 9, p.77.
8) Kahn, O., and Briat, B.: 1976, JCS Faraday II, 72, p. 268.
9) Hay, P. J., Thibeault, J. C., and Hoffmann, R.: 1975, J. Am.
   Chem. Soc., 97, p. 4884.
10) Glerup, J.: 1972, Acta Chem. Scand., 26, p. 3775.
11) Schmidtke, H. H.: 1971, Theoret. Chim. Acta, 20, p. 92.
12) Bencini, A., and Gatteschi, D.: 1978, Inorg. Chim. Acta, 31,
    p. 11.
13) Schaffer, C. E.: 1973, Struct. and Bonding, 14, p. 69.
14) Estes, E. D., Hatfield, W. E., and Hodgson, D. J.: 1974, Inorg.
    Chem. 13, p. 1654.
15) Hodgson, D. J., 1975, Progr. Inorg. Chem., 19, p. 173.
16) Jezowska-Trzebiatowska, B., and Wojciechowski, L.: 1970, Tran-
    sition Metal Chem., 6, p. 1.
17) Kokoszka, G., and Duerst, R. W.: 1970, Coord. Chem. Rev., 5,
    p. 209.
18) Moriya, T.: 1963, in "Magnetism", vol. I, Rado, G. and Suhl,

H., eds., Academic Press, New York.

19) Banci, L., and Dei, A.: 1979, Inorg. Chim. Acta, in press.
20) van Niekerk, J. N., and Shoening, F. R. L. : 1953, Acta Cryst. 6, p. 227.
21) Bleaney, B., and Bowers, K. D., 1952, Proc. Roy. Soc., A214, p. 451.
22) Gregson, A. K., Martin, R. L., and Mitra, S.: 1971, Proc. Roy. Soc., A370, p. 473.
23) Gudel, H. U., Stebler, A., and Furrer, A.: 1979, Inorg. Chem. 18, p. 1021.
24) Wasserman, E., Snyder, L. C., and Yager, W. A.: 1964, J. Chem. Phys., 41, p. 1763; Wasson, J. R., Shyr, C. I., and Trapp, C.: 1968, Inorg. Chem., 7, p. 469.
25) Reedijk, J., Knetsch, D., and Nieuwenhuijse, B.: 1971, Inorg. Chim. Acta, 5, p. 568.
26) de Groot, M. S., and van der Waals, J. H.: 1959, Mol. Phys., 3, p. 190.
27) Gatteschi, D., Mealli, C., and Sacconi, L.: 1976, Inorg. Chem., 15, p. 2774.
28) Bencini, A., Gatteschi, D.: Transition Metal Chem., in press.
29) Slichter, C. P.: 1955, Phys. Rev., 99, p. 479.
30) Sacconi, L., Mealli, C., and Gatteschi, D.: 1974, Inorg. Chem., 13, p. 985.
31) Bencini, A., Gatteschi, D., and Sacconi, L.: 1978, Inorg. Chem., 17, p. 2670.
32) Blumberg, W. E., and Peisach, J.: 1965, J. Biol. Chem., 240, p. 870.
33) Abraham, J. R., Burbidge, P. A., Jáckson, A. H., and Macdonald, D. B.: 1966, J. Chem. Soc. B, p. 620.
34) Boyd, P. D. W., Smith, T. D., Price, J. H., and Pilbrow, J. R.: 1972, J. Chem. Phys., 56, p. 1253.
35) Stratton, W. J.: 1970, Inorg. Chem., 9, p. 517.
36) Dei, A., Gatteschi, D., and Piergentili, E.: 1979, Inorg. Chem., 18, p. 89.
37) Bertini, I., and Gatteschi, D.: 1973, Inorg. Chem., 12, p. 2740.
38) Saraswat, R. S., and Upreti, G. C.: 1978, J. Phys. Soc. Japan, 44, p. 1142.

IRON–SULFUR PROTEINS: COMBINED MÖSSBAUER AND EPR STUDIES |

Eckard Münck and B.H. Huynh

Gray Freshwater Biological Institute/Department of
Biochemistry, University of Minnesota, Navarre,
Minnesota, 55392, USA.

   This article discusses some correlations between electron
paramagnetic resonance and Mössbauer spectroscopy which can be
utilized to unravel some of the complexities of proteins containing many metal atoms. After describing the qualitative
features of Mössbauer spectra observed for Kramers and non-
Kramers systems recent findings on the MoFe protein of nitrogenase (the biological nitrogen fixing system) will be discussed.
Mössbauer and EPR studies of the MoFe protein (30 Fe atoms) have
lead to a detailed spectroscopic characterization of two novel
spin-coupled metal centers.

A.   Introduction

      For the past decade much of the research activity on iron-
sulfur proteins has been centered on characterizing some basic
structures, the 2Fe-2S and the cubane 4Fe-4S centers, in considerable detail. Some of the most important clues regarding
the nature of these centers were furnished by electron paramagnetic resonance (EPR) and Mössbauer spectroscopy. At present
efforts are underway to characterize more complex proteins, such
as enzymes containing more than one type of metallo center, among
them perhaps structures which we have not encountered before.
Without selective isotopic enrichments or clever biochemical
tricks to simplify the data analyses it might not be possible
to describe each center in detail. However, by rigorously
using certain correlations between EPR and Mössbauer spectroscopy
much information can be elicited in a rather straightforward
way. In the following we will discuss these correlations in
simple terms and discuss their application by using nitrogenase

275

*I. Bertini and R.S. Drago (eds.), ESR and NMR of Paramagnetic Species in Biological and
Related Systems, 275–288.*
Copyright © 1979 by D. Reidel Publishing Company

(30 Fe and 2 Mo atoms) as an example.

B.   Some Basic Features of ^{57}Fe Mössbauer Spectroscopy

    In this section we will discuss the orientation dependence
of the intensities of the absorption lines of simple magnetic
Mössbauer spectra (we will assume that the reader is familiar
with the basic principles of the Mössbauer effect. Simple
introductions have been given by Frauenfelder and Debrunner (1)
and by Moss (2). More advanced treatments, relevant to the
topics discussed here, have been published by Lang (3) and
Münck (4)).

    The simplest type of magnetic Mössbauer spectrum results
when an (effective) magnetic field B_{eff} acts on the ^{57}Fe nucleus.
The resulting splittings of the nuclear ground and excited states
are easily calculated from the Hamiltonian

$$\hat{H}_m = -g_N \mu_N \hat{I} \cdot \underline{B}_{eff} \tag{1}$$

where $g_N = 0.18$ and $I = 1/2$ for the nuclear ground state and
$g_N = -0.10$ and $I = 3/2$ for the excited state.

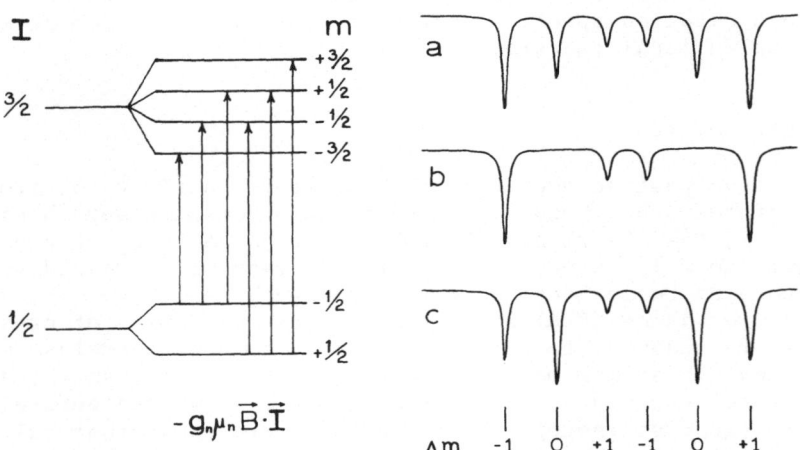

Figure 1.   Energy level diagram of ^{57}Fe showing the nuclear
Zeeman splittings. a) Mössbauer spectrum resulting from a sample
with randomly distributed internal magnetic fields. b) and c)
The internal field of each molecule in the sample is aligned
parallel (b) or transverse (c) to the observed 14.4 KeV radiation.

The ordering of the magnetic sublevels in Fig. 1 indicates the fact that the ground state moment is positive, while the excited state moment is negative. The multipolarity of the 14.4 KeV radiation connecting the ground with excited state levels is magnetic dipole (M1) in character, that is we have the selection rule $\Delta m = 0, \pm 1$, where Δm is the difference of the ground state and excited state magnetic quantum numbers, $\Delta m = m_e - m_g$. Fig. 1 shows the allowed transitions leading to a symmetric six-line pattern (Fig. 1a). In this article we will be particularly concerned with the angular dependence of the allowed transitions of the nuclear Zeeman pattern. The angular dependence can be computed from standard radiation theory (a particularly lucid discussion has been given by Frauenfelder and Steffen (5)) with the result,

$$\text{Intensity } (\Delta m = 0) \varpropto \sin^2 \theta$$
$$\text{Intensity } (\Delta m = \pm 1) \varpropto 1 + \cos^2 \theta \qquad (2)$$

where θ is the angle between the quantization axis (z-axis) of the nucleus and the progation direction of the gamma-radiation. In the simplest case θ is the angle between $\underset{\sim}{B}_{eff}$ and the direction of observation. Note that the intensities of the $\Delta m = 0$ lines are zero if $\underset{\sim}{B}_{eff}$ is parallel to the propagation direction of the 14.4 KeV Mössbauer radiation, while these transitions are maximized for $\theta = 90°$. These situations are depicted in Figures 1b and 1c, respectively.

The excited state of the ^{57}Fe nucleus has a nuclear quadrupole moment Q which may interact with an electric field gradient tensor (EFG). In the principal axis system of the EFG tensor (principal components V_{xx}, V_{yy}, V_{zz}, $\eta = (V_{xx} - V_{yy})/V_{zz}$) the quadrupole interaction is described by

$$\hat{H}_Q = \frac{eQV_{zz}}{12} [3\hat{I}_z^2 - 15/4 + \eta (\hat{I}_x^2 - \hat{I}_y^2)] \qquad (3)$$

In the absence of magnetic effects ($B_{eff} = 0$) the quadrupole interaction causes the nuclear excited state level to split into two states, separated in energy by

$$\Delta E_Q = \frac{eQV_{zz}}{2} (1 + \frac{1}{3} \eta^2)^{1/2}$$

Figure 2 shows the resultant Mössbauer spectrum, a quadrupole doublet consisting of two absorption lines (the lines have equal intensities for samples with randomly oriented molecules; this result is obtained by averaging the intensities of Eqs (2) over all molecular orientations).

In the following section we will discuss the origin of $\underset{\sim}{B}_{eff}$ in the framework of the spin Hamiltonian formalism. To

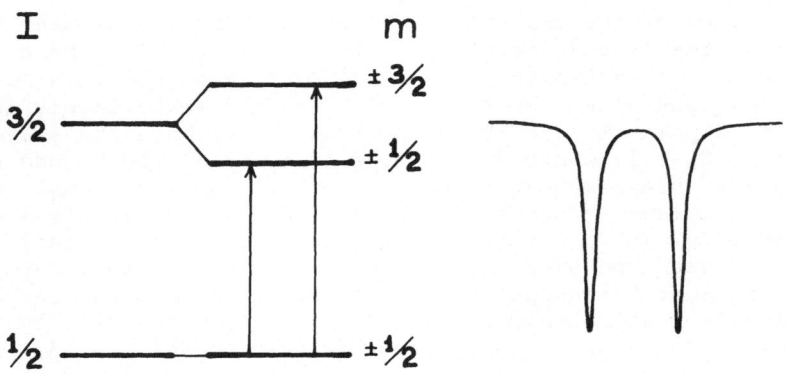

Figure 2. Quadrupole splitting in ^{57}Fe. The I = 3/2 level at
14.4 KeV is split into two sublevels by electric quadrupole
interaction. The nuclear ground state is not split because
a nucleus in a state with I = 1/2 has no spectroscopic quadrupole
moment. On the right a typical Mössbauer spectrum, a quadrupole
doublet, is sketched.

keep matters as simple as possible we will ignore for our
discussion the presence of quadrupolar effects.

C.   PARAMAGNETIC HYPERFINE STRUCTURE AND CONNECTION WITH EPR DATA

Within the framework of the spin Hamiltonian formalism the
physics of the Mössbauer spectra can be understood in a rather
straightforward way. A Hamiltonian sufficient for most applications can be written as

$$\hat{H} = D[\hat{S}_z^2 - \tfrac{1}{3} S(S + 1)] + E(\hat{S}_x^2 - \hat{S}_y^2) + \mu_B \, \hat{\underset{\sim}{S}} \cdot \underset{\approx}{g} \cdot \underset{\sim}{B}$$

$$+ \hat{\underset{\sim}{S}} \cdot \underset{\approx}{A} \cdot \hat{\underset{\sim}{I}} - g_N \, \mu_N \, \hat{\underset{\sim}{I}} \cdot \underset{\sim}{B} \tag{4}$$

For ^{57}Fe the first two terms are much larger than the magnetic
hyperfine interactions. Therefore the first two terms determine
the quantization axis of the electronic spin S. Thus we may
write down a nuclear Hamiltonian by replacing the operator $\hat{S}$
in the magnetic hyperfine term by an appropriately taken
expectation value <S>.

$$\hat{H} = <\underset{\sim}{S}> \cdot \underset{\approx}{A} \cdot \underset{\sim}{\hat{I}} - g_N \; \mu_N \; \underset{\sim}{\hat{I}} \cdot \underset{\sim}{B} \qquad (5)$$

Eq. (5) can be written as

$$\hat{H} = -g_N \; \mu_N \; \underset{\sim}{\hat{I}} \cdot (-<\underset{\sim}{S}> \cdot \underset{\approx}{A}/g_N \; \mu_N + \underset{\sim}{B})$$

$$= -g_N \; \mu_N \; \underset{\sim}{\hat{I}} \cdot (\underset{\sim}{B}_{int} + \underset{\sim}{B}) = -g_N \; \mu_N \; \underset{\sim}{\hat{I}} \cdot \underset{\sim}{B}_{eff} \qquad (6)$$

Eq. (6) connects the spin Hamiltonian with Eq. (1). We see that the internal magnetic field $\underset{\sim}{B}_{int}$ controls the magnetic hyperfine splitting (B_{int} is of order 100 kG). $\underset{\sim}{B}_{int}$ depends in a rather complicated way on the zero-field splitting parameters D and E, on the electronic g-tensor and the A-tensor, and on the orientation of the molecules relative to applied field. We do not want to discuss such intricate matters here. Rather we like to point out some general features which follow from

$$\underset{\sim}{B}_{int} = -<\underset{\sim}{S}> \cdot \underset{\approx}{A}/g_N \; \mu_N \qquad (7)$$

Obviously, $B_{int} = 0$ for a diamagnetic (S = 0) compound. Thus in zero applied field $B_{eff} = 0$ and the Mössbauer spectrum consists of a simple quadrupole doublet (or a single line if $\Delta E_Q = 0$). For complexes with integer electronic spin S the Hamiltonian of Eq. (4) produces in low symmetry non-degenerate electronic levels. It can be shown that singlet states can have no permanent magnetic moment (see ref. 6), i.e. $<\underset{\sim}{S}> = 0$. Thus we have the situation that compounds or spin-coupled systems containing an even number of electrons, i.e. systems with integer or zero electronic spin, produce only quadrupole doublets in zero applied field. The same compounds are in general also not amenable to EPR studies, primarily because the magnetic anisotropies and the large zero-field splittings impose very unfavorable conditions.

For compounds with half-integer electronic spin (Kramers ions) we have to distinguish between situations characterized by fast or slow electronic relaxation rates. In the fast relaxation limit the appropriately taken expectation value $<\underset{\sim}{S}>$ is obtained by averaging over all thermally accessible electronic states, i.e. $<\underset{\sim}{S}>$ is proportional to the magnetization of the material. At low fields the magnetization is proportional to the applied field, and thus vanishes in zero field. Hence quadrupole doublets are observed for Kramers ions in the fast relaxation limit.

In practice we can observe the Mössbauer spectra in the slow relaxation limit if we study the samples at low temperature. (Liquid helium temperature is sufficient in most cases.) In order to discuss this situation we can proceed as follows. For Kramers ions (for instance Fe^{3+}) the applied magnetic field

in Eq. (4) is typically in the order of hundred gauss (just
enough to decouple transferred hyperfine interactions with ligand
nuclei, see ref. 4). Under these conditions the electronic
levels of Kramers ions consist of well separated doublets. It
is customary to describe the magnetic properties of each
doublet by the Hamiltonian

$$\hat{H} = \mu_B \, \hat{\underset{\sim}{S}}' \cdot \underset{\sim}{g}' \cdot \underset{\sim}{B} + \hat{\underset{\sim}{S}}' \cdot \underset{\sim}{A}' \cdot \hat{\underset{\sim}{I}} \tag{8}$$

In Eq. (8) $S' = 1/2$ is an effective spin and g' is an effective
g-tensor describing the Zeeman splitting of the doublet. The
principal axes values of $\underset{\sim}{g}'$ are obtained from EPR experiments.
Some frequently occurring g-values are $g_x' = g_y' = 6$ and $g_z' = 2$
observed for the $\pm 1/2$ doublet of an $S = 5/2$ system with
$E = 0$ in Eq. (4). For $E/D = 1/3$ the $S = 5/2$ manifold produces
a doublet with an isotropic signal at $g' = 4.29$ (in the following
we will always refer to quantities of Eq. (8) and therefore
we will drop the prime).

    In order to keep our discussion simple we will assume an
isotropic A-tensor in Eq. (8). With this assumption the internal
magnetic field $\underset{\sim}{B}_{int}$ is parallel to $\langle \underset{\sim}{S} \rangle$, i.e.

$$\underset{\sim}{B}_{int} = -A\langle \underset{\sim}{S} \rangle / g_N \mu_N \tag{9}$$

We will now discuss two situations. First let us assume that
$\underset{\sim}{g}$ in Eq. (8) is isotropic. Since the Zeeman term in Eq. (8)
is assumed to be much larger than the magnetic hyperfine
interaction it follows that the electronic spin is quantized
along the applied magnetic field $\underset{\sim}{B}$, i.e. $\langle S_z \rangle = \pm 1/2$ and
$\langle S_x \rangle = \langle S_y \rangle = 0$ if we identify the z-axis with the direction of
$\underset{\sim}{B}$. Hence $\underset{\sim}{B}_{int}$ is parallel to $\underset{\sim}{B}$. Consider now a sample of
randomly oriented molecules such as a protein in frozen solution.
No matter how a molecule is oriented relative to the applied
field, the internal field $\underset{\sim}{B}_{int}$ is parallel to the applied field.
Thus by applying a field of a few hundred gauss parallel to
the Mössbauer radiation we force $\underset{\sim}{B}_{int}$, and therefore the nuclear
quantization axis, to be parallel to the observation direction.
Consequently, $\theta = 0$ in Eqs. (2) for the entire molecular population, yielding the pattern of Fig. 1b with suppressed $\Delta m = 0$
transitions. A rotation of the applied field by 90 degrees
aligns $\underset{\sim}{B}_{int}$ transverse to the observed Mössbauer radiation; we
observe now a pattern as indicated in Fig. 1c.

    The situation just described occurs quite frequently. For
$E/D = 1/3$ the high-spin ferric system yields a doublet with
isotropic g-values at $g = 4.29$. Iron sulfur proteins with
reduced 2Fe-2S centers exhibit g-values around $g_x \approx g_y = 1.94$
and $g_z = 2.02$. Also cubane 4Fe-4S clusters have almost
isotropic g-values in the -1 and -3 oxidation states (the states

comparable to oxidized H P F and reduced ferredoxin). Even high-spin ferric hemes with g-values at $g_x = g_y = 6$ and $g_z = 2$ fall into the "isotropic" category. We will address this case briefly below.

Let us now consider the case of an <u>uniaxial</u> system characterized by an extremely anisotropic g-tensor such that $g_x \simeq g_y \simeq 0$ and $g_z \gg g_x$, g_y. In order to discuss this case we write the electronic Zeeman term in Eq. (8) as

$$\hat{H}_z = \mu_B \, \hat{\underset{\sim}{S}} \cdot \underset{\sim}{g} \cdot \underset{\sim}{B} = \mu_B \, \hat{\underset{\sim}{S}} \cdot \underset{\sim}{B}' \tag{10}$$

with $\underset{\sim}{B}' = \underset{\sim}{g} \cdot \underset{\sim}{B}$. From the formal structure of the Zeeman term it is clear that the electronic spin is now quantized along the direction of $\underset{\sim}{B}'$. If $g_x = g_y = 0$ the direction of $\underset{\sim}{B}'$ will coincide with the direction of g_z, i.e. it is determined by and fixed to the molecular geometry. Hence $\underset{\sim}{B}_{int}$ is fixed to the molecule and therefore randomly oriented as the molecular population. Crudely speaking $\underset{\sim}{B}_{int}$ does not care about the direction of the applied field. Consequently the intensities of the Mössbauer spectrum are independent of the direction of applied field. In the simplest case a pattern with intensities 3:2:1:1:2:3 results, as shown in Fig. 1a.

The uniaxial case occurs quite frequently also. The high-spin ferric system for $|D| \gg \mu_B B$ and $E = 0$ yields two doublets with g-values at 0, 0, 6 (the $\pm$ 3/2 doublet) and 0, 0, 10 (the $\pm$ 5/2 doublet), respectively.

We mentioned above that high-spin ferric hemes with g-values at $g_x = g_y = 6$ and $g_z = 2$ will yield Mössbauer spectra whose intensities depend strongly on the direction of the applied field. With the above g-values the $\underset{\sim}{B}'$ field will have a strong tendency to align closely with the x-y plane. (Recall the spring device discussed in the lecture.) Since the system has axial symmetry the $\underset{\sim}{B}'$-field, and therefore also $\underset{\sim}{B}_{int}$, is easily rotatable around the molecular z-axis. Thus $\underset{\sim}{B}_{int}$ can be controlled quite effectively by the applied field, resulting in a field-dependent Mössbauer spectrum.

Isotropic and uniaxial Kramers doublets yield not only qualitatively different Mössbauer spectra, they also have quite different EPR properties. In particular, the uniaxial doublets are EPR-silent. This is readily understood, when we consider the conditions for their occurrence. Extremely anisotropic doublets are in general subdoublets of spin multiplets with $S \geq 3/2$. They result for instance for $|D| \gg \mu_B B$ and $E = 0$ in Eq. (4). The high-spin ferric ion for example has two such doublets, the $\pm$ 3/2 and the $\pm$ 5/2 states. Both doublets have $g_x = g_y = 0$, and $g_z = 6$ and $g_z = 10$, respectively. A spin-flip

at g_z would require a $\Delta m = 3$ or a $\Delta m = 5$ transition, respectively.
Such transitions, however, are highly forbidden since the
operator for the EPR transition probability is linear in S.
For small values of E/D we will get finite values for g_x and g_y,
and weak EPR transition will be observed. Also, the Mössbauer
spectrum will display a weak field dependence.

To summarize, compounds with _integer_ electronic spin S are
in general not amenable to EPR spectroscopy. In zero applied
magnetic fields such compounds yield Mössbauer spectra consisting
of quadrupole doublets. Compounds with _half-integer_ spin yield
at low temperatures Mössbauer spectra exhibiting paramagnetic
hyperfine structure (B_{int}). If the intensities of the magnetic
pattern can be changed easily in a small applied field, we can
predict that an EPR signal can be observed. On the other hand,
a field-independent Mössbauer spectrum will indicate to us that
the electronic ground state is an uniaxial Kramers doublet which
will be EPR-silent.

Although the arguments presented in this section have been
simple and qualitative they allow us to draw quite powerful
conclusions in complex systems. We will demonstrate this in
the next section.

D.   MÖSSBAUER AND EPR STUDIES OF NITROGENASE

In this section we will apply the simple arguments
developed above to a rather complex problem. It is impossible
here to discuss in any detail a protein which contains (most
likely) 30 iron atoms and 2 molybdenum atoms. Therefore we
refer the interested reader to the literature for a detailed
discussion of the physics of the M-centers (7,8,9) and P-clusters
(10).

The molybdenum and iron containing protein (MoFe protein)
of nitrogenase has MW $\simeq$ 220,000. It consists of four subunits
in an $\alpha_2\beta_2$ structure and contains, according to present
evidence, 30 Fe and 2 Mo atoms (10). As isolated the MoFe
protein exhibits a sharp and clean EPR spectrum with g-values
at $g = 4.32$, 3.65 and 2.01 (for the protein from A. vinelandii;
MoFe proteins from other organisms have rather similar g-values).
A typical spectrum is shown in Figure 3. The observed signals
have been shown (7) to result from the $\pm 1/2$ doublet of an
$S = 3/2$ system; the observed g-values are explained by assuming
$|D| \gg \mu_B B$, $E/D = 0.055$ and $g = g_e = 2.01$ in Eq. (4). By
evaluating the temperature dependence of the EPR signals the
zero-field splitting parameter D was determined to be $D = +6$ cm^{-1}
(7,9). Quantitation of the EPR spectrum finally yielded a spin

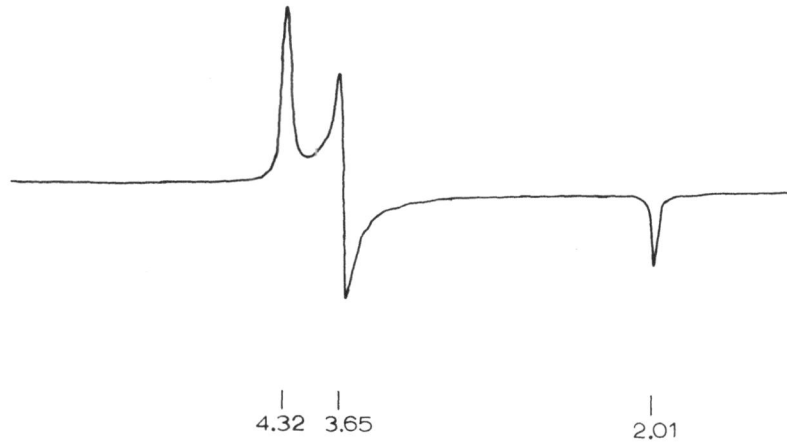

4.32 3.65   2.01

Figure 3.   EPR spectrum of the MoFe protein from <u>Azotobacter</u>
<u>vinelandii</u> taken at X-band at 12 K.

concentration close to 2 spins/molecule, i.e. the MoFe protein
contains 2 centers with spin S = 3/2.

     Is the S = 3/2 signal due to molybdenum or does it result
from an iron containing complex? This question can unambiguously
be answered by studying the protein with Mössbauer spectroscopy.
If the S = 3/2 center is associated with iron we should observe
at low temperatures a magnetic Mössbauer spectrum, the intensities
of which should depend on the orientation of a weak applied
magnetic field. If no such spectrum is found the signal must
result from molybdenum atoms.

     Figure 4c shows a Mössbauer spectrum taken at 4.2 K in a
parallel applied field of 600 gauss. The spectrum is a superposition of two quadrupole doublets, labeled D and Fe^{2+}, and
a magnetic spectrum with features not quite as simple as these
shown in Figure 1. For simplification we have divided the
spectrum into two parts. Figure 4d shows the doublets D and
Fe^{2+}; they represent iron residing in environments of integer
spin (quadrupole doublets at low temperature in weak fields!).
Figure 4b shows the magnetic spectrum"taken"in parallel field,
while Figure 4a shows a spectrum obtained in transverse field.
Note that the intensities depend on the orientation of the
applied field, i.e. the S = 3/2 centers are associated with
iron atoms.

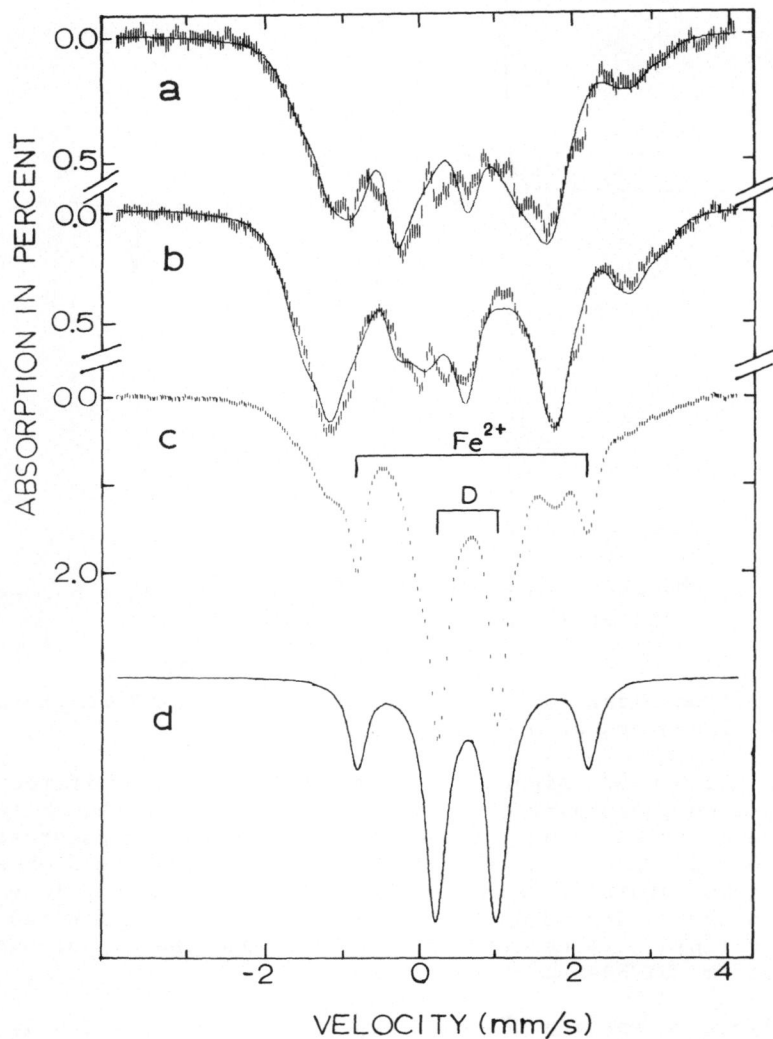

Figure 4. Mössbauer spectra of the MoFe protein from A. vinelandii taken at 4.2 K in a parallel applied magnetic field of 600 gauss. The spectrum shown in c) consists of a superposition of two quadrupole doublets labeled D and Fe^{2+} and a magnetic spectrum resulting from the S = 3/2 centers. For clarity the quadrupole doublets (d) and the magnetic spectrum (b) are shown separately. In a) the magnetic spectrum as obtained in transverse field is shown.

It is fairly straightforward to assess how many iron
atoms are associated with the magnetic spectrum in Fig. 4b.
Assuming that the MoFe protein contains a total of 30 Fe atoms,
we find that 4 belong to doublet Fe^{2+}, 12 belong to component D,
and 12 Fe atoms are associated with the magnetic spectrum.
Taken together, the EPR and Mössbauer data imply that the MoFe
protein contains two S = 3/2 centers, each containing 12/2 = 6
iron atoms! Note that we arrived at a very important conclusion
without relying on any numerical results for the hyperfine
parameters. Further studies (8,9) showed that the S = 3/2
centers belong to the molybdenum and iron containing cofactor
(8) of nitrogenase. Moreover, these centers can be stabilized
in three distinct oxidation states (9). The solid lines in
Fig. 4a and b are an attempt to decompose the magnetic spectrum
into six subsites (9).

Components D and Fe^{2+} provide us with a further opportunity
to apply the considerations of section C. The quadrupole
splitting and the isomeric shift of doublet Fe^{2+} are strongly
suggestive of high-spin ferrous iron (S = 2) in a tetrahedral
environment of sulfur atoms. Surprisingly, the irons of
component Fe^{2+} reside in a diamagnetic (S = 0) environment; this
was proven by studying samples in strong applied fields (7).
These studies showed that component D is diamagnetic also.
Since components D and Fe^{2+} occur in the ratio of 3:1 these
findings lead to the suspicion that the D and Fe^{2+} irons are
sites of a 4Fe-4S cluster (7), with three iron sites yielding
the spectral signature labeled component D while the fourth
site would yield the doublet Fe^{2+}. This suggestion was proven
by studying higher oxidation states of the MoFe protein (10).

Figure 5A shows a low-temperature Mössbauer spectrum of
the MoFe protein after 6 electrons were removed from each
molecule. By following the redox titration it was easy to
show that the central doublet in Fig. 5A represents the oxidized
S = 3/2 centers. The remainder of the sample exhibits a complex
magnetic pattern with absorption stretching from −4 mm/s to
+5 mm/s. The intensities of this pattern are independent of
the orientation of a weak applied field. This suggests very
strongly that this spectrum is associated with a Kramers
doublet for which $g_z \gg g_x$, g_y and $g_z/g_x > 15$. As discussed
above such doublets are in general subdoublets of a spin
multiplet with S ≥ 3/2 when E ≈ 0. The Mössbauer studies show
clearly that this doublet is the electronic ground state (10),
i.e. D < 0. In general spin multiplets with S ≥ 3/2 have at
least one EPR-active Kramers doublet (for E/D = 0 the EPR
active doublet is the ± 1/2 state which is highest in energy
for D < 0). So far, EPR studies have not yielded any signal
attributable to the ± 1/2 state. We believe that the missing
EPR signal can be rationalized as follows: At 4.2 K the

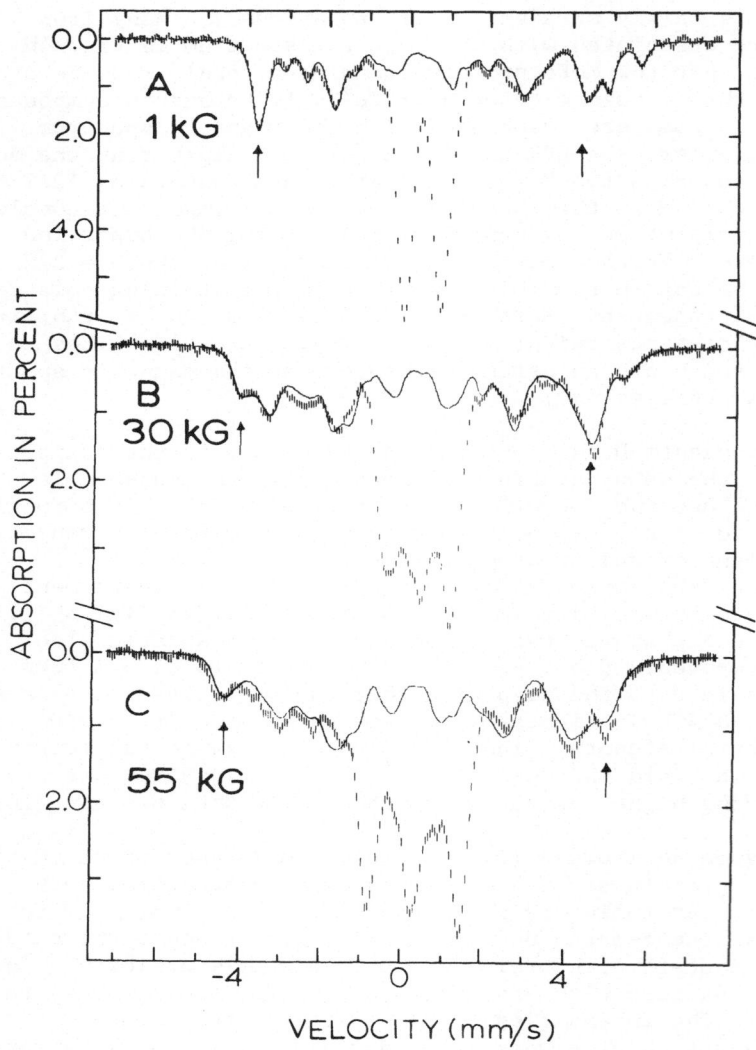

Figure 5. Mössbauer spectra of thionine oxidized MoFe protein
from C. pasteurianum. The data were taken in parallel applied
fields at 4.2 K. A. The central quadrupole doublet results
from the oxidized S = 3/2 centers, now in a state with S = 0.
The remainder is a complex, field-independent magnetic spectrum.
A detailed data analysis shows it to be a superposition of
spectra from 8 subsites (the P-clusters occur in slightly
unequivalent pairs). The spectra in B. and C. (not discussed
in the text) demonstrate spin-coupling.

electronic spin relaxation is long, the EPR active ± 1/2 doublet, however, is not populated at this temperature (Mössbauer studies in strong fields suggest that the ± 1/2 doublet is at least 30 cm^{-1} above the ground state). At higher temperatures the ± 1/2 doublet becomes populated; the electronic spin relaxation rate, however, becomes very rapid and the EPR signal is lost due to life-time broadening.

It took more than two years of hard work to produce the theoretical curve shown in Fig. 5A. The final data analysis supports the idea that components D and Fe^{2+} result from subsites of 4Fe–4S clusters. These clusters, labeled P-clusters, are probably related to familiar 4Fe–4S centers. Their spectroscopic properties, however, are quite unique (10).

The authors have enjoyed a long and fruitful cooperation with Dr. W.H. Orme-Johnson and his coworkers of the University of Wisconsin. This work was supported by a grant from the National Science Foundation PCA 08522.

References

1. Debrunner, P.G., and Frauenfelder, H.:   An Introduction To Mössbauer Spectroscopy, L. May, ed., Plenum Press, New York, 1971.

2. Moss, T.H.: 1973, Methods in Enzymology, Vol. 27, 912, Academic Press, New York.

3. Lang, G.: 1970, Quart. Rev. Biophys. 3, 1.

4. Münck, E.: 1978, Methods in Enzymology, Vol. 54, 346, Academic Press.

5. Frauenfelder, H., and Steffen, R.M.: 1965, Alpha-, Beta-, and Gamma Spectroscopy, K. Siegbahn, ed., North Holland, Amsterdam, 1965.

6. Abragam, A., and Bleaney, B.: 1970, Electron Paramagnetic Resonance of Transition Ions, Chapter 15, Oxford University Press.

7. Münck, E., Rhodes, H., Orme-Johnson, W.H., Davis, L.C., Brill, W.J., and Shah, V.K.: 1975, Biochim. Biophys. Acta 400, 32–53.

8. Rawlings, J., Shah, V.K., Chisnell, J.R., Brill, W.J., Zimmermann, R., Münck, E., and Orme-Johnson, W.J.: 1978, J. Biol. Chem. 253, 1001–1004.

9. Huynh, B.H., Münck, E., and Orme-Johnson, W.H.: 1979,
    Biochim. Biophys. Acta, 527, 192-203.

10. Zimmermann, R., Münck, E., Brill, W.J., Shah, V.K., Henzl,
    M.T., Rawlings, J., and Orme-Johnson, W.H.: 1978, Biochim.
    Biophys. Acta 537, 185-207.

THE SPIN-PAIRING MODEL FOR THE BINDING OF DIOXYGEN TO TRANSITION
METAL COMPLEXES

Russell S. Drago

School of Chemical Sciences
University of Illinois
Urbana, Illinois 61801 (U.S.A.)

INTRODUCTION

The binding and activation of dioxygen is essential to many
biological functions. Cobalt(II) complexes reversibly bind O_2
and the resulting adducts have one unpaired electron. The presence
of an odd electron in these complexes affords a ready probe of the
electronic structure of the O_2 adduct by electron paramagnetic
resonance spectroscopy. This feature and the relative stability
of the cobalt(II) complexes has stimulated considerable research
on the cobalt(II) system. Early interpretations of epr, X-ray
diffraction and infrared results on these dioxygen adducts of
cobalt(II) are classic cases of over-interpretation of data. The
topic will be developed historically and the objective of this
school will be to learn the lessons from the past so they will
not be repeated.

Most dioxygen adducts of cobalt are formed with five coordinate, low-spin cobalt(II) complexes. The analysis of the epr
spectra of low spin cobalt(II) complexes has been treated in
detail (1). Most five coordinate, low spin cobalt(II) complexes
have $g_\parallel$ values of about 2.0 and $g_\perp$ values of 2.3; $g_\perp > g_\parallel$. The
values of $A_\parallel$ and $A_\perp$ are around 80×10^{-4} cm^{-1} and -10×10^{-4} cm^{-1}
respectively.

The epr spectra of the dioxygen adducts have $g_\parallel > g_\perp$;
typical values are 2.10 and 2.00. The values of $A_\parallel$ and $A_\perp$ for
the Coacacen·pyridine complex, originally (2) studied, are -19.1
and -10.0 respectively. (We shall discuss the minus signs later.
Only the absolute value is obtained from the spectrum.)

I. Bertini and R.S. Drago (eds.), ESR and NMR of Paramagnetic Species in Biological and
Related Systems, 289–301.
Copyright © 1979 by D. Reidel Publishing Company

A_{aniso}, which is defined as

$$A_{aniso} = A_{\parallel} - <A>$$
$$(\text{where } <A> = \frac{1}{3}A_1 + \frac{1}{3}A_2 + \frac{1}{3}A_3),$$

can be used to estimate the fractional occupation of d_{z^2}, α^2, from

$$A_{aniso} = \frac{4}{7}P\alpha^2 \qquad (P = k<r^{-3}>_{3d}).$$

Using the A_{aniso} value for the five coordinate complex to calculate P (by assuming $\alpha^2 = 1$) the A_{aniso} for the O_2 adduct was used to calculate an α^2 value of 0.1. It was concluded (2) that nearly complete transfer of the cobalt(II) d_{z^2} electron to dioxygen occurred and the complex was described as a cobalt(III) complex containing ionically bound superoxide ion, $Co(III)O_2^-$.

In a subsequent study (3), the ^{17}O epr was investigated and an A_{iso} (<A>) value for ^{17}O of 21.6 G was reported. A_{iso} for a full electron in a 2s orbital is 1659 G so these authors concurred with the earlier conclusion that the dioxygen adduct was an ionically bound superoxide ion with the unpaired electron occupying a molecular orbital consisting essentially of oxygen 2p orbitals. At $-30°C$ the two oxygens were equivalent, suggesting a dynamic system.

The interpretation of the epr in terms of an ionically bound superoxide ion reportedly (4) received confirmation from infrared and x-ray diffraction studies. In a recent article the problems with these analyses have been discussed (5). Our concern here will be with the epr interpretations.

Though the epr analysis described above is a viable one, it is by no means unique. There are several aspects of the problem that are disconcerting and should not have been overlooked. First, by a no less rigorous analysis (6) of the isotropic cobalt hyperfine coupling constant, it was concluded that the orbital containing the odd electron is 50% cobalt d-orbital. Secondly, six coordinate cobalt(II) complexes typically have $A_{\parallel}$ and $A_{\perp}$ values of 65×10^{-4} and 50×10^{-4} leading to an A_{aniso} value of 55×10^{-4}. Calculating a P value from this data leads to an α^2 value of 0.5 for the dioxygen adduct of cobalt(II). Thus the conclusion that the extent of electron transfer into the bound O_2 of the Coacacen·pyridine adduct is extensive enough to be considered a bound superoxide ion is very much in doubt. The ^{17}O isotropic hyperfine coupling constant, A_{iso}, is not definitive in this regard. One unpaired electron in a 2s orbital gives rise to an A_{iso} value of -1659 G, so only 0.013 2s character is needed for a 21.6 G coupling. There are a variety of mechanisms that could

account for this coupling including a singlet oxygen molecule
coordinated to cobalt(II). It is clear that more information is
needed.

THE SPIN-PAIRING MODEL

As we were initiating experiments to evaluate the ^{17}O anisotropic hyperfine coupling constant, these important results were
published (7). At -100° two $A_{||}$ hyperfine coupling constants of
60 G and 88 G were reported for the non-equivalent directly bound
and terminal oxygen atoms, respectively. Estimating the isotropic
component with $A_{iso} = -41\rho_{O\pi}$ the expression $A_{||} = -144\rho_{O\pi}$ results
for spin density on an oxygen atom. This leads to 0.4 e^- and
0.6 e^- on the directly bound and terminal oxygen atoms. The
authors incorrectly (vide infra) concluded that this confirmed
the formulation $Co(III)O_2^-$; that is, an ionically bound superoxide
ion.

Two very important conclusions can be drawn from these new
results (8). First, the sum of the hyperfine couplings for ^{59}Co
and ^{17}O add up to a spin density greater than one even though
there is only one unpaired electron in the system. This indicates that the cobalt hyperfine is arising via a spin polarization
mechanism. Since all earlier epr interpretations were based on a
direct delocalization mechanism (4), they are incorrect and the
ionically bound superoxide model loses its most convincing experimental support. Second, since the unpaired electron is almost
completely localized on dioxygen, the model describing the bound
O_2 as coordinated singlet oxygen is eliminated.

Having demonstrated what the coordinated dioxygen is not,
we shall now concentrate on describing what it is by using the
spin-pairing model for dioxygen bonding. The spin-pairing model
of binding dioxygen to metals can be viewed as arising from a
typical free radical reaction in which metal spins pair up with
either one or two of the unpaired π* electrons of dioxyen. Five
coordinate cobalt(II) complexes spin pair the unpaired electron in
d_{z^2} with one of the π* electrons of dioxygen forming a sigma bond.
This interaction is illustrated in Figure 1 where for simplicity
the molecular orbitals for the O_2 adduct are generated by combining the essential orbitals of the five coordinate cobalt(II)
complex with the π* orbitals of dioxygen.

The π* orbital of O_2 overlaps with the metal a_1 (d_{z^2}) orbital
to form a σ molecular orbital, ψ_1, containing two electrons and a
corresponding σ* molecular orbital, ψ_3. A second π* orbital of
O_2 is orthogonal to the above σ and σ* MOs. This molecular
orbital, ψ_2, is essentially oxygen based and contains the unpaired
electron. Neglecting, for convenience, the minor overlap with

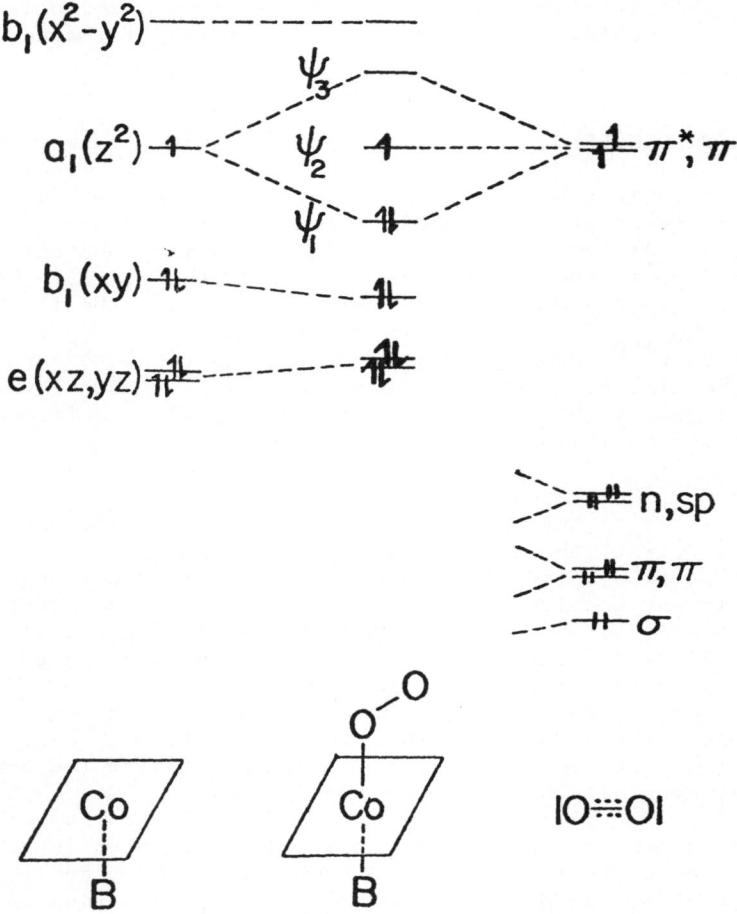

Figure 1. Molecular orbital diagram for the dioxygen adduct of Co(II). One electron from an O_2 π^* orbital spin-pairs with one unpaired electron from Co(II) forming the ψ_1 molecular orbital; the remaining unpaired electron is in an O_2 π^* orbital.

other orbitals, ψ_1, ψ_2 and ψ_3 have the form:

$$\psi_3 = \beta(d_{z^2}) - \alpha(\pi_1^*) \tag{1}$$

$$\psi_2 = \pi_2^*(a2p_1 + b2p_2) \tag{2}$$

$$\psi_1 = \alpha(d_{z^2}) + \beta(\pi_1^*) \tag{3}$$
$$\text{where } \beta = (1 - \alpha^2)^{\frac{1}{2}} \tag{4}$$

In this model, when $\alpha = 0$, the complex is formulated as an ionic superoxide, $Co-O_2^-$; when $\beta = 0$, $Co-O_2^+$; and when $\alpha = \beta$, $Co-O_2$. Whether the bound oxygen resembles O_2, O_2^+ or O_2^- depends on the coefficients of a molecular orbital which does not contain unpaired electrons. Thus, contrary to the reported (7) interpretation of the anisotropic hyperfine coupling constant, one cannot formulate a $Co(III)O_2^-$ description from the fact that the odd electron is localized on the oxygen atoms.

With the only unpaired electron in the system residing essentially on O_2, one might next ask how the cobalt hyperfine coupling arises. If one interprets the cobalt hyperfine as arising from a direct delocalization of the electron onto cobalt, a summation of the anisotropic ^{17}O and Co hyperfine coupling constants would indicate more than one unpaired electron in the system, even though only one exists. An indirect, rather than direct, mechanism must lead to negative spin density on the metal center. The spin polarization of the paired electrons in ψ_1 by the unpaired electron in ψ_2 produces negative spin density at the cobalt nucleus and gives rise to a cobalt hyperfine coupling in the epr.

The application of the spin pairing model to other metal-dioxygen complexes is straightforward. When the electron configuration of the metal can provide two electrons to spin-pair with dioxygen, a metal-oxygen double bond results. Some illustrative examples are presented in Figure 2. This topic has been discussed in greater detail in the literature (9). Note that the result of application of this model to the systems in Figure 2 predicts the correct susceptibility and metal character of the orbitals containing the unpaired electrons.

ELECTRON TRANSFER ANALYSIS

A pronounced variation in the cobalt hyperfine coupling constant in a series of O_2 adducts has been observed (8), indicating extensive variation in the cobalt contribution to ψ_2 and thus in the nature of the bound O_2. Next we shall indicate how an analysis of the cobalt hyperfine coupling constant permits an evaluation of the charge on the bound O_2 fragment. We shall cast these results in terms of the electron transfer into the bound O_2. Referring back to equation 3 and recalling ψ_1^2 is density, we would obtain from a zero overlap, Mulliken population analysis, $2\alpha^2$ for the cobalt electron density (the two arises from the fact that there are two electrons in ψ_1). The remaining density $2(1 - \alpha^2)$ is assigned to oxygen. Since oxygen had one of these electrons before bonding, the electron transfer, E.T., into O_2 is given by

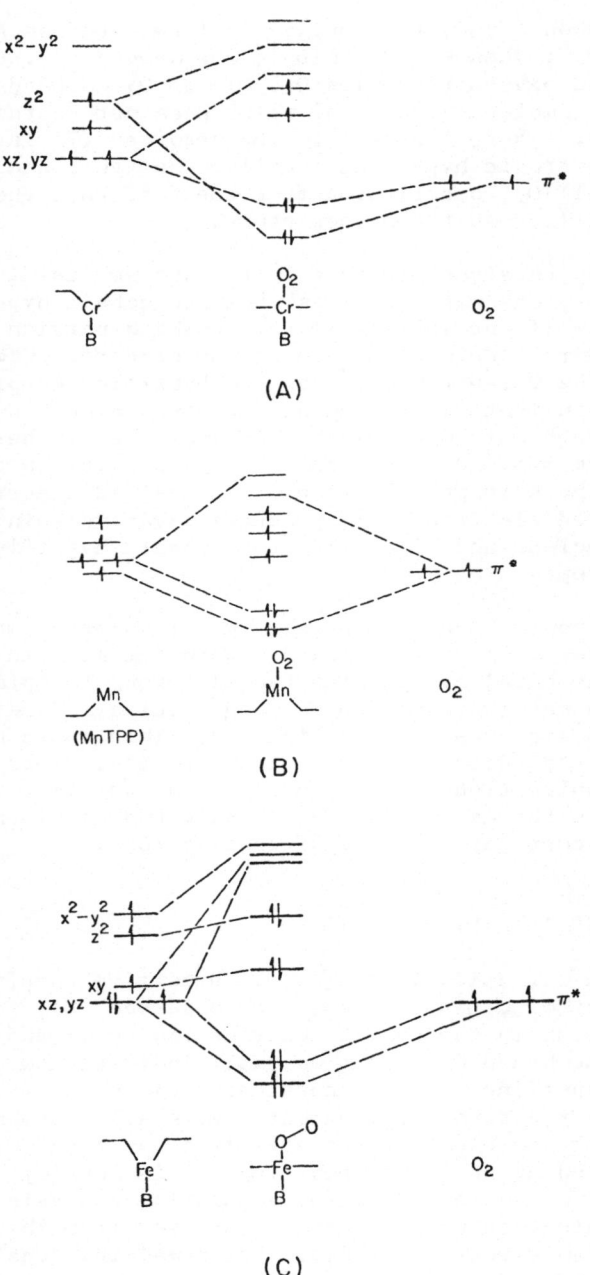

Figure 2. Molecular orbital diagram for O_2 adducts of Cr(II), Mn(II) and Fe(II) showing spin-pairing upon adduct formation.

$$E.T. = 2(1 - \alpha^2) - 1 \qquad (5)$$

Our next concern is to obtain the cobalt contribution to ψ_1 by an analysis of the indirect cobalt hyperfine coupling constant, that is, the spin polarization.

Spin polarization has been studied in detail for hydrocarbon radicals and their derivatives. The four line, isotropic proton hyperfine coupling of the methyl radical (23 G) is used to evaluate Q in the equation:

$$a_H = Q\rho_C \qquad (6)$$

The measured, indirect, isotropic proton hyperfine coupling constant in a radical, a_H, is used to calculate the spin density on the directly bonded carbon, ρ_C, using a value of Q equal to 23. The scaling constant Q is given by (10):

$$Q = \left(\frac{1}{1 - S_o^4}\right) \left(\frac{J_{ph} - J_{ps}}{\Delta E_{2,1}}\right) a_{Ho} \qquad (7)$$

where a_{Ho} represents 506 G for a hydrogen atom, J_{ph} represents the exchange integral $<p(1)h(2)|\frac{1}{r_{12}}|h(1)p(2)>$, p refers to a carbon p_z orbital, s to a hydrogen 1s, h to a carbon sp^2 orbital, S_o to the h,s overlap integral $<h|s>$ and $\Delta E_{2,1}$ to the energy difference between p_z and the σ bonding orbital. Though all of these factors can influence the magnitude of Q, the value remains remarkably constant in a large series of hydrocarbon radicals and their derivatives (11). In the series PO_4^{2-}, CO_3^- and OH^- (an oxygen lone pair electron removed from the diamagnetic anion in each case), Q for A_{iso} is proportional to the sp^n hybridization suggesting

$$Q = U\alpha^2$$

where U is the scaling factor and α the s character in sigma bond. ΔE is apparently not effecting U.

For the dioxygen adducts of cobalt(II), we are concerned with the anisotropic cobalt hyperfine coupling and we can write

$$\frac{A'_{aniso} \, obs}{A_{aniso} \, 3d} = U_{3d}\rho_o\alpha^2 \qquad (8)$$

From the ^{17}O epr we know that ρ_o the spin density on the adjacent oxygen is 0.4 and will assume that it remains constant in the O_2 adducts used in our analysis. If we can determine U_{3d}, all quantities in equation 8 can be determined except α^2, the square

of the cobalt coefficient in ψ_1. We can solve for this and calculate the electron transfer by using equation 5. The A'_{aniso}
value used in equation 8 should be corrected for the dipolar
contribution to the coupling constant from the 0.6 of an electron
on the terminal oxygen using

$$A_{x,y} = \frac{-g\beta_e g_N \beta_N}{r^3} \qquad \text{and} \qquad A_z = \frac{2g\beta_e g_N \beta_N}{r^3} .$$

$(A'_{aniso} = A_{aniso} - A_{dipolar})$. In a typical dioxygen adduct of
cobalt(II), the value of $A_{dipolar}$ is 1.0×10^{-4}.

The constant, U_{3d}, is evaluated from the epr spectra of some
(12) free radical alkyl nitroso complexes of cobalt(III);
$(CN)_5Co-N\overset{O}{\underset{R}{\diagdown}}{}^{2-}$. The nitrogen hyperfine coupling constant indicates that 0.7 of an electron is localized on nitrogen. Correcting
for $A_{dipolar}$ from the 0.3 of an electron on oxygen and assuming*
15% cobalt character in the cobalt-nitrogen sigma bond, the constant U_{3d} can be determined. Substituting this value into equation
8 and using experimentally determined values for A'_{aniso}-obs leads
to the electron transfer values reported in Table I.

The data in Table I show an increasing extent of electron
transfer as the average ligand field strength (N > 0) of the coordinated groups is increased. This is predicted on the basis
of Figure 1. As the d_{z^2} energy increases (from, for example, an
increased spherical field), its contribution to ψ_1 will decrease
and that of oxygen will increase. This transfers the bonding
electron density to oxygen.

The values of E.T. reported in Table I are only semi-
quantitative. If the value for α^2 in the cobalt-nitrogen bond
were 0.20, the E.T. values would all be about 0.2 lower but the
trends would be the same. If the energy difference between ψ_2
and ψ_1 were to decrease with increasing electron transfer, ΔE
of equation 7 would change, α^2 would decrease and the E.T. values
for the large electron transfer systems would increase most. If
the odd electron delocalizations on the directly bonded oxygen
were to decrease with increasing E.T., this would tend to make
the α value larger than that calculated and this would tend to
decrease the E.T. values on the high end of the scale most. The
calculations in Table I were made for the d_{z^2} contribution to ψ_1
and there is some cobalt 4s orbital mixed in. The neglect of 4s

---

*In six coordinate cobalt(II) complexes 7 to 10% nitrogen
character was found in the essentially metal d_{z^2} orbital. Since
Co(III) is more acidic, 15% is a reasonable value. Assuming 10%
and 20% caused some E.T. values to differ by 0.2 but the same
trends were observed.

Table I
Electron Transfer Values for Various O_2
Adducts of Cobalt(II) Complexes

Compound[a]	E.T.[b]
$Co(acacen) \cdot py \cdot O_2$[c]	0.4
$Co(acacen) \cdot H_2O \cdot O_2$	0.1
$Co(p-OCH_3TPP) \cdot py \cdot O_2$[c]	0.6
$Co(p-OCH_3TPP)(1-MeIm)O_2$	0.5
$Co(PPIXDME)(py)O_2$	0.5
$Co(salen) \cdot py \cdot O_2$	0.5
$Co(SMDPT) \cdot O_2$	0.5
$Co(DMGH) \cdot py \cdot O_2$	0.7
$Co(DPGB)_2 \cdot CH_3CN \cdot O_2$	0.8
$Co(DPGB)_2 \cdot acetone \cdot O_2$	0.8
$Co(DPGB)_2 \cdot HMPA \cdot O_2$	0.7
$Co(X-SalDAPE) \cdot O_2$	
X = 5-H	0.2
= 5-OMe	0.3
= 5-Br	0.3

[a]Abbreviations:
salen, N,N'-ethylenebis(salicyliden-
    imine)
SMDPT, bis(salicylidene-γ-imino-
    propyl)methylamine)
DMGH, d-methylglyoxime
DPGB, BF_2-capped diphenylglyoxime
X-SalDAPE, bis(salicylidene)di-
    aminopropylether
p-OCH$_3$TPP, p-methoxyphenyl-meso-
    tetraphenylporphyrin
PPIXDME, protoporphyrin(IX) di-
    methyl ester
HMPA, hexamethylphosphoramide

[b]Electron transfer from Co(II) to
O_2. The results are semiquanti-
tative but trends are accurate.

[c]M.O. calculations on these systems
agree with our E.T. values (13).

leads to low values of α^2 and large values of E.T. in the series.
Finally, since the electron resides on dioxygen and in view of
the low symmetry of the complexes, the g and A tensors are
probably not aligned with the molecular axes as has been shown
in a single crystal study of oxycobaltmyoglobin. The single

crystal study does show that the frozen solution spectra provide
a good gauge of the actual anisotropy. In view of the semi-
quantitative nature of these results, the second order corrections
described by McGarvey (1) were not made. What with all the words
of caution listed above, the encouraging aspect is the agreement
between the E.T. values determined above and those reported independently from molecular orbital calculations (13,14).

THE ISOTROPIC COBALT HYPERFINE COUPLING CONSTANT

The isotropic cobalt hyperfine coupling constant could provide information about the 4s character in ψ_1 from the relation:

$$\frac{A_{isoCo}}{A_{4sCo}} = U_{O-Co}\rho_O$$

A_{4s1} for a full electron in a cobalt 4s orbital is $+1232 \times 10^{-4}$
cm^{-1}. When the amount of 4s character calculated in this way is
compared with the 3d character calculated from A_{aniso}, a large
apparent 4s character (20 to 65%) is obtained. In contrast to
this finding for O_2 adducts and the nitroso complexes, typical
values for five coordinate cobalt(II) complexes show about 4%
4s orbital mixed into the molecular orbital containing the unpaired electron. The complicating factor for the radical ligands
is attributed to a non-differential overlap of the dioxygen (or
nitroso) π^* orbital and the filled cobalt 3s orbital (8). ($A_{F.C.}$
for 1 electron in 3s is 3,666 G.) As the Co-O-O angle increases,
the differential overlap will increase monotonically. This complicating feature makes the interpretation of A_{isoCo} difficult.

RELEVANCE OF THE SPIN-PAIRING MODEL TO HEMOGLOBIN COOPERATIVITY

A series of thermodynamic studies of axial base binding
($\Delta H_{1:1}$) and dioxygen coordination (ΔH_{O_2}) to cobalt(II) protoporphyrin IX dimethyl ester has been carried out in the poorly
solvating solvent, toluene (15). These results show that as the
strength of axial base binding increases, the strength of dioxygen binding to the five coordinate cobalt adduct increases.
As the axial base interaction increases in this series of complexes, the d_{z^2} orbital is driven higher in energy relative to
the oxygen π^* orbital (see Figure 1). The higher the initial
energy of the d_{z^2} electron in the five coordinate adduct, the
more it is stabilized upon forming the O_2 adduct and falling
back to occupy ψ_1.

The application of the spin-pairing model to iron(II) complexes suggests that the effects of axial base variation on the

d-orbital energies will result in the same trends in dioxygen
binding strengths as observed in the cobalt(II) protoporphyrin
IX dimethyl ester adducts. These considerations have implications regarding the mechanism of cooperativity in the hemoglobin
system (15).

In the restraint theory of cooperativity in hemoglobin (16),
it is proposed that Hb has two alternative quaternary structures:
the R state, whose O_2 affinity is essentially that of the isolated
subunits; and the deoxy T state whose O_2 affinity is diminished.
In the deoxy T state the proximal histidine is held in place by
the protein and restrains the five-coordinate iron(II) from
picking up O_2 and becoming six-coordinate due to movement of iron
toward the mean porphyrin plane. Weakened O_2 binding results.
According to this model, no restraint is present in the deoxy
form of native hemoglobin. Conformation change to the R state
removes the restraint for O_2 binding and a higher affinity for
O_2 results for the R conformation. Various triggering mechanisms
for the conformation change have been proposed. Cooperativity
is observed for both carbon monoxide and dioxygen binding in
both Hb and CoHb (17).

The spin-pairing model predicts, generally, that dioxygen
binding will become weaker as the strength of the axial base bond
becomes weaker. This relationship suggests an alternative description for cooperativity. In the T form of hemoglobin the
protein restrains the histidine from interaction with the high
spin iron as strongly as it might in the absence of the restraint.
This restraint is manifested upon O_2 coordination and is illustrated in Figure 3. The restraint could bend the histidine,
preventing direct overlap of nitrogen with the iron d_{z^2} orbital,
or simply prevent the two from as close an approach as desired.

According to the enthalpy relationship between $\Delta H_{1:1}$ and
$-\Delta H_{O_2}$, the oxygen affinity for the restrained conformation is
lowered relative to what it would be in the absence of restraint.
This effect causes both the metal histidine and the metal-dioxygen
bonds to be weaker in the T form than they would be in the absence
of the restraint. Coordination of dioxygen in this form builds
up potential energy in the system that could be released if the
restraint could be removed by a protein conformation change. If
enough energy for the change cannot be generated upon binding the
first dioxygen, multiples of this potential energy arise on
binding the second, etc., dioxygens. Note that potential energy
is to be gained upon removal of restraint from a stronger metal-
base interaction as well as a stronger dioxygen interaction for
each oxygenation step that occurs in the T form. Upon coordination of the third dioxygen more than enough potential energy is
available from the increased iron-nitrogen and iron-O_2 bond
strengths in the R form to effect the endothermic T to R protein

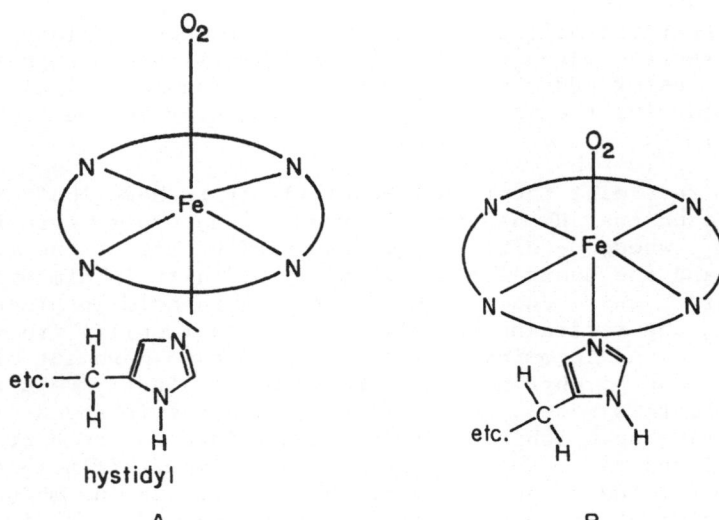

A   B

Figure 3. 3A corresponds to the T state. The Fe interaction
with the hystidyl residue is being restrained by the protein.
A weaker Fe-O_2 interaction occurs, denoted by a longer Fe-O_2
bond. 3B corresponds to the R state. The Fe hystidyl interaction
is not being restrained. A stronger Fe-O_2 interaction occurs,
denoted by a shorter Fe-O_2 bond.

transformation. With the protein in the R-form, addition of the
fourth dioxygen is expected to be the most exothermic of all the
O_2 binding steps (see Figure 4).

According to this spin-pairing description of cooperativity,
a further understanding of the mechanistic details of cooperativity

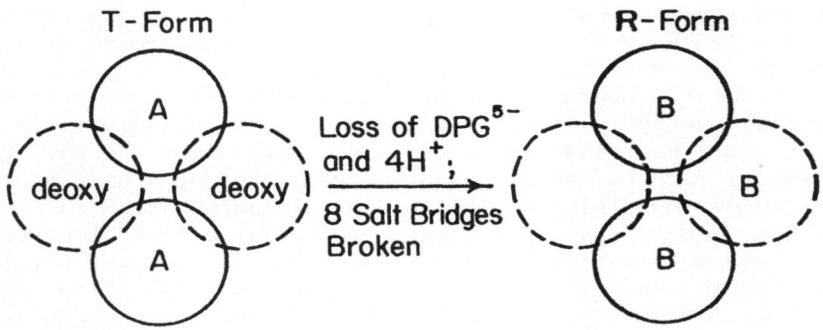

Figure 4. Schematic diagram of hemoglobin conformation change.

is an enormously complicated problem which requires understanding all of the protein-protein interactions that make the T-form more stable than the R-form. As discussed in the literature (15), more new problems were introduced by the study of coboglobin, than there were answers provided. A more complete understanding of cooperativity than offered above will be difficult to obtain but details about changes in the iron environment in the various protein conformations will be of importance in substantiating or rejecting this proposal.

(1)   McGarvey, B.R.: 1975, Can. J. Chem.53, p. 2498.
(2)   Hoffman, B.M., Diemente, D.L., and Basolo, F.: 1970, J. Am. Chem. Soc.92, p. 61.
(3)   Melamud, E., Silver, B.L., and Dori, Z.: 1974, J. Am. Chem. Soc.96, p. 4689.
(4)   Basolo, F., Hoffman, B.M., and Ibers, J.A.: 1975, Acc. Chem. Res.8, p. 384.
(5)   Drago, R.S.: 1979, Inorg. Chem.18, p. 1408.
(6)   Chien, J.C.W. and Dickinson, L.C.: 1972, Proc. Nat. Acad. Sci.69, p. 2783.
(7)   Getz, D., Melamud, E., Silver, B.L., and Dori, Z.: 1975, J. Am. Chem. Soc.97, p. 3846.
(8)   Tovrog, B.S., Kitko, D.J., and Drago, R.S.: 1975, J. Am. Chem. Soc.98, p. 5144.
(9)   Drago, R.S. and Corden, B.B.: 1980, submitted, p. 000.
(10) McConnell, H.M. and Chestnut, D.B.: 1958, J. Chem. Phys.28, p. 107.
(11) Cramer, R.E. and Drago, R.S.: 1968, J. Am. Chem. Soc.90, p. 4790; Drago, R.S. and Petersen, H.: 1967, J. Am. Chem. Soc.89, p. 3978.
(12) Symons, M.C.R. and Wilkinson, J.G.: 1972, J. Chem. Soc.68, Faraday Trans. II, p. 1265.
(13) Fantucci, P. and Valenti, V.: 1976, J. Am. Chem. Soc.98, p. 3832.
(14) Redieu, A., Rohmer, M.-M., Bernard, M., and Veillard, A: 1976, J. Am. Chem. Soc.98, p. 3717.
(15) Drago, R.S., Beugelsdijk, T., Breese, J.A., and Cannady, J.P.: 1978, J. Am. Chem. Soc.100, p. 5374.
(16) Perutz, M.F.: 1976, Br. Med. Bull.32, p. 195 and references therein.
(17) Imai, K., Yonetani, T., and Ikeda-Saito, M: 1977, J. Mol. Biol.109, p. 83 and references therein.

EPR STUDIES OF COPPER CENTERS IN PROTEINS

Giuseppe ROTILIO

Institute of Biological Chemistry,University
of Rome, Rome, Italy

The phenomenological criterion of the fraction of
total copper that is detected by EPR classifies copper
proteins in three groups, with well defined correlations
of EPR properties with their structure and function.
Proteins having only EPR-invisible copper contain
either Cu(I)-thiolate centers acting in copper-storage
or binuclear copper pairs able to bind oxygen. Proteins
showing all their copper by EPR have mononuclear copper
centers that display distinct EPR parameters, depending
on solvent access to the metal site and related to spe
cific protein ligands and mechanisms of electron transfer.
The presence of multiple copper centers results in
approximately 50% EPR-detectability and in ability to
reduce O_2 to H_2O.

1. INTRODUCTION.

Copper was among the earliest transition metals to be
detected and characterized in biological system by EPR.
This is due to its wide occurrence as a prosthetic
group of various functional classes of protein and
enzymes, that began to be obtained in large amounts as
purified molecules in the fifties, thus permitting to
start an actual physico-chemical and spectroscopic
analysis at the molecular and submolecular level. An
additional factor that led to the extensive use of EPR
spectroscopy in copper biochemistry is that it is diagnostic for one of the valence states of concern in
biological systems, namely Cu(II), which, having only

303

*I. Bertini and R.S. Drago (eds.), ESR and NMR of Paramagnetic Species in Biological and
Related Systems, 303–325.*
Copyright © 1979 by D. Reidel Publishing Company

one unpaired electron (d^9), is particularly simple to
study by EPR.

Up to date, 20 years since the appearance of the
first EPR report on biological copper (1), a large
amount of data is available, as well as a number of
comprehensive reviews (2). The exceeding majority of
studies have been carried out on proteins, mainly enzymes,
carrying copper as an essential component of the active
site. Only in proteins copper is present as a definite
constituent, apart from some rare pigments (3). Therefore, my survey will be limited to copper-protein systems.

Some general guide lines are apparent, both in
the history of the whole topic and in the strategy of
every single experimental undertaking in this research
field.

a) The first issue is to quantify how much of the
copper that is present in a protein is detected by EPR,
in an attempt to establish the valence state of the
metal ion. This is currently done by comparing the integrated intensity of the EPR signal with the total
copper as determined by chemical reaction or atomic
absorption. In many cases, copper is not seen by EPR,
or is seen only in part. However, as it will be seen
below, it is not straight forward to establish the valence
state of the EPR invisible copper on the sole basis of
absence of EPR signals. In this concern, optical spectroscopy and magnetic susceptibility of the system
supply additional elements for the correct establishment
of the valence state.

b) The next problem to face is the EPR characteri
zation of the species under study: g values, hyperfine
splitting constants of both copper and liganding nuclei
and relaxation properties are the essential items of
study. In this context, electronic and circular dichroic
spectra, and NMR investigations of effects on ligand
and solvent nuclei by the paramagnetic copper (T_1 para
and T_2 para) are of primary aid. ENDOR (4) and pulsed
EPR (5) measurements have been sporedically used for
ligand investigation. Artificial insertion of Co(II)
in the place occupied by copper in the native protein,
without modification of protein structure and function,
can give useful information, in view of the peculiar
magnetic and optical properties of the replacing ion.
Replacement of natural copper by isotopically pure
copper is often useful, and in most such cases obtaining
a copper - free protein capable of recombining with
the metal as to give a copper protein with properties
typical of the native state is essential. The aim of
these studies is to clarify the protein environment

around the copper, with special attention to the parti
cular geometry imposed by the macromolecular ligand to
the metal ion and to the nature of the specific copper
coordinating aminoacid side groups. It is not infre-
quent, in this kind of approach, to reach the conclusion
than more than one type of paramagnetic copper is present
in the protein under study, each one with distinct
electronic features. When X-ray crystallography, X-ray
absorption or photoelectron spectroscopy data are avai
lable, as they actually begin to be in recent years,
interpretation of EPR results will be greatly facilitated.
          c) Further insight into copper-proteins by EPR is
usually concerned with investigation of the role of the
copper ion in the protein function. All of the copper
proteins, apart from the storage proteins called thioneins, are involved, either directly or indirectly,
in the electron transfer to oxygen. Table I includes
the functional classes of copper proteins with reference
to this specialized biological role, which is shared
only by iron in such extensive and comprehensive terms.
However the precise mechanism by which this general
function is fulfilled in each case, has to be defined
in every copper protein and EPR has been of great impor
tance in this respect. Redox titrations, anaerobic
techniques of substrate addition, rapid freezing (6)
procedures are of current application in this area, and
complementary data from optical fast kinetics, steady
state enzyme kinetics, establishment of specific inhibi
tion patterns are essential as well. In special cases
selective removal of individual copper types in proteins
having more than a single copper species is important
to define the essentiality of a copper type in a parti
cular reaction.
          In the following discussion, background optical
and magnetic protein of copper complexes will be omitted,
as well as technical peculiarities of copper EPR, with
reference to the introductory section of the review
articles listed under reference 2. The matter is subdi
vided according to the phenomenological criterion of
EPR - detectability of the protein-bound copper. In
spite of its empirical character, this criterion gives
rise to quite homogeneous classes and subclasses of
copper protein and turns out to be related to functional
aspects of various proteins more closely than immediately apparent. Cytochrome oxidase will be excluded from
this review, firstly because its functional and structural complexity is far higher than any other system
discussed here, secondly because it contains a second
EPR-active prosthetic group, namely heme iron. Being
intermediate between copper and heme enzymes and being

TABLE 1

FUNCTIONAL CLASSES OF COPPER PROTEINS

CLASSES OF FUNCTIONS	Cu-PROTEINS	ANALOGOUS Fe - PROTEINS	
		HEME	NON HEME
I. Electron transfer from protein to O_2 without separation of charge (oxygen binding).	hemocyanin	hemoglobin	hemerythrin
II. Electron transfer from protein to O_2 with separation of charge (oxygen reduction) 1) with liberation of all O_2 as:			
1a) $H_2O_2(H_2O_2)$-forming oxidases)	amine ox., galactose ox.		
1b) $H_2O(H_2O$-forming oxidases).	laccase, ceruloplasmin, ascorbate ox.	Cytochrome oxidase	
2) with insertion of at least one atom of O_2 into an organic mole	tyrosinase, dopamine -hydroxylase.	Try-oxygenase, cyt P-450	Many

TABLE 1 (Continued)

FUNCTIONAL CLASSES OF COPPER PROTEINS

CLASSES OF FUNCTION	Cu-PROTEINS	ANALOGOUS Fe - PROTEINS	
		HEME	NON HEME
cule (oxygenases).			
III. Electron transfer between reduced O_2 species (O_2^- or H_2O_2-dismutases, peroxidases)	superoxide dismutase	hydroperoxidases	superoxide dismutase
IV. Electron transfer between proteins (non-autoxidizable e^- transferase, involved in e^- transport chains with O_2 at one end)	azurin,plastocyanin, stellacyanin,umecyanin, rusticyanin.	cytochromes	ferredoxins

still confronted with basic problems of preparation
and biochemical characterization, it actually constitutes
an independent class among biological systems susceptible
to EPR investigation.

2. COPPER - PROTEINS CONTAINING ALL THE COPPER IN A FORM THAT IS NOT DETECTED BY EPR IN THE NATIVE STATES.

These proteins belong to three functional classes,
namely the copper storing proteins (copper-thioneins),
the copper oxidase - oxygenase known as tyrosinase and
the oxygen carriers hemocyanins. From the structural
standpoint, it is now well established that two different
types of copper center are present in these systems,
that is a diamagnetic mononuclear copper -thiolate
center in thioneins and a binuclear copper pair in tyrosinase and hemocyanins.

2.1. Copper-thioneins.

These proteins represent the only case where $Cu(I)$
is the permanent valence state of the total protein-
bound copper in the native molecule. They belong to a
class of sulfur-rich low molecular weights proteins
which are able to bind several metals (i.e. Zn, Cu, Cd,
Hg) at very high metal-protein ratios (10 g atoms
metal/10,000 g protein). Yeast contains a more specific
copper thionein which has recently been subjected to
intensive investigation leading to the conclusion that
the absence of copper EPR signal is due to the presence
of $Cu(I)$ and not to spin-coupled $Cu(II)$ pairs (7). The
strategy adopted to solve this problem provides a good
example of an approach for copper valence determination
that combines EPR with specialized biochemical techniques.
The protein as prepared has no d-d transitions and shows
a very low percentage of the total copper as $Cu(II)$ in
the EPR spectrum: denaturing conditions, such as low or
high pH values or boiling, fail to increase the EPR
signal intensity in the absence of oxygen; in the presence of EDTA, at 100°C, approximately 50% as the copper
become detectable, such as for the disproportionation
$2 Cu(I) = Cu(0) + Cu(II)$. Oxygenation of denatured
samples led to slow appearance of practical all the
copper as $Cu(II)$. Treatment with thiol reagents or high
potential oxidizing agents released the copper as $Cu(II)$,
while reducing agents had no affect, such as expected
if the copper were present as $Cu(II)$. Dissociation of
the copper by H^+ did not result in changes of the U.V.
spectrum of the protein which could be attributed to

oxidation of thiol groups to disulfides.

Functionally, these results point to cysteine -
Cu(I) complexes as a way of tightly binding copper for
physiological storage and to oxidation/reduction processes
as a possible mechanism for controlling copper mobiliza
tion from its cellular storing sites.

2.2. Hemocyanin and tyrosinase: the proteins containing a single binuclear copper center.

These two proteins have well distinct biochemical
function, but represent a single system from the experimental standpoint. All the major chemical and spectro
scopic approaches give comparable results in either case,
suggesting substantial similarity of the active sites.
Both proteins exist in three states, i.e. oxidized,
reduced and oxygenated, all EPR silent and diamagnetic
at all temperatures. The relation between the three
states in shown below, though it should be kept in mind
that this general scheme does not take into account
specific differences related to the source of the protein.

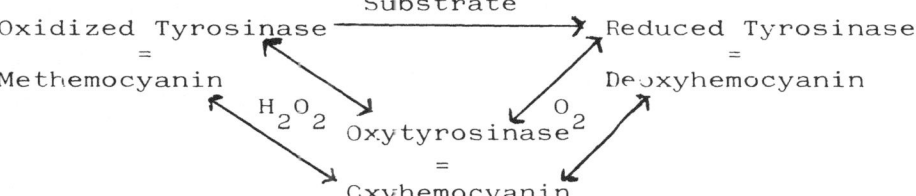

The two proteins differ essentially in the relative
stability of the three states. Hemocyanin, being an
oxygen-carrying protein is isolated in the stable oxyge
nated state, naturally shuttles between the oxy-and
deoxy-forms and is functionally inactive in the oxidized
derivative; tyrosinase, being a hydroxylating enzyme,
is isolated in the oxidized form, is reduced by substrate,
and is likely to occur as an oxygenated transient form
in the catalytic cycle. These differences may arise from
differences in protein structure but the prosthetic
moiety, that gives rise to the three states appear to
be the same in both proteins. It behaves like a pair of
adjacent copper ions, which is bicupric in the oxidized
state, bicuprons in the reduced state, and a bicupric-
peroxo unit in the oxygenated state. The protein struc
ture is likely to play a role in the reconstitution of
the copper-free protein, which requires Cu(I) in the case
of hemocyanin, while is obtained with Cu(II) in the case
of tyrosinase. In the latter protein, EPR has been used
to follow Cu(II) binding to the apoenzyme (8), and it
has been shown that the Cu(II) EPR signal disappears

with a relatively slow and complex time course, which
is in fact indicative of slow conformational rearrangement
of the protein to bring the metal ions in the proper
mutual position. In spite of these differences the presence of a binuclear copper cluster in both prcteins
has been established on the basis of some fundamental
criteria. In particular binding of O_2 (in hemocyanin)
or H_2O_2 (in tyrosinase) has a Cu/ligand = 2 stoichiometry
and gives rise to absorption bands that are very similar
in both position (345 and 600 nm) and extinction coeffi
cient (9000 M^{-1} cm^{-1} and 600 M^{-1} cm^{-1} per copper,
respectively). These bands are typical of cupric copper
(9), though no paramagnetism is detectable at any tempe
rature by either EPR or magnetic susceptibility (10-
12). Peroxidic oxygen bridging a bicupric center in the
oxy forms is also suggested by laser resonance Raman
scattering measurements (13).

Apparently, most of the available information on
this class of proteins comes from other techniques than
EPR. However it is possible to obtain EPR spectra of
hemocyanin and tyrosinase in particular conditions,
cbviously related to the binuclear nature of the copper
center. In this light, these studies have special me-
thodological relevance and must be discussed here.

Reaction with NO. NO is isoelectronic to O_2 and
CO, which are specific ligands of hemocyanin or tyrosi
nase. Reaction with NO of both proteins (14, 15) gives
rise to a mixed EPR signal, consisting of one signal,
easily saturable, which is a typical axial Cu(II) signal
accounting for approximately half the copper, and of
another signal, non saturable, with apparent   m=1
(g=2) and   m=2 (g=4) transitions that clearly arise
from very loosely (dipole - dipole) coupled Cu(II)
pairs, as demonstrated from the 7 line hyperfine struc
ture on the g=4 signal and the normal Curie behaviour
of the signal intensity as the function of temperature.
The mono - and dinitrosyl complexes of the binuclear
cluster (with one copper in the cuprous state in the
moncnitrosyl one) are responsible of the spectra, which
appear to be characteristic of 2 contiguous cupric ions
approximately 6 Å apart (16). These complexes represent a particular type of met forms, as methemocyanin
and oxidized tyrosinase (17) are diamagnetic (Cu-Cu
distance   4 Å).

Half-apo and half-oxidized derivatives. In these
cases the approach of monomerization of the binuclear
cluster, already exploited in the mononitrosyl derivative, is furtherly expanded, using either singly oxidized,
Cu(I)... Cu(II),or apo-derivatives where only one copper
had been removed and then oxidized,(...)... Cu(II).

Mononuclear Cu(II) complexes of the active site could thus be obtained from hemocyanin (18, 19). They exhibit EPR signals of the type discussed in 3.2, and thus offer the opportunity of carrying out ligand perturbation studies as widely experienced with that class of copper sites.

Reaction with β-mercaptoethanol. The reaction with this thiol compound has been detected in Neurospora tyrosinase (20), and then found also in other tyrosinases. An EPR spectrum of rather unusual feature is obtained, with very close g values (g = 2.19, g = 2.05) and no resolved $A_\parallel$ and $A_\perp$, accounting for 30% of the total copper. Again, one-electron reduced form of the binuclear Cu(II) cluster is suggested, and its relevance lies in its resemblance to the EPR signal attributed to the copper of cytochrome-oxidase. In both cases the unusual EPR parameters can be interpreted as to be due to delocalization of the Cu(II) spin on a sulfur ligand, giving the system the partial character of a sulfur radical (2 d).

Functional implications. The presence of a single binuclear copper cluster appears to be utilized for either oxygen transport (hemocyanin) or substrate hydroxylation (tyrosinase). In this respect, this type of copper center bears considerable functional analogy (see Table I) to monuclear heme Fe(II) centers that bind oxygen either for oxygen transport (e.g. hemoglobin) or for substrate hydroxylation (e.g. cytochrome P-450).

3. COPPER PROTEINS WHERE ALL THE COPPER EXISTS AS AN EPR - DETECTABLE FORM: THE MONONUCLEAR Cu(II)-PROTEINS.

There are two distinct subclasses of such proteins, easily differentiated by a single property, which all other properties are consequent on. This property is the presence of at least one water molecule in the first coordination sphere of the cupric ion. Proteins that lack this feature, i.e. proteins where Cu(II) is surrounded by only protein ligands, are the single-copper"blue" proteins, with unique optical and EPR spectra, and low solvent nuclear magnetic resonance rates in native conditions: they are endowed with the specialized capability of fast outer sphere electron transfer. Proteins having water as a ligand have more regular, but still peculiar, EPR and optical spectra, show high rates of water proton nuclear relaxation in conditions of fast exchange with the solvent, and are, when redox-active, specialized in fast inner sphere electron transfer. In the latter class, nuclear magnetic relaxation rate of solvent protons is

preferentially measured as a complement of EPR spectroscopy in a variety of solvent perturbation experiments.
Proteins of both types are among the best examples of
applications of theoretical chemistry of copper to the
aim of geometry assignments. They also are the only
known examples of copper proteins, the structure of which
has been investigated at the molecular and atomic level
by the aid of X-ray crystallography analysis and amino
acid sequence determination. A third subclass should be
mentioned in this context, that is the artificial copper
proteins, i.e. proteins that naturally contain other
metal ions as prosthetic groups, but accept copper in
the metal binding site without altering their structure.
They usually belong to the water-coordinating class, but
there is a recent report of copper substitution in a
zinc enzyme producing a "blue" copper site (21). Metal
substitution is in any case a widely used approach in
the investigation of this class of protein, also in the
opposite situation of introducing other metals in the
place naturally occupied by copper.

3.1. Proteins with a single mononuclear copper center
isolated from the solvent: "blue" electron transferases
not reoxidizable by oxygen.

They are small proteins ($1-2 \times 10^4$ M.W), capable of
fast electron transfer to and from the copper ion, which
however excludes oxygen. They have non systematic names,
namely azurin, plastocyanin, stellacyanin, umecyanin,
rusticyanin, and are of bacterial or plant origin. The
electronic spectrum is very similar in all cases, and
is dominated by high extinction coefficient around 600
nm. The EPR spectra is uniquely characterized by very
narrow, and some times unresolved $A_{\|}$. The EPR parame-
ters are in general not affected by solvent perturbation
(e.g. addition of anions, pH changes), slight modification of them being imputable to minimal conformational
protein rearrangements. This is in agreement with the
lack of large solvent NMR relaxation effects (in solutions of native single blue copper proteins, $T_{1 \, para} \approx
T_{1 \, dia}$) and indicates a rigid protein environment around
the copper, such as that can not be altered by ligand
exchange without irreversibly destroying the specificity
of the copper binding site. In other work, EPR spectroscopy of the native protein has given an outsanding
contribution to the field because of the peculiarity of
the EPR parameters, which have stimulated plenty of
theoretical and model work (22- 26), but have not been
optimally responsive to biochemical challenge, which
only can offer hints for structural information. Never-

theless, some definite approaches can be outlined, and
a few paradigmatic result from our laboratory are sum
marized in Table 2.

T A B L E   2

Copper environment in single copper "blue" proteins as
monitored by EPR.

Experimental per turbation.	Observed Magnetic Re sonance phenomena.	Interpretation
Raising pH	Bleaching of the blue color with reduction of the copper in azu rin and plastocyanin reversed by ferricya nide (27,28)	Sulfur coordi- nation
External ligands	The mercurial p-chlo romercuribenzoate re versibly changes the EPR spectrum of pla- stocyanin to that of simple nitrogenous complexes of Cu(II) (28)	Sulfur+nitro- gen coordina- tion
Metal removal and recombination	Cu(II)recombines to apo-stellacyanin in two steps characte- rized by different EPR and PRR parame- ters:fast binding as a water relaxing- non blue Cu(II), fol lowed by slow trans- formation into a non- relaxing "blue" Cu(II) (29)	Sulfur coordi nation in a site excluding water
	High resolution ^{1}H- NMR of copper-free azurin reveals reso nances attributable to cysteine and hi- sticine protons (30)	Cysteine+histi dine as protein ligands

EPR analysis has therefore been decisive, before
the recent information obtained by direct structural

investigation, in evidencing sulfur coordination and
lack of solvent interactions in the copper centers of
these protein. The peculiarity of their coordination
geometry was also supported by the spectral properties
of Co(II) derivatives of stellacyanin plastocyanin and
azurin, pointing to tetrahedral simmetry (31-33). X-
ray diffraction pattern of plastocyanin (34) and azurin
(35) have recently confirmed histidine and sulfur
(cysteine plus methionine) coordination in a substantially
tetrahedral site isolated from the solvent by four protein
ligands. The location of the copper is, however, rather
superficial; thus explaining how, in some cases, excess
of external reagents can reversibly affect the EPR spec
trum, as consequence of limited ligand exchange (28).
However, aminoacid composition of stellacyanin indicates
the absence of methionine and prevents generalization of
this model to all the single "blue" copper proteins.
Stellacyanin and rusticyanin also have a distinct EPR
spectrum in so far that they show a strong departure from
axial simmetry (36,37). A pentacoordinate Cu(II) center
has in fact been suggested for stellacyanin (38).
     Functionally, these proteins belong to the copper
proteins, together with the inert metal-storing thioneins
and with superoxide dismutase, that do not react with
oxygen in the Cu(I) state. They are specialized in protein-protein electron transfer like the cytochromes, and
share with cytochromes a buried metal center, probably
connected with the protein-protein contact areas on the
surface <u>via</u>   protein residues especially conductive
to electrons. The nature of their copper center (sulfur
coordination, solvent exclusion, tetrahedral geometry)
appears to be properly poised for outer sphere fast
electron transfer.

3.2. Proteins with a solvent accessible mononuclear
copper center.

     This class is not homogenous. It includes rather
poorly characterized oxidases (galactose oxidase, amine
oxidases) and one intensively studied and very well
known enzyme (superoxide dismutase). It is convenient
to include the artificial Cu-protein, such as Cu-carbo-
nic anhydrase and Cu-transferrin, in this context, since
their EPR spectra have the same general characters as
those of the present class of copper proteins.
     The oxidases belonging to this class produce H_2O_2.
While amine oxidases contain a second prosthetic group
of carbonylic nature, galactose oxidase is unique among
oxidases in having just a single copper per protein
molecule as the only non-protein component. In both

types of oxidases the copper is accessible to molecules
from the solvent (39, 40) in the resting enzymes and
exists largely (70-100% depending on the preparation)
as Cu(II). The major EPR results obtained with these
enzymes is that the signal intensity and line shape are
not significantly affected by addition of substrate
(41, 42), indicating that redox cycling of copper in the
catalytic mechanism in unlikely. Though, copper is essen
tial to the enzyme activity, as demonstrated by copper
removal and readdition experiments (41, 43). In amine
oxidases there is preliminary evidence that the active
enzyme unit, probably a dimer, bears two EPR-distinct
copper ions (40). Nitrogen coordination has been demonstrated in both types of enzymes from superhyperfine
structure (41, 44), but no indication is available on
the coordination number and geometry.
  Superoxide dismutase, on the other hand, offers the
best example of application of EPR studies to obtain
structural information. Table 3 shows that knowledge
gained by magnetic resonance methods was substantially
confirmed by the more recent X-ray crystallographic
analysis (45).

T A B L E   3

Correlation between magnetic resonance observations and
X-ray crystallography data for copper, zinc superoxide
dismutase.

Magnetic resonance	X-ray crystallography
The two copper centers of the di mer (33,000 M.W.) are EPR equiva- lent and non interacting (46,47)	Identical metal sites on identical subunits. The copper in the di- mers are 33 Å apart
Deuterium effect on EPR line width (46). Enhancement of water T_1 (48,49).Binding of anions to copper at the site of interaction with the solvent (46 -48)	One side of the copper is open to solvent access
Superhyperfine structure from 3 magnetically equivalent nitrogens (47). Line broadening of the C_2- protons from 3 His in the ^{1}H-NMR of reconstituted apoprotein (50)	The copper has His 44, 46,61 and 118 as li- gands
Rhombic EPR line shape (47)	The copper is in a distorted square pla nar arrangement

T A B L E   3 (Continued)

Magnetic resonance	X-ray cristallography
Exchange coupling between the copper and the cobalt introduced in the place of the zinc, as only possible if a common ligand bridges the two metals (51 - 53)	The copper and the zinc are approximately 6 Å apart with His 61 positioned so that apparently each of its ring nitrogens is liganded to one of the metals

It also shows the utility of complementing EPR spectroscopy with a number of biochemical perturbation, such as anion addition, pH changes and metal substitution as well as of supplemtary magnetic resonance observations such as higher frequency measurements, ligand isotopic effects on line width and hyperfine structure, proton NMR and water proton nuclear relaxation. From these studies and parallel investigations on the enzyme mechanism, the copper binding site of superoxide dismutase appears as a unique bioinorganic complex, in so far that it is specifically associated to zinc through a bridging histidine imidazole group and shuttles between the Cu(II) and Cu(I) state with very high efficiency. The two features are probably functionally related, and it has been suggested that the structure of the zinc-copper cluster is particularly suited for a change of the copper coordination number in the redox process (54, 55). In this way a tetrahedral Cu(I) and a pentacoordinate Cu(II) sites would be created in the protein by minimal conformational rearrangements and the same kinetic advantage of the rigid and preformed tetrahedral geometry of the blue copper center would be obtained. These two types of highly efficient electron-transferring copper centers differs in a second major point: at variance with the "blue" electron carriers, the superoxide dismutase copper coordinates a water molecule which is displaced by substrate or inhibiting anions, such as CN , in the process of enzymic activity (56). The superoxide dismutase copper is therefore poised for inner sphere redox reactions while the blue centers are specific for outer sphere electron transfer. The different protein ligands in the two cases - all histidines in the former one, histidine plus sulfur aminoacids in the blue centers - are certainly related to the fact that water coordination is the electron gate in the case of inner sphere mechanisms, while sulfur coordination is essential for conveying electrons to the copper from the protein surface. In both cases, the reduced protein is rather inert to

reaction with oxygen, thus confirming the rather general rule - with the only possible exception of galactose
oxidase - that a single mononuclear copper center is not
capable of oxygen activation.

The EPR characterization of the copper binding
site of superoxide dismutase has been supported by comparison with copper-carbonic anhydrase, an inactive
derivative of the native Zn-containing enzyme that is
also structurally well known. Table 4 presents a general overview of the significant correlations between
tha magnetic rescnance properties of the two systems.

T A B L E   4

Correlation between the copper binding sites of carbonic anhydrase and superoxide dismutase as studied by
magnetic resonance

	Superoxide dismutase	Carbonic anhydrase
EPR line shape	Rhombic (47) Axial at pH 5 (57) or 10 (46)	Axial (58)
Reaction of the with ferrocya- nide	Rhombic copper redu ced (59) Axial copper not reduced:at pH 5 complex with peculiar EPR spectrum (60)	Not reduced: complex with pe- culiar spectrum (60)
Effect of mono- valent anions on the EPR line shape	Change to axial (47)	Change to rhom- bic (58)
Effect of mono- valent anions on the enhance- ment of the sol vent proton T_1^{-1}	Inhibition (56)	No effect (61)
Interpretation	Pentacoordinate nati ve site (45) Four- coordinate on reduc- tion (55) Anions di- splace copper-bound water (56)	Four coordinate native site (55) Pentacoordinate anion complexes (38) Anions do not displace cop- per-bound water (61)

Notably the opposite effect of monovalent anions on
the copper EPR and T_1 parameters in the two proteins
illustrates particular binding situations, which may be
interpreted on the basis of different symmetries and
coordination numbers (55) and make up a sort of reference
system to be utilized for studies of copper centers
lacking such a high structural information (see **4.1.c**).
        Studies with Cu-transferrin are less significant
in this context, mainly because the structural details
of the metal binding sites of this protein are less
known. However, they have special relevance to the problem of multiple copper binding sites in the same protein.
This circumstance has been already mentioned for amine
oxidases, and will again been discussed in the case of
blue oxidases. The Cu-transferrin system illustrates
how fine changes of the ligand environment, in substantially similar binding sites, produce clearly distingui-
shable EPR spectra with specific differences in anion
binding and pH sensitivity (62).

4. COPPER PROTEINS WITH PARTIALLY EPR – DETECTABLE
COPPER: THE PROTEINS WITH MULTIPLE COPPER CENTERS THAT
REDUCE O_2 TO H_2O.

        The property of forming water from oxygen is shared
by some oxidases and all oxygenases, and is a privilege
of copper and iron enzymes (see Table I). Among the
copper enzymes, only one can not been located in this
section, namely tyrosinase, all the copper of which is
in the state of an EPR – silent pair (see 2.2.). The
exceeding majority of the H_2O – forming copper enzymes
consists of the "blue" oxidases, which also represent
the extreme case of copper heterogeneity, as they apparently contain in each protein molecule the binuclear
the "blue" and the water – relaxing non "blue" copper
center in an at least 1:1:1 ratio. Dopamine   -hydroxy-
lase might be considered analogous to amine oxidase or
galactose oxidase as having only one type of copper
that is partially detected by EPR. However, in dopamine
 -hydroxylase the percentage of EPR silent copper is
approximately 50% and the metal centers undergo definite
redox changes that are absent in the H_2O_2 – forming
oxidases (63, 64). This protein is still poorly characterized, but EPR has been succesfully used to show copper
reduction by the cosubstrate ascorbate, oxidation to the
full complement of the chemically determined copper by
ferricyanide or by oxygen plus substrate, appearance of
superhyperfine lines typical of nitrogenous coordination
by protein ligands on substrate binding (63). Redox

potential of the copper has been also determined by
EPR (64), because of the absence of any visible absorption band as a probe of the extent of its oxidation.
Lacking evidence for the presence of coupled copper
pairs, this enzyme would be a rare case where Cu(I) is
present in the resting state.

4.1. The blue oxidases: laccase, ceruloplasmin, ascorbate
oxidase.

     The contributions by Malmström, Vänngård and their
coworkers dominate the EPR results on these proteins.
They proposed the coexistence in a single blue oxidase
molecule of the three independent types of copper centers
mentioned above, showed the essentiality of all of them
to the enzyme activity and established their redox equilibria and kinetic properties (2b, 2c). Recent work from
the same group also led to EPR characterization of transient intermediates of reactions, by the combined used
of rapid freezing techniques and isotopic labelling
(65, 66). From these and other studies, a picture of
these enzymes emerges, essentially based on the cooperation, through specific mediation by the protein moiety,
of structurally independent copper centers, which bear
considerable analogy to the independent classes described
under 2.2, 3.1 and 3.2 but are strictly interdependent
for their spectral and functional properties. The presen
ce of Cu(I) in the resting enzymes seems to be excluded,
at least in laccases.
     Furthermore, the interdependent three types of
copper center of blue oxidase, have some peculiar EPR
properties with respect to the independent analogous
complexes of single center - containing copper proteins
(see 2.2, 3.1 and 3.2).
     a) In these proteins at least 50% of the total
copper is diamagnetic at all temperatures and is believed
to be Cu(II)-Cu(II) pairs, as it takes up two electrons
cooperatively (67). However reaction with NO does not
give in all cases the same results. In ceruloplasmin a
typical dinitrosyl derivative like that of hemocyanin
or tyrosinase is formed, after the blue copper has been
reduced by NO (68). In ascorbate oxidase the blue copper
is reduced but no dipole-coupled copper signal is formed
(69). In laccase even the reduction of the blue copper
is sluggish and detectable only above 40°C (70). Similarly, treatment of laccase by β-mercaptoethanol did
not give rise to the typical EPR signal observed in
tyrosinase (71). These differences may be due to different distances between the two coupled Cu(II) ion but
do not exclude in the negative cases the presence of a

Cu(I)-Cu(III) pair.

b) The blue copper in these enzymes, except the laccase from Rhus vernicifera, exhibits larger parallel hyperfine splittings than in the single blue copper containing protein (2c). In ceruloplasmin it consists of two components with different EPR parameters and different reactivity (72, 73). The coordination to sulfur ligands of the blue copper in laccase is still rather dubions (74).

c) The third copper center interacts with water, as demonstrated by ^{17}O EPR in laccase (75) and binds anions with some selectivity (for instance F$^-$ in laccases) However, water relaxation is still present after comple te removal or anion saturation of this copper (76, 77). Nitrogenous protein ligands are certainly part of its coordination sphere, as evident from superhyperfine structure detected under non-denaturing conditions (78-80).

A recent report has pointed out the similarity of the EPR parameters in the CN$^-$ and N$_3$ adducts of this copper in laccase and the analogous copper-carbonic anhydrase anion derivatives, suggesting pentacoordinate anion complexes in both proteins (81).

d) In all EPR studies of reactions of copper centers in blue oxidases, functional interdependence of the three types of sites is evident (82).

The role of copper in the mechanisms of action of these oxidases is still uncertain for the various centers and the various catalytic steps in spite of a great deal of mechanistic work, essentially based on the EPR analysis of rapidly frozen mixtures of substrate and enzyme under various conditions. By analogy, one is attempted to think of the water-coordinating copper as the substra te binding site (see the case of superoxide dismutase), the binuclear center as the oxygen activating one (like in hemocyanin and tyrosinase), and the blue copper as the mediator of intramolecular electron transfer (similarly to azurin and plastocyanin). However, these oxidases do not bind CO, as expected if the diamagnetic copper behaved as in hemocyanin or tyrosinase. Furthermore, evidence has been presented from ^{17}O work by the Göteborg group that the water-coordinating copper participates in the oxygen reducing site, probably together with the binuclear cluster (83). In conclusion, these studies on the mechanism confirm the functional interdependence of all the copper sites in these oxidases. If cooperativity between sites is the substantial difficulty in applying the simple models of single copper center-carrying protein to the understanding of the blue oxidases the same property give the blue oxidases

the opportunity of being in turn a good model for more
complicated oxygen-activating systems, such as cytochrome
oxidase.

Acknowledgement - I wish to thank Mrs. Valeria
Di Chiara for typing the manuscript.

R E F E R E N C E S

1. Malmström, B.G., Mosbach, R. and Vänngård, T.:
   1959, Nature 183, pp. 321-322.
2. a) Gould, D. and Ehrenberg, A.: 1968, in'Physiology
   and Biochemistry of Hemocyanins', Ghiretti, F. ed.,
   Academic Press, New York, pp. 95-112. b) Malkin, R.
   and Malmström, B.G.: 1970, Adv. Enzymol. 33, pp.
   177-243. c) Vänngård, T.: 1972, in 'Biological Appli
   cations of ESR', Swartz, H.M., Bolton, J.R., and
   Borg, D.C., eds., Wiley Interscience, New York,
   pp. 411-447. d) Peisach, J. and Blumberg, W.E.:
   1974, Arch. Biochem. Biophys. 165, pp. 691-708.
3. Blumberg, W.E. and Peisach, J.: 1965, J. Biol. Chem.
   240, pp. 870-876.
4. Rist, G., Hyde, J. and Vänngård, T.: 1970, Proc.
   Natl. Acad. Sci U.S.A. 67, pp. 79-86.
5. Mondovì, B., Graziani, M.T., Mims, W.B., Oltzik,
   R. and Peisach, J.: 1977, Biochem. 16, pp. 4198-4202.
6. Palmer, G.: 1967, Methods in Enzymol. 10, pp. 594-609.
7. Rupp, H., Cammack, R., Hartmann, H.J. and Weser, U.:
   1979, Biochim. Biophys. Acta, in press.
8. Kertesz, D., Rotilio, G., Brunori, M., Zito, R. and
   Antonini, E.: 1972, Biochim. Biophys. Res. Commun.
   49, pp. 1208-1215.
9. Van Holde, K.E.: 1967, Biochem. 6, pp. 93-99.
10. Nakamura, T. and Mason, H.S.: 1960, Biochem. Biophys
    Res. Commun. 3, pp. 297-299.
11. Moss, T.M., Gould, D.C., Ehrenberg, A., Loehr, J.S.
    and Mason, H.S.: 1973, Biochem. 12, pp. 2444-2449.
12. Dooley, D.M., Scott, R.A.,Ellinghaus, J., Solomon,
    E.I. and Gray, H.B.: 1978, Proc. Natl. Acad. Sci
    U.S.A. 75, pp. 3019-3022.
13. Loehr, J.S., Freedman, T.B., and Loehr, T.M.: 1974,
    Biochim. Biophys. Res. Commun. 56, pp. 510-515.
14. Schoot-Uiterkamp, A.J.M.: 1972, FEBS Letters 20,
    pp. 93-96.
15. Schoot-Uiterkamp, A.J.M., and Mason, H.S.: 1973,
    Proc. Natl. Acad. Sci. U.S.A. 70, pp. 993-996.
16. Schoot-Uiterkamp, A.J.M., Van Cer Deen, H., Berendsen,
    H.C.J., and Boas, J.F.: 1974, Biochim. Biophys.
    Acta 372, pp. 407-425.
17. Makino, N., MacMahill, P., Mason, H.S., and Moss,

T.H.: 1974, J. Biol. Chem. 249, pp. 237-242.

18.   Himmelwright, R.S., Eickman, N.C. and Solomon,
      E.I.: 1978, Biochem. Biophys. Res. Commun. 81,
      pp. 237-242.
19.   a) Himmelwright, R.S., Eickman, N.C. and Solomon,
      E.I.: 1978, Biochem. Biophys. Res. Commun. 81,
      pp. 243-247. b) Himmelwright, R.S., Eickman, N.C.
      and Solomon, E.I., J. Am. Chem. Soc. 101, pp.
      1576-1586.
20.   Deinum, J., Lerch, K. and Reinhammar, B.: 1976,
      FEBS Letters 69, pp. 161-169.
21.   Maret, W., Dietrich, H., Ruf, H.H. and Zeppezauer,
      M.: 1979, J. Inorg. Biochem., in press.
22.   Malmström, B.G. and Vänngård, T.: 1960, J. Mol.
      Biol. 2, pp. 118-124.
23.   Brill, A., Martin, R., and Williams, R.J.P.: 1964,
      In 'Electronic Aspects of Biochemistry', Pullman,
      B. ed., Academic Press, New York, pp.519-557.
24.   Blumberg, W.: 1966, in 'The Biochemistry of Copper',
      Peisach, J., Aisen, P. and Blumberg, W., eds.
      Academic Press, New York, pp. 49-66.
25.   a) Gould, D. and Ehrenberg, A.: 1968, Europ. J.
      Biochem. 5, pp. 451-455. b) Hyrayama, Y. and
      Sugiura, Y.: 1979, Biochem. Biophys. Res. Commun.
      86, pp. 40-47.
26.   Brill, A., and Bryce, G.: 1968, J. Chem. Phys.48,
      pp. 4398-4404.
27.   Finazzi Agrò, A., Rotilio, G., Avigliano, L., Guer
      rieri, P., Boffi, B. and Mondovì, B.: 1970,Bio-
      chem. 9, pp. 2009-2014.
28.   Graziani, M.T., Finazzi Agrò, A., Rotilio, G.,
      Barra, D. and Mondovì, B.: 1974, Biochem. 13, pp.
      804-809.
29.   Morpurgo, L., Rotilio, G., Finazzi Agrò, A. and
      Mondovì, B.: 1974, Arch. Biochem. Biophys. 161,
      pp. 291-296.
30.   Rotilio, G., Finazzi Agrò, A., Avigliano, L.,Lai,
      A., Conti, F., Franconi, C. and Mondovì, B.: 1970,
      FEBS Letters 12, pp. 114-117.
31.   Mc Millin, D.R., Holwerda, R.A. and Gray H.B.:
      1974, Proc. Natl. Acad. Sci. USA, 71, pp.1339-1341.
32.   Mc Millin, D.R., Rosenberg, R.C. and Gray, H.B.:
      1974, Proc. Natl. Acad. Sci USA 71, pp. 4760-4762.
33.   Solomon, E.I., Rawlings, J., Mc Millin, D.R.,
      Stephens, P.J. and Gray, H.B.: 1976, J. Am. Chem.
      Soc. 98, pp. 8046-8048.
34.   Colman, P.M., Freeman, H.C., Guss, J.M., Murata,
      M., Norris, V.A., Ramshaw, J.A.M. and Ventakappa,
      M.P.: 1978, Nature 272, pp. 319-324.
35.   Adman, E.T., Stenkamp, R.E., Sieker, L.C. and

and Jensen, L.H.: 1978, J. Mol. Biol. 123, pp.35-47.

36. Peisach, J., Levine, W.G. and Blumberg, W.E.: 1964, J. Biol. Chem., 242, pp. 2847-2858.
37. Cox, J.C., Aasa, R. and Malmström, B.G.: 1978, FEBS Letters 93, pp. 157-160.
38. Morpurgo, L., Finazzi Agrò, A., Rotilio, G. and Mondovì, B.: 1976, Europ. J. Biochem. 64, pp.453-457.
39. Marwedel, B.J., Kurland, R.J., Kosman, D.J. and Ettinger, M.J.: 1975, Biochem. Biophys. Res. Commun. 63, pp. 773-779.
40. Barker, R., Boden, N., Cayley, G., Charlton, S.C., Henson, R., Holmes, M.C., Kelly, I.D. and Knowles, P.F.: 1979, Biochem. J. 177, pp. 289-302.
41. Mondovì, B., Rotilio, G., Costa, M.T., Finazzi Agrò, A., Chiancone, E., Hansen, R. and Beinert, H.: 1967, J. Biol. Chem. 242, pp. 1160-1166.
42. Kosman, D.J., Bereman, R.D., Ettinger, M.T. and Giordano, R.S.: 1973, Biochem. Biophys. Res. Commun. 54, pp. 856-861.
43. Amaral, D., Bernstein, L., Morse, D. and Horecker, B.L.: 1963, J. Biol. Chem. 238, pp. 2281-2284.
44. Cleveland, L., Coffman, R.L., Coon, P. and Davis, L.: 1975, Biochem. 14, pp. 1108-1115.
45. Richardson, J.S., Thomas, K.A., Rubin, B.H., and Richardson, D.C.: 1975, Proc. Natl. Acad. Sci. USA 72, pp. 1349-1353.
46. Rotilio, G., Finazzi Agrò, A., Calabrese, L., Bossa, F., Guerrieri, P., and Mondovì, B.: 1971, Biochem. 10, pp. 616-621.
47. Rotilio, G., Morpurgo, L., Giovagnoli, C., Calabrese, L. and Mondovì, B.: 1972, Biochem. 11, pp. 2187-2192.
48. Fee, J.A. and Gaber, B.P.: 1972, J. Biol. Chem. 247, pp. 60-65.
49. Terenzi, M., Rigo, A., Franconi, C., Mondovì, B., Calabrese, L., and Rotilio, G.: 1974, Biochim. Biophys. Acta 351, pp. 230-236.
50. Stokes, A.M., Hill, H.A.O., Bannister, W.H. and Bannister, J.V.: 1973, FEBS Letters 32, pp. 119-123.
51. Calabrese, L., Rotilio, G. and Mondovì, B.: 1972, Biochim. Biophys. Acta 263, pp. 827-829.
52. Rigo, A., Terenzi, M., Franconi, C., Mondovì, B., Calabrese, L. and Rotilio, G.: 1974, FEBS Letters 39, pp. 154.
53. Rotilio, G., Calabrese, L., Mondovì, B., and Blumberg, W.E.: 1974, J. Biol. Chem. 249, pp.3157-3160.

54. Mc Adam, M.E., Fielden, E.M., Lavelle, F., Cala-
    brese, L., Cocco, D. and Rotilio, G.: 1977, Bio-
    chem. J. 167, pp. 271-274.
55. Rotilio, G., Morpurgo, L., Calabrese, L., Finazzi
    Agrò, A., and Mondovì, B.: 1977, in 'Metal-Ligand
    Interactions in Organic Chemistry and Biochemistry',
    Pullman, B. and Goldblum, N. eds. Reidel, Dordre-
    cht, pp. 243-253.
56. Rigo, A., Viglino, P. and Rotilio, G.: 1975, Bio-
    chem. Biophys. Res. Commun. 63, pp. 1013-1018.
57. Rotilio, G., Calabrese, L., Bossa, F., Barra, D.,
    Finazzi Agrò, A. and Mondovì, B.: 1972, Biochem.
    11, pp. 2182-2187.
58. Morpurgo, L., Rotilio, G., Finazzi Agrò, A., and
    Mondovì, B.: 1975, Arch. Biochem. Biophys. 170,
    pp. 360-367.
59. Rotilio, G., Morpurgo, L., Calabrese, L. and Mon-
    dovì, B.: 1973, Biochim.Biophys.Acta 302, pp.229-
    235.
60. Morpurgo, L., Mavelli, I., Calabrese, L., Finazzi
    Agrò, A., and Rotilio, G.: 1976, Biochem.Biophys.
    Res. Commun. 70, pp. 607-614.
61. Bertini, I., Conti, G., Luchinat, C. and Scozza-
    fava, A.: 1977, Inorg. Chim. Acta 23, pp.L15-L16.
62. Zweier, J.L.: 1978, J. Biol. Chem. 253 pp.7616-
    7621.
63. Walker, G.A., Kon, H. and Lowenberg, W.: 1977,
    Biochem. Biophys. Acta 482, pp. 309-322.
64. Ljones, T., Flatmark, T., Skotland, T., Petersson,
    L., Backstrom, D. and Ehrenberg, A.: 1978, FEBS
    Letters 92, pp. 81-84.
65. Aase   R., Brändén, R., Deinum, J., Malmström,B.G.,
    Reinhammar, B. and Vänngård, T.: 1976, FEBS Let-
    ters 61, pp. 115-119.
66. Aasa, R., Brändén, R., Deinum, J., Malmström, B.G.,
    Reinhammar, B. and Vänngård, T.: 1976, Biochem.
    Biophys. Res. Commun. 70, pp. 1204-1209.
67. Fee, J.A., Malkin, R., Malmström, B.G., and Vänn-
    gård, T.: 1969, J. Biol. Chem. 244, pp. 4200-4207.
68. Van Leeuwen, F.R., Wever, R. and Van Gelder, B.F.:
    1973, Biochim. Biophys. Acta, 315, pp. 200-204.
69. Van Leeuwen, F.X.R., Wever, R., Van Gelder, B.F.,
    Avigliano, L., and Mondovì, B.: 1975, Biochim.
    Biophys. Acta 403, pp. 285-291.
70. Rotilio, G., Morpurgo, L., Graziani, M.T. and
    Brunori, M.: 1975, FEBS Letters 54, pp. 163-166.
71. Deinum, J.: 1977, Ph. D. Thesis, University of
    Göteborg, pp. 1-15.
72. Deinum, J. and Vänngard, T.: 1973, Biochim. Bio-
    phys. Acta 310, pp. 321-330.

73. Falk, K.E., and Reinhammar, B.: 1972, Biochim.
    Biophys. Acta 285, pp. 84-90.
74. Briving, C. and Deinum, J.: 1975, FEBS Letters 51,
    pp. 43-46.   o
75. Deinum, J. and Vänngard, T.: 1975, FEBS Letters
    58, pp. 62-65.
76. Rigo, A., unpublished results.   o
77. Malmström, B.G., Reinhammar, B. and Vänngard, T.:
    1968, Biochim. Biophys. Acta 156, pp. 67-76.
78. Malkin, R., Malmström, B.G. and Vänngard, T.:1968,
    FEBS Letters 1, pp. 50-54.
79. Brändén, R. and Reinhammar, B.: 1976, Biochim.
    Biophys. Acta 405, pp. 236-242.
80. Dawson, J.H., Dooley, D.M. and Gray, H.B.: 1978,
    Proc. Natl. Acad. Sci. USA 75, pp. 4078-4081.
81. Desideri, A., Morpurgo, L., Rotilio, G., and Mon-
    dovì, B.: 1979, FEBS Letters 98, pp. 339-341.
82. Morpurgo, L., Rotilio, G., Finazzi Agrò, A. and
    Mondovì, B.: 1974, Biochim. Biophys. Acta 336,
    pp. 324-328.
83. Brändén, R. and Deinum, J.: 1977, FEBS Letters 73,
    pp. 144-146.

THE BINDING OF DIVALENT IONS TO TRANSFER NUCLEIC ACIDS

M. Guéron and J.L. Leroy

Groupe de Biophysique, Ecole Polytechnique,
91128 Palaiseau, France.

1. INTRODUCTION

Transfer ribonucleic acids (tRNA) are the adaptor molecules which translate the codons of messenger RNA into the sequence of amino-acids which make up the protein specified by the messenger. Each transfer RNA (molecular weight around 27000) is composed of approximately 80 nucleotides linked by 3'-5' phosphodiester bonds. The tRNAs which code for the various aminoacids seem to have similar structure. tRNA^{Phe} from yeast has been investigated in most detail. Its secondary and tertiary structure are shown in Fig. 1.

For protein synthesis, tRNA is first charged with its cognate amino-acid at the 3' end (-CCA), by means of a specific amino-acid ligase. The charged tRNA then complexes with a protein called the elongation factor (EF-Tu) and is transferred to the ribosome, where its anticodon recognises the codon on the messenger, following which the aminoacid is incorporated in the nascent protein. These processes require that tRNA be in its proper structure, and for this reason among others (such as the requirement for ATP-Mg) they only take place in certain salt conditions.

Transfer RNA like other polynucleotides is a polyelectrolyte because each of the backbone phosphodiester groups is negatively charged (pK $\approx$ 2). In distilled water the repulsion between phosphates disrupts the secondary and tertiary structure. As one adds salt, the secondary structure (clover leaf, Watson-Crick double helical stems) forms. This can be followed by the UV hypochromism due to the stacking of the bases. In a solution containing 10 mM NaCl, this process is already complete. Addition of

327

I. Bertini and R.S. Drago (eds.), ESR and NMR of Paramagnetic Species in Biological and

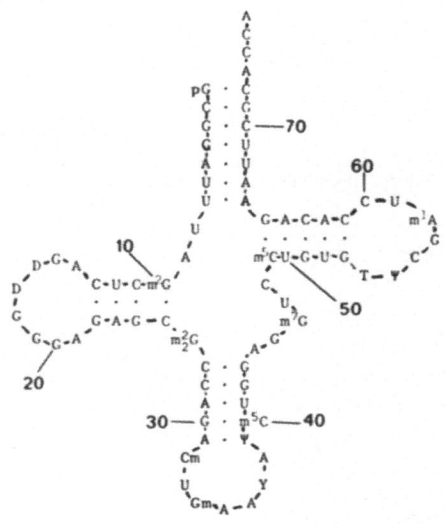

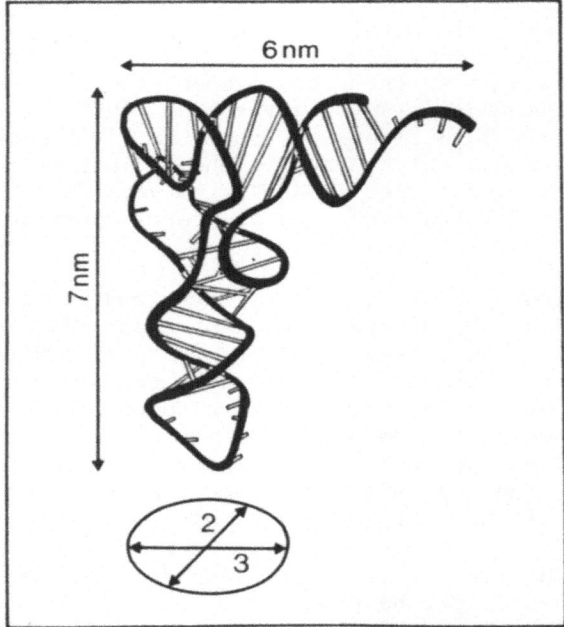

Fig. 1 : a) The clover-leaf structure of yeast tRNA^Phe.
          b) The three-dimensional structure of tRNA^Phe.
The black line is the phosphate-sugar backbone and the
rungs represent the bases. There are 75 singly charged
phosphodiesters, plus one terminal phosphate. (Re-
drawn from Rich A. and Kim S.H. (1978) Scientific
American 238, 52-62).

more salt (0.1 M to 1 M NaCl) induces the formation of the
tertiary structure, as shown by biological activity or by the
study of spectroscopic probes sensitive to tertiary structure.
Among such probes is the naturally occuring modified base 4-thio-
uracil in position 8 of many E. coli tRNAs.

Instead of sodium, the structural changes can also be brought
about by the addition of divalent cations such as Mg (whose
concentration in cells is around 10 mM). It was observed at an
early date that Mg is effective at much lower concentrations than
Na. The secondary and tertiary structures are formed for a Mg
concentration as low as 0.1 mM, and they seem to appear nearly at
the same Mg concentrations. Correlatively, the affinity of tRNA
for Mg was found to be much higher than for sodium.

Another interesting observation has been made more recently.
Whereas the structure of tRNA in 0.1 mM Mg is similar to that
obtained in high NaCl (0.1 to 1 M), it appears from the measurement of diffusion coefficients that a conformational change may
occur as the Mg concentration increases from 1 to 10 mM.

The marked effects of magnesium on tRNA structure have lead to
the suggestion that there are special binding sites for Mg on
tRNA, and that the incorporation of Mg in these sites is
necessary in order to form the tertiary structure. For instance,
it is sometimes thought that as Mg is added to tRNA in a salt
solution (for instance 50 mM NaCl) the first four or five Mg bind
selectively to as many sites. However the arguments in favor of
such an interpretation are not convincing. Among the most often
quoted arguments is the recent observation of four or five Mg
ions in the X-ray structure of tRNA[Phe]. However the crystals used
in these studies contain, apart from two spermines per tRNA, only
Mg as a counterion. Thus the number of Mg per tRNA is not four or
five, but around 35, by charge neutrality. The other Mg are not
seen because they are less rigidly held and have therefore a
smaller Debye-Waller factor, but they must be there nevertheless,
and their mobility is no indication that their binding equilibrium constant is smaller. Other arguments for specific and
strong binding sites derive in general from early and incorrect
interpretations of the binding curves.

It should be noted that the Mg concentrations necessary for
binding 5 Mg to tRNA are below 1 mM, even in the presence of
100 mM Na. Hence the afore mentioned effects of large magnesium
concentrations are independent of the problem of strong binding
sites, and will not be studied here.

One should also note that since most or all of the tertiary
structure (including the tertiary base pairs which are seen in

nmr of exchangeable protons) is formed in 100 mM Na in the absence of Mg, this ion cannot play an essential role in the secondary or tertiary structure of tRNA.

2. EPR TITRATION OF MANGANESE BINDING TO tRNA

Manganese substitutes well for magnesium as regards the structure and function of tRNA. It can therefore be used as a probe of these properties. In this section we show how one can titrate the association of Mn to tRNA by using the EPR of free Mn. In section 3 the EPR of bound Mn is used to study the binding sites. In section 4, the distribution of sites is studied by the effect of Mn on the ^{31}P nmr of the phosphate groups.

2.1. EPR of free Mn (hexaaquoion) at room temperature. The exchange regime between free and bound Mn.

Mn^{2+} is a d^5 ion. The crystal field is low, so that the five electron spins align (Hund's rule) giving rise to L=0, that is an S-state ion. Hence the effects of the spin-orbit coupling are small. They are in fact much smaller than for the isoelectronic Fe^{3+}. Due to the weak spin-orbit coupling, the ion exhibits a long relaxation time and the EPR can be observed even in solution at room temperature. For the same reason, the zero-field splitting is often small. In the hexaaquoion, the five ($\Delta m_S=1$) transitions corresponding to the possible values of m_S are all found together at g=2, i.e. the distorsions from cubic symmetry are negligible, and D can be set approximately equal to zero, on the scale of the Zeeman energies. The system thus behaves as an S=5/2 ion in cubic symmetry.

Due to the nuclear spin of Mn (I=5/2), each $\Delta m_S=1$ line is split into 2I+1=6 hyperfine lines (Fig. 2).

Estimates of the longitudinal relaxation time τ_{1s} obtained from water proton NMR are about 10^{-8} sec at room temperature at X band. (τ_{1s} increases with temperature and frequency). The line-width is due in part to the spread between the transitions corresponding to different m_S. Second-order terms in the hyperfine interaction affect this spread, in such a way that the spread is minimum for the 4th line starting from low field. The intrinsic line-width then corresponds to $3\times10^{-9}s$.

The EPR of Mn bound to tRNA will be discussed below. At this point, we note only that the linewidth is 4 times larger than for the hexaaquoion (80 G instead of 20) and that the spectrum is not shifted.

Hence its amplitude in the derivative mode is sixteen times less. Thus the EPR signal is a measure of the free Mn, apart from a

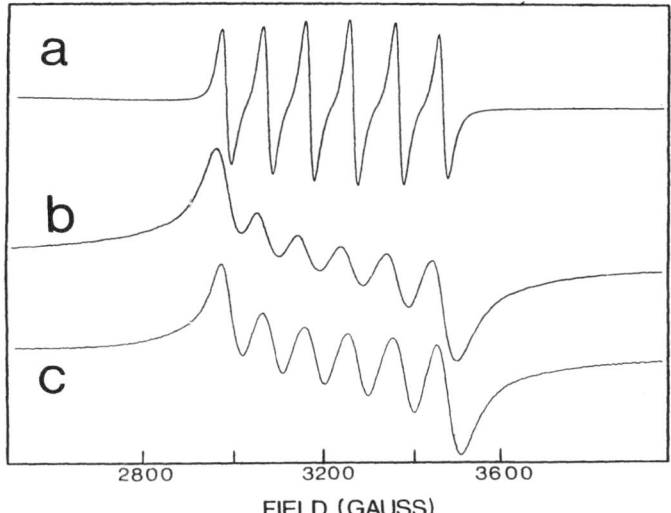

Fig. 2 : X-band EPR spectra of manganese, T= 298 K.
a) hexaaquoion, $[Mn^{2+}]$=0.5 mM.
b) Mn-ATP complex, [ATP]=10 mM, $[Mn^{2+}]$=0.5 mM.
c) Mn-tRNA complex, Baker's yeast tRNA^{unfrac}=0.5 mM,
$[Mn^{2+}]$=0.5 mM, $[Na^+]$=50 mM, pH=7.0.

small contribution from bound Mn which can often be neglected, or, if necessary, corrected for. The total quantity of Mn in the sample being known, the bound Mn is then found by difference, and the titration can thus be performed.

This interpretation is valid under the assumption of distinct signals for the free and bound species, that is under conditions of slow exchange. The inverse situation would be one where Mn exchanges between the hexaaquo- and the tRNA-bound species at a rate large enough for only one spectrum to be obtained. This situation cannot be ruled out a priori since the bound signal is in most cases too weak to be observed. However two independent arguments demonstrate that the situation is that of slow exchange.

The spectra of free and bound Mn (Fig. 2) consist of a group of 5 unresolved lines for each hyperfine transition. On going from one species to the other the width changes from 20 to 80 G. If two such lines are to collapse in a single one, the chemical exchange time τ_m has to be smaller than $(\gamma_e \Delta B)^{-1}$ where ΔB is a number of the order of 80 G. Already for ΔB=10 G this corresponds to $\tau_m < 5 \times 10^{-9}$ s, which is much smaller than the chemical exchange time for manganese binding to phosphates, which is in the range of 10^{-6} to 10^{-7} s.

The second argument is based on an experimental study of the line shape and its variation as a function of bound and free species. In a solution containing enough tRNA that nearly all ions are bound, only the bound signal is observed, with its 80 G width. One now doubles the Mn concentration. Some of the added ions will also be bound and others, at most 50%, will be free. If fast exchange prevails, this will give a signal whose width is at least as large as the average of the bound and free widths, or 50 gauss. But if exchange is slow, the free signal will be only 20 G wide, and being 16 times more intense than the bound signal in the derivative presentation, it will be easily observed. A 20 G-wide signal is indeed observed, demonstrating slow exchange.

In conclusion, the narrow EPR signal is due only to free Mn and is thus a valid and convenient tool for the study of Mn binding to tRNA.

2.2. The titration experiments

The Scatchard representation is commonly used in biochemical titrations. Let M be the metal, T the tRNA, n the number of binding sites for Mn. Let b,f and t stand for bound, free and total respectively. For independent and identical binding sites, the mass-action law is :

$$\frac{[M_b]}{[M_f](n\,T_t - [M_b])} = K \tag{1}$$

If one sets $y=[M_b]/([M_f]T_t)$ and $x=[M_b]/T_t$ this reads $y=K(n-x)$. The (x,y) plot is then a straight line which gives directly nK (value of y for $x=0$) and n (value of x for $y=0$) (Fig. 3a).

Other shapes can also be found. The plot of Fig. 3b corresponds to increased affinity as successive ions bind. This means that the binding is cooperative, and is unambiguous evidence that the structure is being modified upon ion binding. Such a plot has indeed been found for tRNA in certain conditions (e.g. 30 mM NaCl) where the addition of Mn brings about the formation of the more compact tertiary structure which exhibits increased elec-trostatic effects, and where the resulting change in affinity can be observed. Indeed such plots provided early evidence for the existence of tertiary structure.

The plot of Fig. 3c is typical of plots observed for Mn-tRNA titrations. It corresponds to decreased affinity for successively bound ions. This can be explained in two ways. Either one has two types of sites, independent and with differing affinities. The strong sites are first populated, then the weak sites.

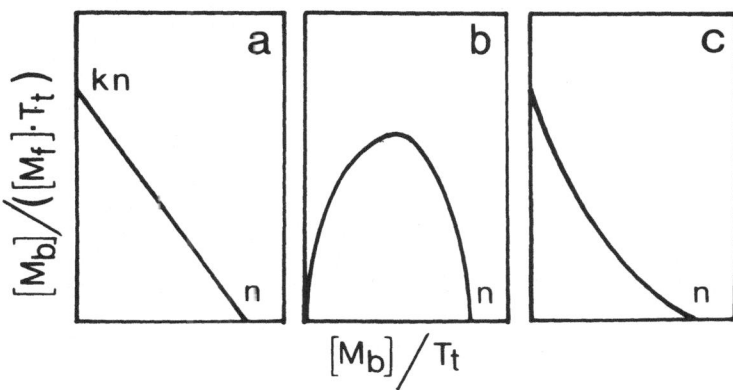

Fig. 3 : Examples of Scatchard binding plots.
a) Independent and identical binding sites.
b) Cooperative binding : the affinity increases as
successive ions bind.
c) Anticooperative binding : the affinity decreases as
successive ions bind. For two types of independent
sites, the plot looks similar.

Or, there is only one type of sites, but they interact in such a
way that the first bound ions decrease the affinity for other
ions (anticooperativity).

As we mentioned above, a common interpretation of the anticooperative plots has been that there are a small number (4 to 6)
of special sites on tRNA, to which Mn (or Mg) ions bind strongly.
Only after these sites are occupied would the ions go to the
other, weak, sites, which are usually assumed to be phosphates.

However, this interpretation of the Scatchard plots is untenable,
for several reasons :

a) the affinity for Mn varies strongly as a function of sodium
concentration, and the number of "strong sites" derived from the
plots of Fig. 4 changes with Na concentration.

b) the analysis in terms of independent "strong" and "weak" sites
ignores the electrostatic effects which occur upon the binding of
ions to polyelectrolytes, and which must by themselves lead to
curved anticooperative plots, even if the sites are all identical.

c) anticooperative plots, quite similar to those for tRNA, are
also observed for DNA, where special sites would not be expected.

The polyelectrolyte theory will be considered later. For the

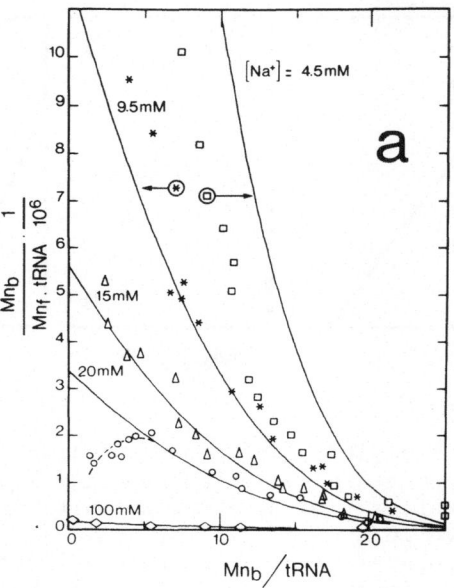

Fig. 4a

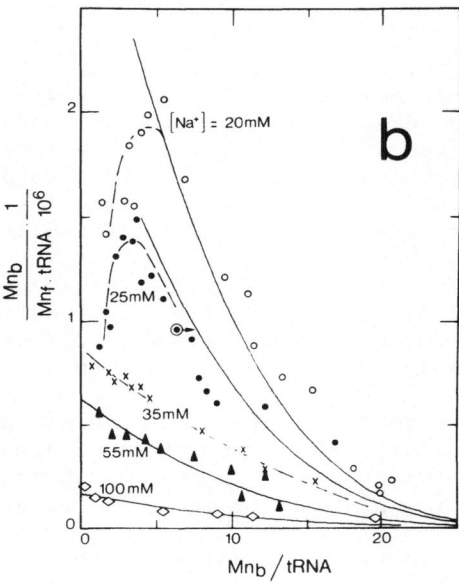

Fig. 4b

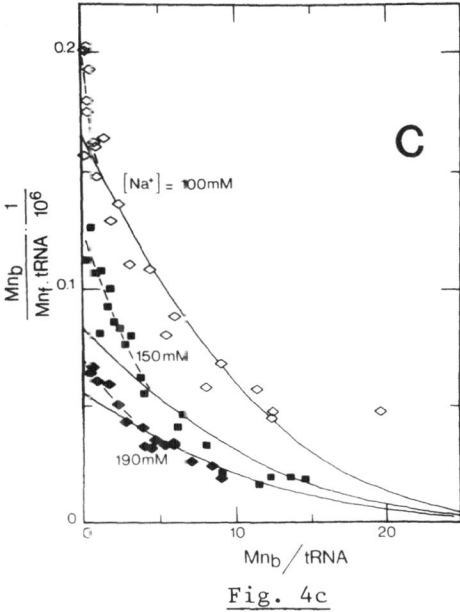

Fig. 4c

<u>Fig. 4</u> : Scatchard plots of Mn^{2+} binding to yeast
tRNA^{Phe} at 25°C for various sodium concentrations. The affinity decreases as the sodium concentration is increased, showing the important role of
electrostatic interactions. Full curves correspond to
the model mentioned in the text. Dashed curves are
traced through the experimental points.

a) The low-salt curves (up to 15 mM Na$^+$) are easily
interpreted by the model : the plot is dominated by
electrostatic interactions, swamping the effect of any
special binding sites or of tRNA folding.

b) In intermediate salt (20 and 25 mM Na$^+$) the
electrostatic interactions are smaller, and the folding
of tRNA gives rise to the cooperative plots.

c) In high salt (50 mM Na$^+$ and more) tRNA is folded
even in the absence of divalent ions and behaves as a
rigid system. The fit of the electrostatic model, (with
the same parameters as in low salt) is fair, except for
the first points, for which the large ordinate and
steep slope (dotted curves) may suggest specific
binding sites.

The curves are tagged by sodium concentration.

present, we conclude that the binding results do not require the existence of special binding sites. It can be shown also that because of the curvature due to polyelectrolyte effects, the plots cannot be very sensitive to special sites. The situation is different in high salt were the electrostatic effects are reduced. Evidence for one special site has been reported in these conditions, at a temperature of 4°C.

In view of the insensitivity of binding measurements to the presence of special sites, we have attempted to investigate their occurence by other methods.

3. EPR OF BOUND MANGANESE

We have seen that the spectrum of the aquoion can be interpreted as due to a $S=5/2$, $g=2$ spin in an environment with cubic (octahedral) symmetry, taking into account the hyperfine interaction to second order. The relaxation rates can be explained qualitatively by the fluctuations of the ZFS (whose average is zero) due to collisions of water molecules which distort the hexaaquoion. The spectra of Mn bound to adenosine monophosphate or adenosine triphosphate are very similar to those of the aquoion, from which they can be derived simply by homogeneous broadening. (This was verified by computer simulations). Since there is reasonable evidence from numerous NMR experiments that Mn is directly bound to phosphate in these systems, this indicates that the ZFS of Mn remains very small even in some cases where the first coordination sphere is asymmetric. The case of Mn in highly concentrated imidazole also gives a similar spectrum. However the coordination to nitrogen rather than oxygen spreads the metal electrons out more and the hyperfine splitting due to the Mn nuclear spin is therefore reduced (by about 6%). The relaxation mechanism of these systems may be the same as for the hexaaquoion, the greater broadening being explained possibly by the longer correlation time of the fluctuations responsible for the instantaneous ZFS.

If one now considers the same systems in an ethylene glycol glass (40% ethylene glycol, 60% water) at 77°K, one is again faced with a collection of spectra derivable from each other simply by a change in linewidth. The spectra can be interpreted as presenting a small, non-zero ZFS whose principal effect is to allow the appearance of "forbidden lines" corresponding to $\Delta m_S=1$, $\Delta m_I=1$. Each point midway between hyperfine lines is thus flanked by two forbidden lines. Their transition probability relative to the $\Delta m_I=0$ lines is proportional to $(D/B)^2$ where D is the zero-field splitting and B the magnetic field. It has been verified that these lines are indeed much weaker at higher fields (70 GHz EPR).

From the intensity of the forbidden lines, one can estimate that

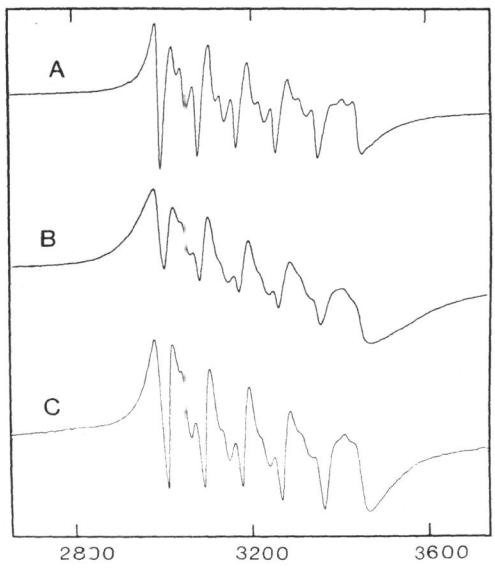

FIELD (GAUSS)

Fig. 5 : X-band EPR spectra of manganese in an
ethylene glycol glass (40% ethylene glycol, 60% water)
at 77°K.
A : hexaaquoion ; [Mn^{2+}]=0.85 mM.
B : Mn-ATP complex ; [ATP]=10 mM, [Mn^{2+}]=0.85 mM.
C : Mn-tRNA complex ; Baker's yeast tRNA^{unfrac}=1 mM,
[Mn^{2+}]=0.5 mM.

the static ZFS is in the range of 100 G.

Before turning to tRNA, let us see what one finds for Mn bound to
proteins. On the time scale of EPR, a protein is static, since
its rotational correlation time is in the order of 10^{-9}-10^{-8}
seconds (τ_R=4$\pi\eta$a^3/3 kT, where η is the viscosity, and $\underline{a}$ the
molecular radius). Hence even in the liquid, a powder spectrum is
found. Depending on the protein, one finds spectra corresponding
to different values of the ZFS, from values smaller than the
hyperfine structure A ($\approx$100 gauss) to values in the range of
1000 gauss giving significant effects on the spectrum.

We can now turn to manganese bound to tRNA, and consider whether
the EPR gives evidence for special, strong binding sites.

To understand what might be expected, we look at the metal sites
on the crystal structure of tRNA^{Phe}. Among these we find the
following :

- One magnesium coordinates directly to phosphate 19 and binds
  through water molecules to N7 of G20, N3 of C60 and to riboses
  of G20 and U59. Manganese can displace magnesium from this
  site. It binds directly to N7 of G20, and its water ligands can
  make hydrogen bonds to N3 of C60, to phosphates 19 and 10, and
  to riboses of G20 and U59.

- Two other magnesiums have two direct metal to phosphate
  coordination. One binds to phosphates 8 and 9, another to
  phosphates 20 and 21.

The last two ions are 1.7 nm apart, and their distances to the
first ion mentioned are 1 and 0.8 nm respectively.

If these sites were the main ones to which manganese goes at
first, there is the possibility that :

a) the EPR of the first manganese bound would show a reduced
hfs because of the coordination to N7 of G20 and/or a large
ZFS.

b) the sites located closest together would be populated, leading
to a large dipolar broadening (the field $\mu/(4\pi\epsilon_0 c^2 r^3)$ of a
manganese ion is 60 G at 1 nm).

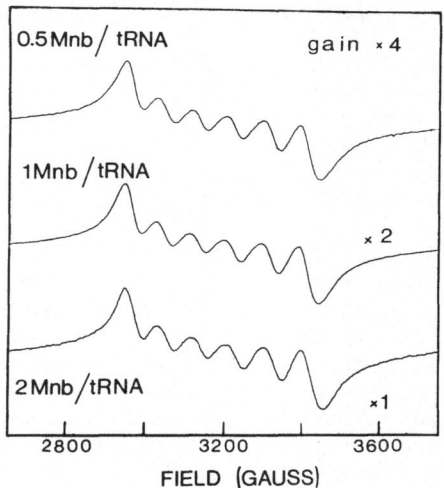

Fig. 6 : X-band EPR spectra of manganese bound to tRNA.
No change in signal shape is observed for the first
ions bound to tRNA, and the intensity is proportional
to the number of ions. T=285 K, Baker's yeast
$tRNA^{unfrac}$=1 mM, Na^+:0.15 M, pH=7.0.

However the observations do not show these effects. As Mn is
added, from 0.5 to 2 per tRNA, no change in signal shape
is observed. The intensity is proportional to the number of
bound ions (Fig. 6), and this excludes the possibility that one
ion in a special site is unseen because of excessive broadening.
The spectrum remains at all times very similar to that observed
for the aquoion or the ion bound to nucleotides.

Further addition of manganese broadens the EPR spectrum, due to
dipole–dipole interaction between the electronic magnetic moments
(Fig. 7). The small value of the broadening, together with the

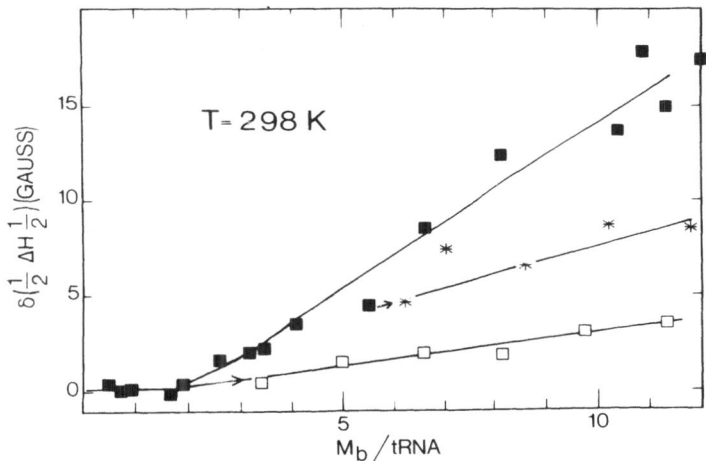

Fig. 7 : Broadening of the EPR spectra of Mn–tRNA due
to the dipole–dipole interaction between bound manganese, versus the number of divalent ions bound.
The broadening observed as manganese is added ($\blacksquare$) is
compatible with a random distribution over the entire
molecule. The effect of added magnesium on the spectrum
observed in the presence of 2 Mn/tRNA ($\square$) and
5.5 Mn/tRNA ($\ast$) indicates however that manganese is
more spread out than a random distribution.
T=298 K, Baker's yeast tRNA^{unfrac}=0.5 mM, Na$^+ \approx$50 mM,
pH=7.0.

proportionality of integrated intensity to number of Mn allows
us to exclude a preferential population of the two neighbouring
sites by any two of the first four bound ions. On the contrary,
the broadening observed is compatible with a distribution of
manganese spread-out over the entire molecule. These results hold
both at room temperature and in low-temperature glass.

An interesting effect occurs when magnesium is added. It is then observed that the broadening increases (Fig. 7). This can be explained if the manganese were originally more spread-out than a random distribution, because of the high charge of the molecule. Upon addition of magnesium electrostatic effects are reduced due to neutralization of phosphate charges, and the manganese, as well as the chemically equivalent Mg, assume a distribution closer to random. Hence their average distance is reduced and the dipolar broadening increased.

In conclusion, the EPR experiments do not favor the view that manganese ions bind preferentially to a few sites on tRNA. The titration studies are interpretable in terms of identical binding sites to a polyelectrolyte (the theory of polyelectrolytes will be discussed later), except possibly for one strong site at 4°C. The bound Mn EPR suggests broadly distributed sites with similar properties. If a site with a large ZFS exists, we exclude the possibility that it be among the first sites to become occupied. It is only the lack of sensitivity of the method which stops us from making stronger statements regarding the equivalence of the binding sites. The EPR provides no evidence whatever of sites with particular properties.

These conclusions hold both in room temperature solutions and in low temperature glass. In the next section we use ^{31}P nmr to investigate the properties of the binding sites.

4. PHOSPHORUS nmr AND RELAXATION OF tRNA

If manganese binds specifically to certain sites, the neighbouring phosphorus nuclei should have their resonance preferentially broadened and relaxed, as manganese is added. From perturbation of the ^{31}P spectrum, one should obtain information on the Mn binding sites.

4.1. Relaxation of phosphorus by manganese

The magnetic coupling between manganese and phosphorus has a dipolar term and a scalar one. The relaxation times for phosphorus bound to manganese are given by

$$1/T_{1M} = DD\left(\frac{6\tau_d}{1 + \omega_p^2\tau_d^2}\right) \tag{2}$$

$$1/T_{2M} = DD\left(4\tau_d + \frac{3\tau_d}{1 + \omega_p^2\tau_d^2}\right) + C\,\tau_{con} \tag{3}$$

where

$$DD = \frac{1}{15} S(S+1) \ g^2\beta^2\gamma^2 \ (4\pi\epsilon_o c^2 r^3)^{-2} \tag{4}$$

$$C = \frac{1}{3} S(S+1) \ A^2/\hbar^2 \tag{5}$$

In these formulae r is the Mn-P distance, ω_p is the phosphorus Larmor frequency, τ_d is the dipolar correlation time which is for our purpose equal to the rotational correlation time τ_R, τ_{con} is the correlation time for the contact interaction and is here equal to the longitudinal spin relaxation time τ_S. The formulae are written for the high field case, that is when $\omega_s\tau_d$ and similar terms are much larger than one. DD and C are respectively the dipolar and contact coefficients. The formulae are in SI units, with $4\pi\epsilon_o c^2 = 10^7$.

In crystals, one finds a Mn-O-P distance r=0.33 nm, and A/h=3.3×10^6 Hz. This gives DD=0.2×10^{14} s^{-1} and C=12.5×10^{14} s^{-1}. Because C is much larger than DD, T_2 will be much shorter than T_1 for typical values of $\tau_d\simeq\tau_R\simeq 10^{-10}$ to 10^{-8} s and τ_s values of 10^{-9} to 10^{-7} s. If on the other hand Mn binds to P indirectly, i.e. through an intermediate water molecule, r is larger and DD is reduced as r^{-6}. The scalar interaction A, being mediated through bonds (in this case through a hydrogen bond Mn-O-H...O-P) would be reduced drastically.

In practice, the relaxation effects of Mn are so great that one works with much less than 1 Mn per binding site. What is then observed is the resonance of the free phosphorus, modified by the intermittent binding of Mn.

Consider a case such as that of AMP where only one type of site is present. The relaxation rates due to Mn are :

$$\frac{1}{T_{ip}} = \frac{f}{T_{iM} + \tau_M} \tag{6}$$

where i=1,2,f is the ratio of Mn per phosphate in the solution, and paramagnetic shifts are neglected ; τ_M is the residence time of Mn on the phosphate.

This leads to two cases :

a) fast exchange $T_{iM} \gg \tau_M \qquad \frac{1}{T_{ip}} = \frac{f}{T_{iM}} \tag{7}$

The description here is that as Mn comes on one phosphate it starts the relaxation process, but leaves before it can finish since $\tau_M \ll T_{iM}$. The relaxation rate is simply T_{iM}^{-1}, reduced by f, the fraction of the time that phosphate is relaxed by Mn.

b) $T_{iM} << \tau_M$. Hence $\dfrac{1}{T_{ip}} = \dfrac{f}{\tau_M}$

This is the situation of slow exchange. Here the relaxation process goes to completion everytime a phosphate encounters Mn. The necessary time is only T_{iM}, whereas the paramagnetic ion stays on for a time τ_M, so a lot of "relaxing power" is wasted.

The slow exchange result may be recast as follows. We observe that τ_M/f is equal to τ_{free}, the mean time between two visits by Mn to a given phosphorus. Thus $T_{ip} = \tau_{free}$. This formulation of slow exchange remains valid when more one type of phosphorus is present, and it will be used for the tRNA problem. One can say that the relaxation corresponds to an uncertainty broadening due to the finite lifetime of the free species.

In the case of phosphorus-Mn interactions, because DD<<C, it often happens that we find a fast exchange situation for T_1 and slow exchange for T_2.

4.2. Phosphorus nmr of tRNA

A phosphorus spectrum of yeast tRNA[Phe] is shown in Fig. 8 showing a number of resolved isolated peaks. The only definite assignment at this time is the peak at lowest field which is assigned by its pH shift to the terminal phosphate. If tRNA is heated to ca. 70°C, all the peaks collapse to a unique position, showing that the shifts are due to the secondary and tertiary structure. The high temperature chemical shift agrees with those found for dinucleotides. The main peak of native tRNA is about 0.5 ppm upfield from this position, and it is ascribed to the phosphates in double-helical configuration. Study of the chemical shifts of other phosphate compounds suggests that the shifts are due to variations in the dihedral $O_5'-P-O_3'$ angle or in the torsional angles around $O_5'-P$ and $P-O_3'$. Addition of magnesium brings changes in the spectrum which are similar to those brought about by reducing the temperature. This suggests that the changes are not due directly to interaction with Mg, but rather to the conformational changes brought about by the divalent ion.

What can we expect on addition of manganese to tRNA? Qualitatively, because of the short range of the magnetic interactions, we hope to learn which phosphates (as described by their nuclear resonance peaks) have affinity for manganese. By analysis of the relaxation we may try to obtain kinetic information, and also to estimate phosphorus to manganese distances. This will help us to define the nature of the binding. By considering the various effects as a function of the number of manganese added, we can also investigate (in fact we shall exclude) the possibility that there exist a small number of strong sites independent of the phosphates.

We observe the following effects :

a) Spectral modifications by manganese (Fig. 8) : the immediate and striking result is that manganese in very small quantities broadens all the resonances.

b) Longitudinal relaxation : manganese relaxes all the resonances.

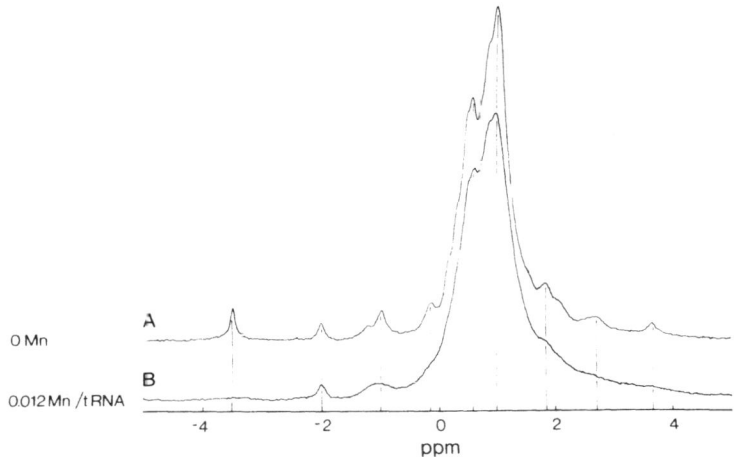

Fig. 8 :
A : Phosphorus NMR spectrum of yeast tRNA^{Phe} (1 mM) in the presence of 10 Mg/tRNA.
B : All the resonances broaden after addition of 0.012 Mn^{2+}/tRNA.
Na$^+$=0.15 M, triethanolamine chloride buffer = 45 mM, pH=7.3, T=303 K. These spectra were measured at 111.7 MHz in the Fourier transform mode, with proton noise decoupling.

These results show right away that tRNA has no highly specific site for Mn, by contrast, say, to the site for zinc in carboxy-peptidase. On the contrary, Mn comes in the vicinity of all the phosphates of the molecules.

In fact this is not surprising. Suppose that there were one site with an affinity 76 times larger than the phosphates ; because there are 76 phosphates, (assumed identical for the purpose of the argument) the manganese ion would still only spend half the time in the strong site, the other half being spent on the phosphates.

We now go onto a more detailed discussion. A first observation is that the spectral broadenings due to Mn increase rapidly with temperature. This shows that for T_2 we are in the slow exchange regime. Hence $T_{2p}=\tau_{free}$, and this represents the time between two successive encounters of a given phosphate with manganese. To determine T_{2p}, a spectrum broadened by Mn is substracted from one without Mn, which has been broadened by computer (Fig. 9). The values of T_{2p}^{-1} for various phosphate peaks are shown in Fig. 10. We observe that :

$$A * \mathcal{L}_{(\Delta)} - B$$

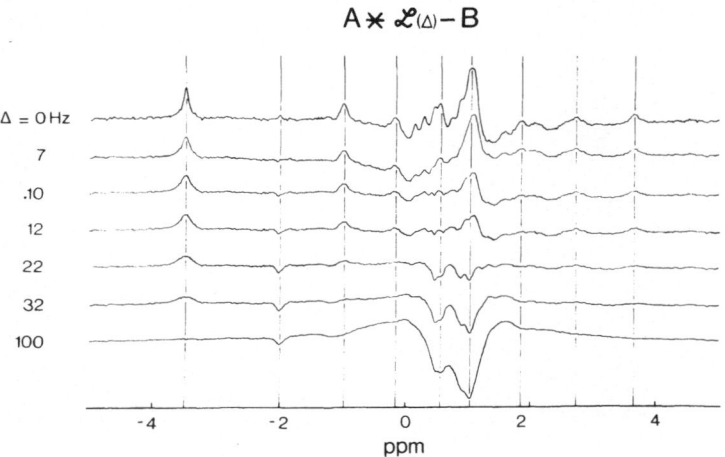

Fig. 9 : Computed difference spectra : The spectrum broadened by Mn (spectrum B in Fig. 8) is substracted from the spectrum without Mn^{2+} (spectrum A in Fig. 8) which has been broadened by convolution with a Lorentzian fonction of linewidth Δ.
A resonance whose broadening by Mn is equal to Δ vanishes in the difference spectrum.

a) the broadenings vary along the spectrum. However the most broadened peaks are only 8 times more broadened than the main cluster of phosphates, the least broadened only 3 times less. (The extent of broadening is shown by the slope of the lines on the plot.)

b) the most broadened peak is the terminal phosphate (left-most peak), followed by the peaks at 4 and 2.8 ppm. The least broadened peak is at -2 ppm.

c) the substraction spectra give no evidence of strongly broadened lines in the central region of the spectra.

Turning now to T_{1p}^{-1}, we find that $T_{1p}^{-1} \ll T_{2p}^{-1}$. If T_{1p} were in slow exchange, we would have $T_{1p}=T_{2p}=\tau_{free}$. Since this is not the case we are in rapid exchange.

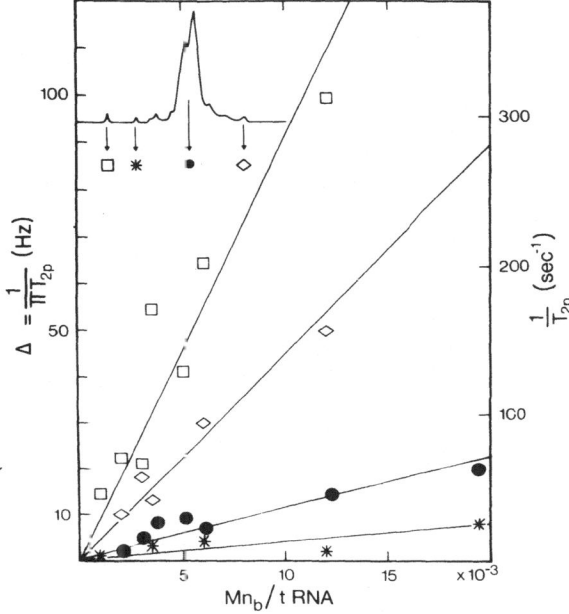

<u>Fig. 10</u> : Measured values of $1/T_{2p}$ for various phosphate
peaks and for the main cluster of phosphates versus the
number of Mn bound per tRNA.
The most broadened peak (terminal phosphate, -3.5 ppm)
is only 7 times more broadened than the main cluster of
phosphates, the least broadened peak only 3 times less.
Yeast tRNA^{Phe}=1 mM, 10 Mg/tRNA, Na$^+$=0.15 M, triethanolamine chloride buffer = 45 mM, pH=7.3, T=303 K.

Fig. 11 shows the values of T_{1p}^{-1} as a function of manganese
concentration. As for T_{2p}^{-1}, the values for different peaks are not
very different, and the plots are linear as a function of Mn/tRNA.

In the fast exchange regime, we have :

$$T_{1p}^{-1} = f \; T_{1M}^{-1} = \frac{\tau_M}{\tau_M + \tau_{free}} \; T_{1M}^{-1} = \frac{\tau_M}{\tau_{free}} \; T_{1M}^{-1} \qquad (8)$$

where f is the fraction of the time that the phosphate harbours
a Mn, and where f and all other quantities may differ depending
on the phosphate considered. The last equality comes from
$\tau_M \ll \tau_{free}$, and is justified by the observation that with very
little manganese, all sites are free most of the time. This
assertion is also supported by the linearity of the plots, as
discussed earlier.

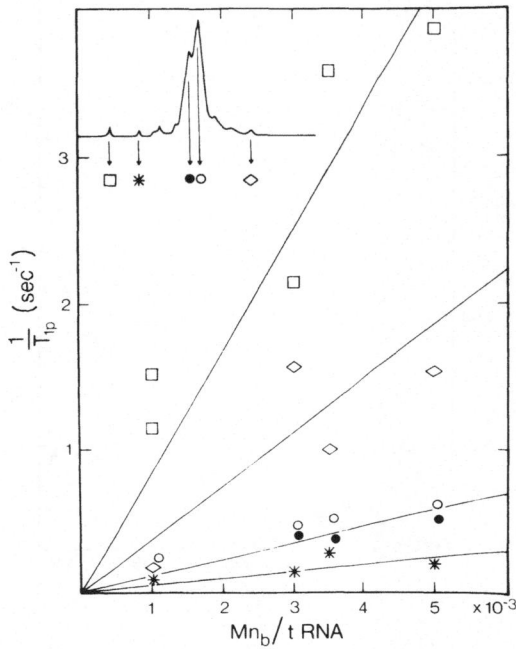

Fig. 11 : Measured values of $1/T_{1p}$ for various phosphate peaks and for the main cluster of phosphates
versus the number of Mn bound per tRNA.
Comparison with Fig. 10 shows that T_{2p}/T_{1p} is about the
same for the different lines. Experimental conditions
same as Fig. 10.

To continue the discussion, we must make assumptions concerning
T_{1M}. Our first assumption is that T_{1M} is the same for all sites.
This is plausible. One notes in particular that differences in
the structure of binding sites could give rise to drastic changes
in T_{1M} and hence T_{1p} because of the r^6 dependence : but no such
thing is seen.

If T_{1M} is equal for all phosphates, T_{1p}^{-1} is proportional to
τ_M/τ_{free}, hence T_{2p}/T_{1p} is proportional to τ_M. Comparison of
figures 10 and 11 shows that T_{2p}/T_{1p} is about the same for all
sites. Therefore all phosphodiester sites have the same τ_M. The
variations of T_{2p} and T_{1p} among the different sites are then
ascribed to τ_{free}.

4.3. The effect of adding magnesium

The experiments just described have been complemented by studies
in different conditions, namely in the presence of Mg, at 6 and
30°C, and for Mn/tRNA ratios smaller or larger than one. The
results are rather complex, and may be described as follows :

a) Mn/tNRA<<1.

In this range, the spectra are satisfactorily resolved. As Mg is
added, the differences in broadening (or T_1) between peaks
decrease. Also the braodening and relaxation of the central
region increase.

b) 0.1<Mn/tRNA<8, T=6°C.

At large Mn/tRNA ratios, it becomes visible that in the absence
of Mg, the main peak has two components, the major one (80%)
being broadened and relaxed six times more than the other.
Broadening and relaxation are proportional to manganese concentration. In the presence of 20 Mg, the effects of manganese are the
same as for the above major component, and the minor component
has nearly vanished.

The interpretation of these observations is too complex to
provide a clear picture. The linearity of Mn effects in the
presence of magnesium, up to 6 Mn/tRNA argues once more against
the presence of a small number of strong binding sites. However
these experiments deal only with the average phosphate, because
of the severe broadening. The possibility that magnesium has
expelled Mn from the presumed strong sites can be refuted because
in the absence of magnesium the Mn dependence is also linear.
Detailed comparison between the two cases is however difficult
because of inhomogeneous broadening in the absence of Mg.

The analysis of the resolved spectra in low Mn must explain
increased broadening and relaxation of the main peak and of some
others as magnesium is added. This would involve, as regards the
main peak, artifacts of measurement caused by differences in the
broadening and relaxation of these phosphates ; and, as regards
resolved peaks, a redistribution of Mn, due to reduced electrostatic interactions upon addition of magnesium.

4.4. The mode of binding

We return to the interpretation of the observed broadening and
relaxation by low manganese concentrations in the presence of
10 Mg/tRNA (Figs. 8-10).

In the conditions of the experiment, we know from the EPR

titration that the vast majority of the ions are bound.
Nevertheless the bound Mn ions are in equilibrium with free Mn
in the immediate vicinity. The free manganese concentration in
the immediate vicinity is designated CIV_2 (where the 2 stands
for divalent), and will be discussed later in the framework of
polyelectrolyte theory. Let K be the equilibrium constant for
binding. Then we have :

$$\tau_M/\tau_{free} = K \times CIV_2$$

The constancy of τ_M for all sites thus suggests that all binding
sites are identical (same K) and that the variation of τ_{free}
reflect only the variation of the CIV_2 in various regions of
tRNA. Such variations are indeed expected because of the variations in the values of the average surface charge density of the
polyelectrolyte. For instance the CIV_2 is expected to be less
towards the anticodon loop than in the central region of the
molecule. At this point one should note that we have neglected
possible competition for the binding sites from other ions such
as Na, or Mg if present. We may mention that if large quantities
of salt are present, the electrostatic effects of the polyelectrolyte charge are reduced. Therefore the CIV should vary
less on the various parts of tRNA, and the relaxation times
should differ less among phosphates. This is indeed observed.

With the above analysis we can now determine the value of τ_M
common to all phosphates. Since nearly all of Mn is bound, we
have :

$$\sum_i f_i = \sum_i \tau_M/(\tau_M + \tau_{free}) = F \qquad (9)$$

where f_i is the proportion of the time that phosphate i harbours
a Mn, and where F is the number of Mn per tRNA. For the common
τ_M, which is much smaller than all τ_{free}, we have :

$$\frac{1}{\tau_M} = F^{-1} \sum_i \frac{1}{\tau_{free}} = F^{-1} \sum_i T_{2p}^{-1} \qquad (10)$$

The value of τ_M varies with temperature. It is approximately
2.5×10^{-6} s at 30°C, a value quite similar to that determined in
early work for ribonucleic acids.

Lastly, we can try to determine the nature of the binding of Mn
to the phosphates, in particular the value of the hyperfine
coupling constant and the Mn to P distance.

At 30°C we know that τ_M is larger than T_{2M} (slow exchange), hence
T_{2M} is shorter than 2.5×10^{-6} s. By Eqs (3) and (5), and assuming
$\tau_S = 10^{-8}$ s, this means that A/h must be larger than 0.6×10^6 Hz, a

value which is 5.6 times smaller than that corresponding to the value for direct Mn—O—P bonding. Since T_{2M}^{-1} is proportional to $A^2\tau_S$, A could even be somewhat smaller, if τ_S is larger than 10^{-8} s, as it could be in the high field (6.3 T) used in the present experiments.

We therefore turn to the study of T_1. For the present purpose, we are not interested in any particular phosphate, but treat them all in common. The experiments are carried out in the presence of magnesium, so that a unique T_1 is observed. We attempt to compute the Mn to P distance r in such conditions. To this effect we need the correlation time. This is estimated by studying the relaxation rate T_{10}^{-1} in the absence of Mn (Fig. 12). At 111 MHz, this is caused mainly by chemical shift anisotropy and varies as $\tau_R/(1 + \omega_P^2\tau_R^2)$.

The relaxation rate shows a maximum at 70°C. At this temperature $\tau_R=\omega_p^{-1}=1.4\times10^{-9}$ s. The values at all temperatures can then be determined. Alternatively, τ_R may be obtained by comparing the relaxation by Mn at two frequencies (Fig. 13).

One can then deduce r from the measured T_{1p} by Eqs (2) and (4). The results are shown in Fig. 14 where it seems that r varies with temperature. This may be due to a change in the mode of binding to phosphates as a function of temperature. However we cannot entirely exclude the interpretation that ions trapped in the native structure become more available to phosphates upon melting. Certainly the dipolar interaction of manganese with the phosphates of melted tRNA, or polyA, or polyU is larger than with the phosphates of native tRNA. Studies on other systems are in progress.

In conclusion, manganese ions semm to show only moderate dis-crimination among the various binding sites of tRNA. These binding sites include all the phosphates and there may be no others. The characteristics of binding are similar (K, τ_M) and it would seem that differences between sites reflect mainly the variations in the concentration of free ions (CIV_2) in the immediate vicinity of the various phosphates. The exact mode of binding is not yet understood. Direct binding occurs when tRNA is melted at high temperature, but the Mn to P distance may be larger in the native structure. It is clear that site-binding occurs, both from the EPR spectra and from the large values of τ_M (microseconds) measured by nmr. The demonstration of site-binding is highly significant for polyelectrolyte theory.

How do the above considerations fit in with the various descrip-tions of specific sites for divalent ions in the literature?

As regards the arguments for 4 to 6 strong sites based on

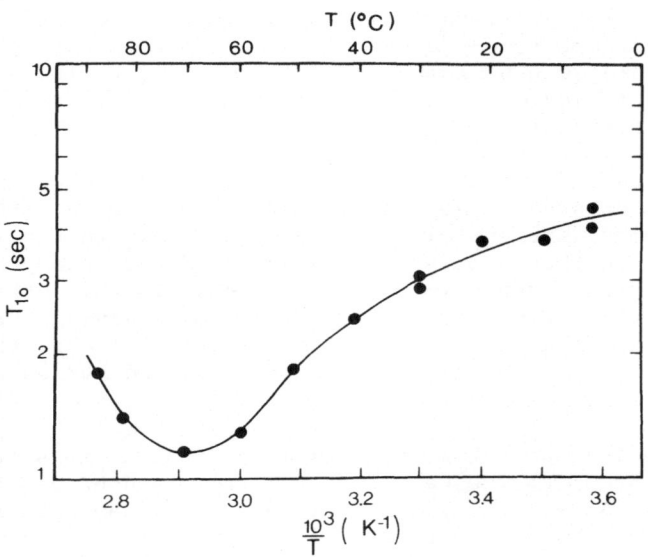

Fig. 12 : Relaxation rate T_{10} measured at 111.7 MHz for
the main cluster of tRNA phosphates versus reciprocal
temperature.
Baker's yeast tRNA^{unfrac}=1.5 mM, 20 Mg^{2+}/tRNA,
Na$^+$=0.15 M, triethanolamine chloride buffer=45 mM, pH=7.3.

titration studies (Scatchard plots), it is clear that they
reflect only a misinterpretation due to the ignorance of polyelectrolyte effects. In fact Scatchard plots for binding to DNA
and tRNA are quite similar.

Concerning spectroscopic evidence (fluorescence, low field
proton nmr) that divalent ions are close to this or that nucleo-
tide, we think that in every case the observations are compatible with a description where the metal does indeed go where
claimed, but also spends part of the time elsewhere on
the molecule. Beyond the demonstration that an ion goes to a
given site, one should now attempt in each case to determine the
proportion of the time spent there.

As for the sites found in crystals (albeit only with extensive
refinement procedures) and assumed to occur in solution, we see
no reason to doubt their existence but consider that their
affinity may be not much higher than that of other sites. It
has been suggested that ions at these sites are necessary for

the native structure of tRNA, but we know from low-field proton nmr that most of the tertiary structure forms even without magnesium. One may wonder on the other hand if full occupation of these sites, necessitating perhaps rather high divalent ion concentrations, is not related to the conformational change which is now thought to occur between 1 and 10 mM Mg.

Other experiments have located rare-earth ions at various places

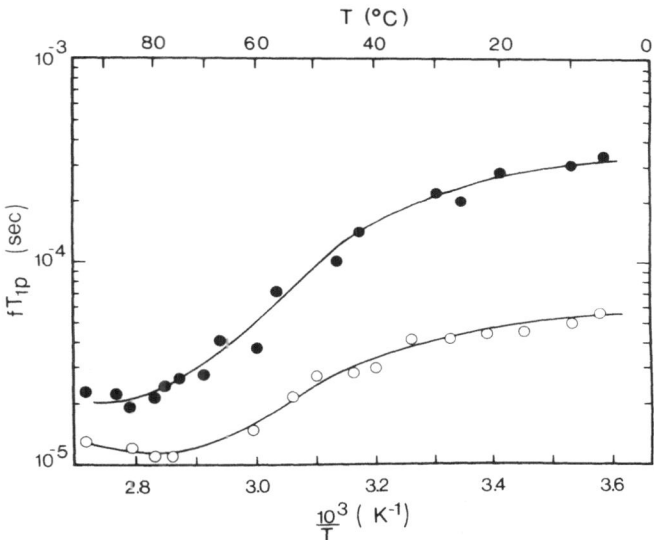

Fig. 13 : Comparison of the measured values of fT_{1p} for the main cluster of tRNA phosphates at ν_a=111.7 MHz (●) and ν_b=40.5 MHz (○). From the relaxation rates measured at 111.7 MHz (T_{1pa}) and 40.5 MHz (T_{1pb}) one calculates the correlation time for interaction between Mn and ^{31}P :

$$\tau_R = \left[\frac{T_{1ap} - T_{1bp}}{T_{1bp}\omega_a^2 - T_{1ap}\omega_b^2} \right]^{1/2}$$

Its values range from 10^{-8}s (T=5°C) to 7×10^{-10} s(T=90°C).

Baker's yeast $tRNA^{unfrac}$=1.5 mM, 10 Mg^{2+}/tRNA, Na^+=0.15 M, triethanolamine chloride buffer=45 mM, pH=7.3, 10^{-2} to 1 Mn^{2+}/tRNA.

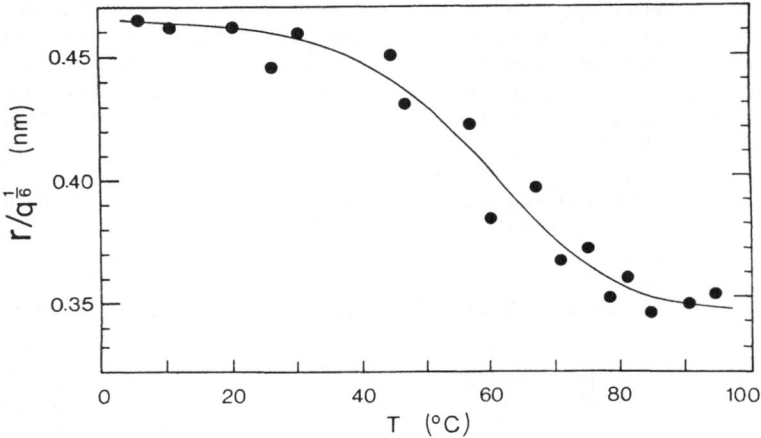

Fig. 14 : Calculated distance between Mn and phosphorus
of tRNA, q being the number of phosphates to which Mn
binds simultaneously.

in tRNA. These ions are trivalent and therefore much more sensitive to the electrostatic potential. Their selectivity for sites
in highly charged areas of the molecule should then indeed be
larger than for divalent ions.

The final word should be of caution. We have not found evidence
of strong sites, and we have shown that evidence from other
sources is weak. But our conclusion is based only on absence of
evidence and must then be tentative.

5. A POLYELECTROLYTE THEORY OF BINDING OF MANGANESE TO tRNA

5.1. Introduction

The reader may have noticed that in the foregoing, mention was
made of an equilibrium constant K for the binding of manganese
to a phosphodiester, but the value of this constant was not
stated. The value of K for the binding of, for instance, Mn to
adenylyl (3'-5') adenosine phosphate, or ApA, has been measured
by EPR. It is in the range of 10 $(M/1)^{-1}$ which makes it seem
at first sight totally irrelevant to the problem of Mn to tRNA
binding, since the apparent affinity which may be read off from
the Scatchard plots of Fig. 4 (albeit loosely, since the
definition of affinity is ambiguous for a curved plot) is in the
range of 10^6 M^{-1} for the first Mn bound in solutions containing
small amounts of sodium chloride. It is precisely this high
apparent affinity, in contrast with the situation found for
solutions of simple salts, which led early workers to the concept
of strong and chemically specific binding sites on tRNA for

divalent ions, in contrast to binding to the phosphate groups.

In the preceding sections, we have seen that on the whole, the various binding sites are rather similar, and we have tried by various methods to evaluate the extent of their differences. But we have not shown how the properties of a "common" site could be understood, or more precisely, how one could account for the observed Scatchard plots and for the binding of manganese to phosphates in terms of the molecular interactions between manganese and an isolated phosphodiester anion. This is what we want to do now : ignoring for the present purpose any differences between binding sites, we want to explain the curvature of the Scatchard plots and their dependence on sodium concentration. Not unexpectedly, this will necessitate an understanding of the distribution of both types of counterions, including how each of them binds and their concentrations in the vicinity of tRNA.

Let us first describe simply what is happening. The negative charge of tRNA attracts counterions in its vicinity, so that their local concentration (CIV) is much higher than in the solution at large. Divalent ions are more attracted than monovalent ions. The equilibrium constant between bound ions and the free ions at their local concentration determines the number of bound ions. The bound ions reduce the net charge of the polyelectrolyte.

As divalent ions are added into the solution, they increase the divalent CIV (CIV_2) and more divalent ions bind. But since the net charge of the polyelectrolyte is being reduced, these effects become weaker and weaker. Thus the apparent affinity decreases as more ions bind because the bound ions reduce the net charge of tRNA. This is why the Scatchard plots are curved (anti-cooperative).

Such effects have been known for a long time. For instance the pH titration of proteins exhibits pK shifts of the titrating groups due to the net charge of the protein. The theory in such a case replaces the protein by a weakly charged sphere of radius R. Typically, the problem is simplified by computing the electrostatic potential under the assumption that the sphere has a uniform superficial distribution of charge whose integral value is the same as that of the protein. Because the superficial charge density is small, one may use the Debye–Hückel approximation to compute the potential ψ. The concentration of protons is then computed as :

$$\rho(r) = \rho(\infty) \exp\phi(r)$$

where $\phi(r) = e\psi(r)/kT$.

Lastly the number of bound protons is determined by the mass-

action law applied between the titrating groups and the free ions
at the surface of the sphere, whose concentration (CIV) is equal
to $\rho(R)$.

It is clear that this type of treatment is very approximate. For
instance it treats the solvent as a continuous fluid and neglects
the reduction of its dielectric constant in the high electric
field of the polyelectrolyte, and it also uses a crude description
of the distribution of charge on the polyelectrolyte. Nevertheless, one can agree that the understanding of this type of model
is an indispensable step in the formulation of polyelectrolyte
theories. And while on the one hand, various improvements of such
models have been attempted, on the other hand it is only recently
that a satisfactory understanding and development of the
simplified models has been achieved. We shall consider polyelectrolyte of simple shapes (cylinder, sphere, plane) coated
with superficial charge at uniform density and we shall first
compute the corresponding counterion distribution.

In a second step the existence of binding sites is introduced and
the counterion distribution is used to compute the binding to the
polyelectrolyte. This treatment is carried out in a self-
consistent way so as to take into account the reduction in
surface charge density by the bound ions.

5.2. The Poisson–Boltzmann equations

The equations of the problem are :

a) the Poisson equation for the potential :

$$\Delta\psi(r) = -\rho(r)/\varepsilon_o D \tag{11}$$

where ρ is the charge density, ε_o the dielectric permittivity
(SI units are used) and D the dielectric constant of the solvent.

b) the Boltzmann expression for ρ, as a sum of contributions of
all ions i, whose charges and concentrations are q_i and n_i :

$$\rho(r) = \sum_i n_i \exp-(q_i\psi(r)/kT) \tag{12}$$

c) the boundary conditions. On the surface of the polyelectrolyte

$$\vec{\nabla\psi} = -\vec{k}\sigma/\varepsilon_o D \tag{13}$$

where σ is the superficial charge density and k is the outwards
normal to the surface. At infinity, $\psi=0$.

Let us first consider a simple case, that of a sphere of radius R
and charge Q in a medium containing monovalent salt only, in a
case where the charge Q is small. We have $n_+=n_-=n$ where n is the

salt concentration, and $q_\pm = \mp e$ where e is the electronic charge. Hence :

$$\rho = - 2ne \sinh(e\psi/kT) \tag{14}$$

Because Q is small, the hyperbolic sine is linearized, leading to

$$\Delta\psi = \frac{2ne^2}{\varepsilon_o D} \frac{\psi}{kT} = \psi/\lambda^2 \tag{15}$$

Here we have introduced the Debye length λ, whose general expression is given by

$$\lambda^{-2} = 4\pi l_B \sum_i n_i z_i^2 \tag{16}$$

where $|e|z_i$ is the ionic charge, and where the useful "Bjerrum length" l_B is :

$$l_B = e^2/4\pi\varepsilon_o DkT \tag{17}$$

l_B is the distance at which the potential energy of two electronic charges is equal to kT. In water, l_B=0.72 nm.

The well-known solution of the present case is

$$\psi(r) = \frac{Q}{1 + Ka} \frac{e^{-(r-R)/\lambda}}{4\pi\varepsilon_o r} \tag{18}$$

It shows that the effect of the charged sphere is screened at a distance comparable to λ. This is due to the accumulation of counterions near the sphere and to the repulsion of the coions according to the linear formulation of Eq. (15).

What is the range of applicability of the linearized Debye-Hückel theory? Roughly speaking, the potential must be small enough in regions where ions are liable to be found. If the salt concentration is low enough, there will be few ions in the vicinity of the sphere (which may itself be simply one of the ions) where the potential is largest. We thus reach the conclusion that the Debye-Hückel law (Eq. (18)) is valid in the limit of infinite dilution, whatever the value of Q. This conclusion is generally studied through one of its consequences : the activity coefficients of ions tend to unity in the limit of zero concentration. This is well verified for simple ions.

With this background, we can understand the surprise of the early workers in the field of polyelectrolytes, when they found that, dilute as you wish, the activity coefficient of the counterions of polyelectrolytes remained smaller than unity! The faith in the Debye-Hückel result was so great that for some time these results were explained by ad-hoc arguments. One would reason that the strong attraction and/or multiple charged sites of the polyelectrolyte would result in some kind of chelation or entrapment of the counterions. This would reduce the number of free counterions, thus resulting in their low activity.

However it was soon found that the assumed chelation did not correlate with the chemical reactivity of the groups on a poly-electrolyte and this made the explanation untenable. It appears that site-binding has ever since carried with some workers the connotation of being the old-fashioned and erroneous way to account for the activity coefficient. This impression is often conforted by the observation that the monomer units of which polyelectrolytes are made do not normally bind to their counter-ions. However this observation is irrelevant since it corresponds to solutions of salts at usual concentrations, whereas the proper comparison is with solutions containing counterions at the much higher CIV concentration.

Why then is the Debye-Hückel limiting law invalid for polyelectro-lytes? The reason is simply that as the charge increases, the concentrations that can be deemed "low" diminish tremendously, and become unattainable. Therefore, the linear approximation is in practice useless for polyelectrolytes. One must use the full Poisson-Boltzmann treatment.

5.3. Poisson-Boltzmann theory of polyelectrolytes

In the early 1950s, the group of Katchalsky and his coworkers treated a cylindrical model of a polyelectrolyte by the Poisson-Boltzmann formalism. They found that the counterions were strongly accumulated close to the polyelectrolyte, and they could explain the reduction of the activity coefficients. The model was a great and justified success and is the basis of much of the later work.

In Fig. 15, we show the Poisson-Boltzmann solution for a cylindrical polyelectrolyte in finite salt. The polyelectrolyte is screened by the accumulated counterions whose integrated charge reaches large values at distances much smaller than the Debye length λ. The counterion concentration close to the poly-electrolyte is very large, and decreases abruptly with distance.

Fig. 15 is taken from a study in which we examined the solutions of the Poisson-Boltzmann equations for polyelectrolytes, in order to see what could be predicted for an irregular shape like that of tRNA. That the shape might be crucial for the counterion

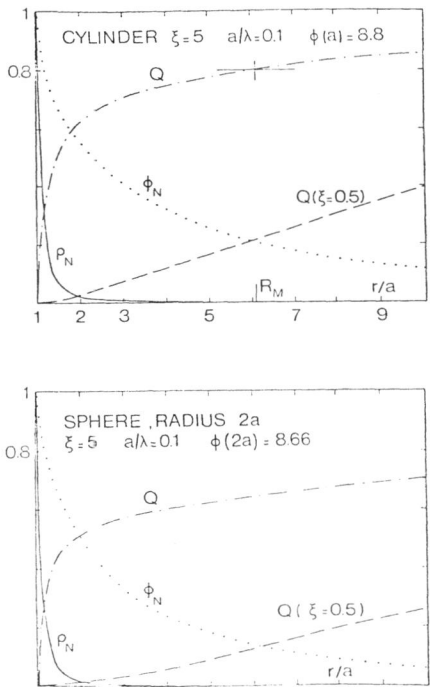

Fig. 15 :
Top : Poisson-Boltzmann solution for a cylinder with
$\bar{\xi}=5$, $a/\lambda=0.1$. If we set $\underline{a}=1$ nm this corresponds to "DNA"
in 1 mM salt. The condensation radius R_M englobes an
integrated charge $(-\cdot-)$ of $1-\xi^{-1}$ per monomer. The full
curve, ρ_N, is the counterion concentration, normalized
to one at the cylinder surface. The normalized potential
$\phi_N(\cdots)$ decays rapidly at first. The value $\phi=1$ is
reached at $\underline{r}=10\underline{a}=\lambda$. The value of the CIV is 6.1 M/l.
Counterion accumulation is obvious by comparison with
the integrated charge for $\xi=0.5$ $(--)$.
Bottom : Poisson-Boltzmann solution for a sphere having
the same superficial charge density and the same radius
of curvature $R_C=2a$ as the cylinder. The value of the
CIV is 5.3.

distribution was suggested by earlier analysis of the equations,
which indicated that at low concentrations of added salt, the
distribution would differ radically for the sphere and the
cylinder : in the zero-salt limit, counterions would accumulate
around a cylinder whose linear charge density was larger than
e/l_B, whereas there was no accumulation around a charged sphere.

With tRNA, which looks as much like a distorted sphere as like a stubby cylinder, these observations had us stranded, and we therefore computed solutions of the Poisson-Boltzmann system for various shapes, at different salt concentrations.

Our analysis yielded the following results :

a) In the low-salt limit, cylinder and sphere indeed behave differently.

b) In higher salt however, the counterion distribution near the polyelectrolyte becomes practically independent of its shape, whether it be cylindrical, spherical or planar. The last case is particularly interesting for its application to micelles and membranes, and also because for the plane we have an analytical solution of the Poisson-Boltzmann equations.

c) With the charge densities commonly encountered, the high-salt condition is satisfied for $R/\lambda \gtrsim 0.2$, where R is the radius of curvature of the polyelectrolyte. For biological polyelectrolytes (e.g. nucleic acids R=2×cylinder radius$\sim$2 nm) the high salt regime already obtains for $\lambda \lesssim 10$ nm, or for monovalent salt concentrations larger than 1 mM. In other words, practically all biological polyelectrolyte systems and experiments fall in the high-salt, shape-independent case. This immediately explains the similarity of the binding curves to tRNA and to (cylindrical) DNA.

5.4. The counterion distribution in the high salt case

In the high-salt case the results for the plane are representative of those for other shapes. We shall now state the most important results as they are found in the analytical treatment of the plane. The accompanying figures and tables give the results of numerical computations for cylinders and spheres.

We consider a highly charged plane, defined in monovalent salt by the condition :

$$|\sigma/e| > (2\pi l_B \lambda)^{-1}$$

which is amply satisfied for usual surface densities and salt concentrations. We then have the following results :

a) The counterion concentration in the immediate vicinity of the plane, or CIV, is given by :

$$CIV = 2\pi(\sigma/e)^2 l_B \tag{19}$$

It is independent of ionic strength and of salt valency, and proportional to the square of the surface charge density σ. We have made a number of computations for $|\sigma/e| = 1.1$ nm^{-2} (This is close to the DNA value, 0.95). For a plane with this surface density, the CIV is 9.2 M, and this enormous value is maintained even if the salt concentration at large tends to zero.

In Table 1 we give the CIV values for spheres and cylinders having the same surface density. The cylinder has a radius of 1 nm like DNA, and with the given surface density, its <u>linear</u>

[salt], (M/l)		10^{-1}	10^{-3}	10^{-5}	10^{-7}
$\xi=2$	cylinder	1.0	0.53	0.44	0.36
	sphere plane : 1.5	1.0	0.28	0.023	3×10^{-4}
$\xi=5$	cylinder	6.8	6.1	5.9	5.8
	sphere plane : 9.2	6.6	5.3	4.1	2.3

Table 1 : Comparison of the CIV, in moles per liter, for sphere and cylinder, for different values of ξ salt concentrations. The radius of curvature is 2 nm (i.e. cylinder radius = 1 nm). The CIVs for the plane are independent of salt concentration. The concentration dependence is also weak for other shapes, above 1 mM salt.

charge density is $\xi=5$, in units of e/l_B. (The value for DNA is $\xi=4.3$). The sphere has the same radius of curvature as the cylinder and the same surface density. The CIV varies very little with salt concentration and its value approximately 6 M/l, is close to that of the plane.

In the top two lines, the CIV is listed for a cylinder and sphere of the same dimensions, but with σ reduced by 2/5, giving $\xi=2$ for the cylinder. The CIV value differs a bit more from that of the plane and is slightly more sensitive to salt. For concentrations below 1 mM, (low salt) the CIV of the sphere falls. One may note that an empirical but commonly used definition of strongly charged polyelectrolyte cylinders is $\xi>1$.

Expressing σ for a cylinder as a function of ξ and the radius a, we have $2\pi\frac{\sigma}{e}al_B=\xi$. Substitution in (19) gives for the CIV of the plane having the same σ :

$$CIV = \xi^2/2\pi a l_B \qquad\qquad (20)$$

An empirical expression for the CIV of the cylinder is :

$$CIV_{cylinder} = (\xi^2/2\pi a l_B) \exp(-2.3/\xi) \qquad\qquad (21)$$

b) As one goes away from the plane, the counterion concentration decays linearly at first. It falls to a quarter of the CIV at a distance T of $|2\pi l_B\sigma/e|^{-1}$. As above, this can be transformed to

$$T = R_c/2\xi = a/\xi \qquad\qquad (22)$$

and these values are also good for sphere and cylinder (Fig. 13).

c) For a mixture of mono and divalent ions, the CIV is the same as in monovalent salt, if the Debye length is kept constant.

This relation is also obeyed approximately by cylinders and spheres. Together with the relations

$$CIV_2/n_2 = \exp(2e\psi/kT) = (CIV_1/n_1)^2 \qquad\qquad (23)$$

it enables us to compute easily the CIVs of the two counterions whose concentrations are n_1 and n_2 respectively. Results obtained from the numerical solution of the Poisson-Boltzmann equation are shown in Fig. 16.

5.5. Application to the binding studies

With the help of the above results, we can try to evaluate the association of ions to a polyelectrolyte.

For a description of the problem, imagine first the polyelectrolyte with no site-bound ions. Its surface density is σ and there is around it the corresponding CIV. Allow the system to evolve towards equilibrium. As some ions from the solution become site-bound, σ is reduced, this reduces the CIV, and the process goes on till both the mass-action and the σ versus CIV relationship are simultaneously satisfied.

The problem is expressed by a system of three equations :

1) the mass-action law expressing the equilibrium between the ions in the solution, whose concentration near the polyelectrolyte is the CIV, and the site-bound ions. The relevant equilibrium constant K is the same that would apply between the counterion and a monomer component of the polyelectrolyte (i.e. $K=10(M/1)^{-1}$

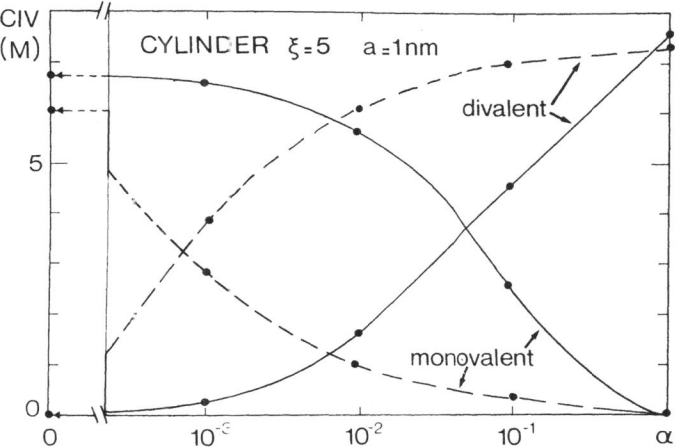

Fig. 16 : The CIV of monovalent and divalent counterions as a function of their relative concentrations at
large.
————————— $\lambda = 0.96$ nm
——————— $\lambda = 9.6$ nm
$\alpha = 2n_2 / (2n_2 + n_1)$

for binding of Mn to ApA).

2) the expression of the surface charge density σ' as a function
of the number of site-bound ions.

3) the relation between the CIV and σ'.

a) Site-binding in monovalent salt

Let P_t be the number of charges. They are carried by the charged
groups which are also presumably the binding sites. Let S_1 be the
number of site-bound ions.

The mass action law is :

$$\frac{S_1}{(CIV)(P_t - S_1)} = K \tag{24}$$

The charge density σ' is related to the original σ by :

$$\sigma' = \left(1 - \frac{S_1}{P_t}\right)\sigma = x\sigma \tag{25}$$

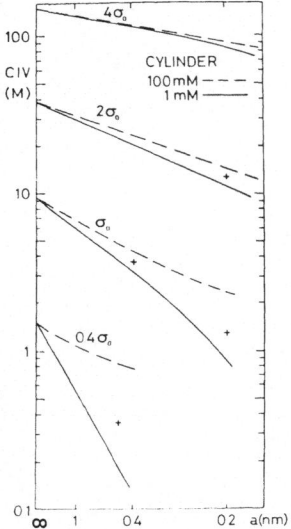 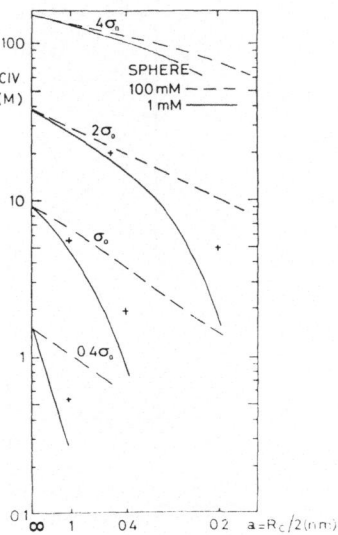

Fig. 17 : The CIV for cylinders and spheres of given surface charge density σ as a function of $1/\underline{a}=2/R_C$. Knowing that $(\sigma_0, \underline{a}=1$ nm) corresponds to $\xi=5$ and that ξ is proportional to $R_C\sigma$, ξ can be determined for any point on the curves. The CIV as a function of σ for given R_C can be obtained by interpolation : It is useful for computing site-binding. The crosses correspond to 10 mM salt.

where x is the reduction factor of the surface charge density. The CIV is a function of the surface charge density, and is supposed to be known. It can be obtained from the numerical solution of the Poisson-Boltzmann equation, for instance by interpolation in Fig. 17.

Here we shall assume for simplicity that the CIV is proportional to σ'^2, the result for the highly charged plane.

Hence :

$$CIV = kx^2 \qquad (26)$$

Equations (24) to (26) give :

$$\frac{1 - x}{kx^3} = K$$

$$kK\, x^3 + x - 1 = 0 \qquad (27)$$

This equation has only one real positive root. As a numerical example we choose $k \approx 5$ (M/1) which is, according to Table 1 an approximate value for our "model DNA". Taking $K=0.1$ $(M/1)^{-1}$, a reasonable guess for Na^+ binding to phosphodiesters, we find for the solution of (27) :

$$x = 0.75$$

This shows that cations are bound to 25% of the sites. By (26) the CIV is then reduced from 6 to 3.4 M/1. We stress that this is simply an example, and is not claimed to be accurate for poly-nucleotides.

In the case of tRNA its irregular structure causes the density of charge to vary along the surface. Hence the CIV and consequently the binding may be larger in some areas (for instance the center part of the molecule) than in others such as the anticodon loop or the aminoacid stem.

b) Site-binding in mixed salts

One must now derive from the Poisson-Boltzmann solution the CIVs of the mono- and divalent cations as described above. One then solves the mass-action equations for competitive binding of the two cations, together with the condition

$$\sigma' = \sigma\left(1 - \frac{S_1 + 2S_2}{P_t}\right) \tag{28}$$

which replaces (25). One sees from Fig. 16 that the divalent CIV can be very sensitive to the monovalent cation concentration at large, and this will influence the site-binding of the divalent ions.

The lines through the experimental points of Fig. 4 are the result of a preliminary calculation of this type, using however a Debye-Hückel solution of the electrostatic problem. The fit is good at all sodium concentrations but this is not too significant. The value of these results is that they show that the data can be fitted by binding to a collection of identical sites, and that the large apparent binding constants observed can be explained on the basis of intrinsic equilibrium constants as low as a few M^{-1}. This is due to the strong accumulation of counterions, which can be expressed as the ratio of the CIV to the concentration at large. The mass-action formalism of binding to polyelectrolytes has been used by several other workers in ways analogous to ours. However the method is not always understood or accepted. A significant example is provided by a recent measurement of the quadrupolar relaxation of Na^{23} in the presence of DNA.[1]

(1) C.F. Anderson, M.T. Record Jr. and P.A. Hart, Biophys. Chem. 7, 301-316 (1978).

This study showed that the relaxation effect of DNA
was well described by a model where the number r of Na relaxed
per phosphate was independent of the sodium concentration at
large. In the interpretation, it was argued that this was
incompatible with a mass-action model.

The present discussion shows that this is a misconception. In
fact, the constancy of the CIV versus salt concentration implies
directly the constancy of the number of bound sodium ions, within
a mass-action model. If these are the relaxed ions, as seems
probable on both theoretical and experimental grounds (such as
relaxation of Na by ApA), the number r of relaxed ions will be
independent of the salt concentration at large, as observed.

5.6. Poisson-Boltzmann versus condensation theory

The analysis given above is rather straightforward. Being based
on work nearly 30 years old, it could have been done much earlier
and should by now have become common knowledge. This however has
not been the case. A possible reason is that the early work,
resting heavily on numerical computations, may have seemed
difficult to apply, despite some efforts to describe in simple
terms the essential properties of the solutions. Nevertheless, it
is only recently that, for example, analytical formulas for
activity coefficients have been provided on the basis of some
general features of the solutions of the Poisson-Boltzmann
equations. It must therefore have come as a relief to many
workers in the field of polyelectrolytes and related fields when
there appeared in the mid-sixties a simple description of the
counterion distribution, from which various properties of polyelectrolytes could be described analytically. This was condensation theory, which states simply that :

a) cylinders with $\xi > 1$ would "condense" in their vicinity $1-\xi^{-1}$
counterions per unit charge, so as to make their net ξ become
unity.

b) the non-condensed ions could be treated in the Debye-Hückel
approximation.

Condensation theory claims to find strong support from experiment,
and it is widely considered as a sound theory of polyelectrolytes.
In our work on tRNA, our first impulse was therefore to try to
explain our results on the basis of condensation theory.

However the tRNA results immediately raised questions that
condensation theory could not solve. In condensation theory,
condensation is a consequence of cylindrical shape, and no
prediction at all is given for other shapes. This makes it
difficult to understand the observed similarity in the binding

to tRNA and to cylindrical DNA. Moreover the theory ignores, negates, or minimizes site-binding, and never provides for its analysis.

An amazing feature of condensation theory, which seems to have worried very few people, is that it is not a complete theory in that it does not provide a complete description of its model. In particular, it is unable to derive the counterion distribution. The closest it comes to a description of the condensed counterions is to assume a uniform concentration in the condensation region. The volume V_p of this region itself has only recently been derived. The value found is independent of salt concentration, in strong contradiction to Poisson-Boltzmann theory ($V_p \to \infty$ in zero-salt). Furthermore the demonstration is incorrect since it is based on the unstated and unproven assumption that V_p is finite in zero-salt. The consequences of the demonstration are also in contradiction with Poisson-Boltzmann, and are in some cases patently incorrect. For instance the value found for the concentration of condensed ions is independent of the cylinder radius a, for given ξ. This would imply that the concentration close to the cylinder would remain constant as $a \to \infty$, which means, for given ξ, as $\sigma \to 0$!

Besides the simplicity of the formalism, there may be another explanation for the popularity of condensation theory. Until recently, much of the work on polyelectrolytes had to do with colligative properties, which are quite insensitive to the properties of the counterion distribution close to the polyelectrolyte (counterion concentration, binding). They can therefore be derived approximately even in theories which fail to provide a correct picture of the counterion distribution. In fact, the postulates of condensation theory are equivalent to a distorted version of some very partial results of Poisson-Boltzmann results, and this explains both the limited successes and the failures of condensation theory.

In order to discriminate between polyelectrolyte theories, experiments sensitive to properties of the immediate neighborhood of the polyelectrolyte are needed. The binding of manganese to tRNA, followed by EPR, or the quadrupolar relaxation of Na by DNA provide data of this sort. More generally the development of spectroscopic studies of polyelectrolytes submits the theories to increasingly serious tests. It is our opinion that, as regards both formal concepts and confrontation with the data, the comparison between Poisson-Boltzmann and condensation theory is strongly in favor of the former.

Acknowledgements

The contribution of Professor G. Weisbuch to the work on poly-electrolytes is acknowledged.

Bibliography

A. "Transfer RNA", edited by S. Altman, MIT Press (1978), is a general reference on tRNA. It contains a chapter by S.H. Kim on the crystal structure of yeast tRNA^{Phe}.

B. On binding of divalent ions to tRNA :

   - "Cooperative Binding of MnII in tRNA", D. Danchin and M. Guéron, Eur. J. Biochem. <u>16</u>, 532-536 (1970).

   - "Electrostatic Effects in Divalent Ion Binding to tRNA", J.L. Leroy and M. Guéron, Biopolymers <u>16</u>, 2429-2446 (1977).

   - "Interactions entre cations divalents et acides nucléiques de transfert". Thèse de doctorat d'état, J.L. Leroy, 1977.

   - "Equilibrium Binding of Magnesium (II) by E. coli tRNA^{fMet}", A. Stein and D.M. Crothers, Biochemistry <u>15</u>, 157-167 (1976).

   - "A Nuclear Magnetic Resonance Study of Secondary and Tertiary Structure in Yeast tRNA^{Phe}", G.T. Robillard, C.E. Tarr, F. Vosman and B.R. Reid, Biochemistry <u>16</u>, 5261-5273 (1977).

C. On nuclear magnetic relaxation by paramagnetic ions in solutions of nucleic acids :

   - The series of articles by R.G. Shulman and collaborators, J. Chem. Phys. <u>43</u>, 3116-3143 ; 3750-3756 (1965).

   - The study of ATP by M. Cohn and T.R. Hugues Jr., J. Biol. Chem. <u>237</u>, 176 (1962).

   - "NMR in Biochemistry" by R.A. Dwek, Clarendon Press, 1973.

D. On EPR of manganese :

   -"Electron paramagnetic relaxation and EPR lineshape of manganous ion complexes in aqueous solutions", G.H. Reed, J.S. Leigh Jr., and J.E. Pearson, J. Chem. Phys. <u>55</u>, 3311-3316 (1971).

E. On polyelectrolytes :

- "Chemical Physics of Ionic Solutions", ed. B.E. Conway and
  R.G. Barradas (Wiley, 1966). See in particular the articles
  by Katchalsky et al., and by Gross and Strauss.

- "Polyelectrolyte Theory. I. Counterion Accumulation, Site-
  Binding and their Insensitivity to Polyelectrolyte Shape in
  Solutions Containing Finite Salt Concentrations", M. Guéron
  and G. Weisbuch, submitted to Biopolymers.

- "Polyelectrolyte Theory. II. Activity Coefficients in Poisson-
  Boltzmann and in Condensation Theory. The polarizability of
  the Counterion Sheath", M. Guéron and G. Weisbuch, J. Phys.
  Chem. in the press.

F. On condensation theory, one may consult :

- "Polyelectrolytes" by F. Oosawa (Dekker, New York, 1971),
  and the review article by G. Manning in Quart. Rev. Biophys.
  11, 179-246 (1978).

PHYSICAL STUDIES OF AZURIN AND SOME METAL REPLACED DERIVATIVES

David R. McMillin and David L. Tennent

Department of Chemistry, Purdue University
West Lafayette, Indiana 57907

The insight into the nature of the blue copper site in azurin
that has been obtained using physical methods is briefly reviewed.
The additional insight that has come from studies of the cobalt(II),
nickel(II) and manganese(II) derivatives of the protein is then
sketched. It is shown that in combination these results have
yielded a resonable description of the electronic and geometric
structure of the metal binding site. In accord with a recent
crystallographic report, it is concluded that the site involves a
cysteine sulfur, a methionine sulfur, and two imidazole nitrogens
disposed in a pseudotetrahedral fashion.

1.   INTRODUCTION

Metalloproteins form a class of coordination compounds, albeit ones with structurally very complex 'ligands'. Including
potential donor groups from the various side chains, e.g. sulfur
from cysteine, cystine or methionine, oxygen from glutamic acid,
aspartic acid or tyrosine, nitrogen from histidine or lysine, as
well as the oxygens and nitrogens derived from the α-carboxyl and
amino functions of each residue, the peptide can present a wide
variety of ligating groups to a metal center. Of course exogenous
donors, e.g. oxygen from water, can also be involved. As variable
as the donor set can be, the peptide structure allows perhaps even
more flexibility in adapting the geometrical arrangement of the
donor atoms, since both the angular *and* the radial distributions
of the donors about the metal can be controlled. The intricate
role that the peptide moiety can play in structuring the metal
binding site in this regard can be appreciated when one considers
that the latter represents only a fraction of the total volume
occupied by the protein. Although crystallographic methods are

369

*I. Bertini and R.S. Drago (eds.), ESR and NMR of Paramagnetic Species in Biological and
Related Systems, 369–379.*

capable of defining protein structures in resonable detail, growing
the requisite single crystals is often difficult.

Consequently, spectral and chemical methods are often used
to study proteins. However, in come cases, e.g. zinc proteins,
there are few spectral properties to examine. In others, e.g.
the blue copper proteins, the spectra of the protein centers are
unique and not easily related to those of structurally characterized
small molecule analogues. As shown below for the case of the blue
copper protein azurin (*Pseudomonas aeruginosa*), in such systems
metal replacement studies in conjunction with physical studies
can facilitate analysis.

2.   SPECTRAL STUDIES OF AZURIN AND ITS METAL REPLACED DERIVATIVES

2.1   Studies of Native Azurin

Azurin is a small bacterial protein (M.W. ~16,000 daltons)
that contains a single type 1 or blue copper site. In reference
1 one finds a recent review of the experimental work carried out
on this and related copper proteins. Here we briefly sketch the
ideas that have emerged from ESR, optical and ^{1}H-NMR studies of
the protein.

The ESR spectrum of a frozen solution of native protein is
presented in Figure 1. The striking feature of the spectrum is
the narrow hyperfine splitting by copper in the g_{11} region. The
coupling constant A_{11} is about 0.006 cm^{-1} and is considerably
smaller (in absolute value) than those found in typical small
molecule systems. (0.012 cm^{-1} < $|A_{11}|$ < 0.022 cm^{-1}) Unfortunately,
the intrepetation of A_{11} is made difficult by the fact that there
are several effects contributing to the hyperfine interaction.

The coupling interaction involves the Fermi contact, the
through space nuclear spin-electron spin dipolar and the nuclear
spin-electron orbit interactions. In a typical complex of Cu(II)
with tetragonal symmetry a d(x^2-y^2) ground state obtains and the
combination of the Fermi contact (via polarization of filled inner
s-orbitals) and the nuclear spin-electron spin dipolar interactions
combine to give a large (negative) A_{11} value. To explain the low
A_{11} value of the blue copper site, as well as the anomalously
intense visible absorption bands, *vide infra*, it has been suggested
that site involves distorted tetrahedral geometry wherein d-p
orbital mixing is permitted.(2) The dipolar contribution from a
spin in a 4p(z) orbital would give a *positive* contribution to A_{11}
and would canal out part of the coupling due to the Fermi contact
interaction involving inner shells. More favorable on energy
grounds, the admixture of 4s orbital character into the ground
state would have a similar effect because unpaired spin density

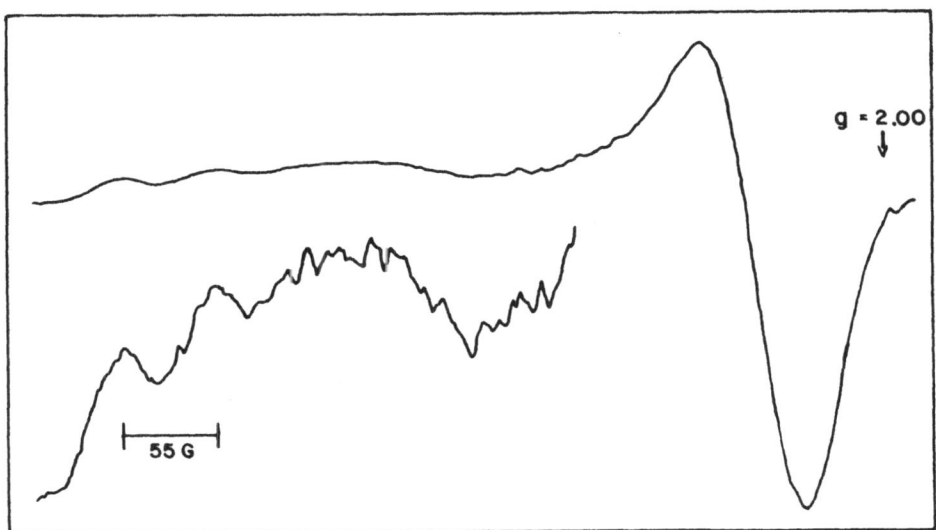

Figure 1.   X-band ESR spectrum of Cu(II)Az in frozen buffer
            solution at -180°C.

in 4s gives a *positive* contribution to the Fermi contact interaction.(3) Alternatively, extensive delocalization of the unpaired
spin onto ligand(s) provides a mechanism for decreasing hyperfine
splitting by the metal nucleus.(4)

    Another distinctive characteristic appears in the electronic
absorption spectrum of blue copper proteins, namely extremely
intense absorption bands ($\epsilon \approx 5000$ $M^{-1}cm^{-1}$) in the region of
~600 nm. (See Figure 2.) Early on, the absorptions were assigned
as "d-d" bands having unusually large intensities because of 4p
orbital admixture into the ground state wave functions.(2) Subsequently it was suggested that the bands could be assigned to
low-lying charge transfer transitions(5), an assignment made more
plausible by the recent identification of additional weaker absorptions in the near infrared region which are presumably the "d-d"
transitions.(6) As should be evident, the detailed assignments of
the optical transitions are not readily deduced from the spectral
data of the native protein.

    Somewhat more definitive information has come from NMR studies
of azurin.(7-10) Azurin contains four histidines, but the ^{1}H-NMR
spectra of the reduced protein as a function of pH reveal only
two titrateable histidines in the usual pK region. Later, ^{13}C-NMR
studies confirmed that two histidines of the reduced protein do
not titrate (8). From these results it could be inferred that two
histidines act as ligands in the copper binding site.

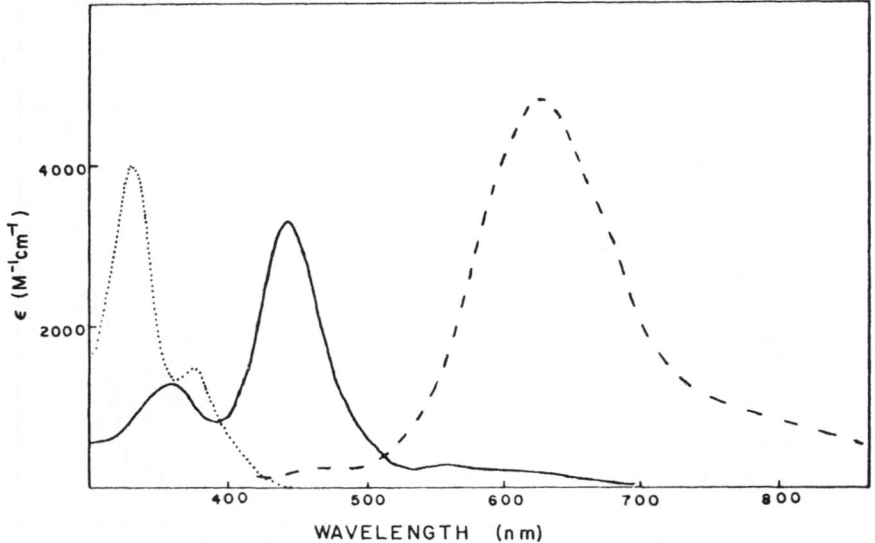

Figure 2.   Charge transfer absorption spectra of Co(II)Az,
            Ni(II)Az and Cu(II)Az.

      In summary, the ESR and optical studies of azurin suggest
the presence of a distorted tetrahedral binding site. The observation of low-lying charge transfer bands implicate a strongly
reducing ligand, possibly cysteine (5). Finally the NMR work
implicates two histidine ligands.

2.2 Optical Studies of Metal Replaced Derivatives of Azurin

      The presence of low-lying charge transfer absorptions can be
extablished by metal replacement studies because the energies of
such absorptions are a very sensitive to the metal center involved.
In Figure 2 the visible and near UV absorption spectra of the
cobalt(II), nickel(II) and copper(II) derivatives of azurin,
denoted Co(II)Az, Ni(II)Az and Cu(II)Az, are presented. The
derivatives were prepared by reacting the metal ions with apoazurin,
and fluorescence data, titrations with a mercurial reagent as well
as competitive binding studies support the postulate that metals
bind at a common site.(11) Based on comparisons with small molecule
systems, it is immediately evident that the intense bands which
shift from ~620 nm to ~440 nm to ~330 nm in going from Cu(II) to
Ni(II) to Co(II) can be identified as ligand to metal charge
transfer transitions. Using the empirical methods developed by
Jørgensen (12), it is even possible to infer the donor types which
may be involved. (11,13) For this analysis it is helpful to include results from the more highly resolved circular dichroic
spectra which are summarized in Table 1.

According to Jørgensen's formulation, the energies of the charge transfer bands should be correlated by the expression

$$\bar{\nu}_{corr} = 3.0(x_L - x_M)$$

where $\bar{\nu}_{corr}$ is a corrected transition energy expressed in μm^{-1}. It is obtained from the experimentally observed transition energy by correcting for certain interelectron repulsion effects and

Table 1. CD Data for Azurin Derivatives

Derivative	Band Energy (μm^{-1})	Sign[a]
Cu(II)Az[b]	1.24	−
	1.61	+
	1.92	+
	2.14	−
Ni(II)Az[c]	1.78	−
	2.00	+
	2.28	+
	2.55	+
	2.82	−
Co(II)Az[c,d]	2.45	−
	2.68	+
	3.00	+

[a]Denotes whether the feature appears as a positive or negative feature in the CD spectrum. [b]Data from Tang, S-P. W., Coleman, J.E., Mayer, Y.P.: 1968, J. Biol. Chem. 243, p. 4286. [c]Data from Reference 13. [d]Bands above $3\mu m^{-1}$ are obscured by absorptions of the peptide moiety.

effects of the ligand field splitting. (12) The symbols x_L and x_M denote the optical electronegativities of the ligand and metal, respectively, involved in the transition. These parameters are empirically determined and are transferable, i.e. for a given ligand or metal center the same parameter is used for all complexes of a given symmetry.

Biological donor groups that could be expected to give rise to such low-lying charge transfer bands are cysteine sulfur and methionine sulfur. And indeed, a very satisfactory correlation obtains using literature values for the optical electronegativities of these two type sulfur donors and the metal centers. (11) The relatively weak bands appearing at lowest energy are associated with the nonbonding levels of a thiolate (cysteine) sulfur. The next higher energy band, which is the most intense transition, can be associated with the sigma bonding level of the thiolate

sulfur. The next higher band is attributable to charge transfer
from a methionine sulfur. In principle two transitions can
originate from a methionine sulfur, and it is possible that the
remaining band to higher energy, which always appears as a negative
feature in the CD, is also attributable to methionine sulfur. In
the few literature reports of charge transfer from methionine type
sulfur, however, only one band seems to be resolved. Therefore
the assignment of the band in question to methionine sulfur (13)
must be viewed as tentative. Recalling the NMR studies, two histidines are involved as ligands and therefore charge transfer from
imidazole moieties is also a possibility. This possibility can
be more carefully analyzed when a x_L value for an imidazole moiety
is available.

The Mn(II)Az derivative has also been prepared, but it gives
a featureless absorption spectrum in the accessible region. Presumably, the combination of a low x_M value and a large spin pairing
energy (12) result in very high energy charge transfer bands in
this case.

2.3 Magnetic Resonance Studies of Metal Replaced Derivatives of Azurin

We searched for an epr signal from Mn(II)Az but met with little
success. We did find a weak signal at times but it very closely
resembled the epr spectrum of Mn(II) in buffer solution. Whereas
in symmetrical environments Mn(II) gives strong epr signals, in
low symmetry environments the signals can spread over a very wide
range spectral range and can become very difficult to observe. (14)
In fact the loss of an Mn(II) epr signal can often be used to
quantitate the uptake of Mn(II) from solution by proteins. (15)
The epr studies of Mn(II)Az do, however, illustrate the thermodynamic selectivity that the azurin binding site exhibits toward
copper. The stability constant for the binding of Cu(II) is
very large compared to that for Mn(II). In the case of native
azurin even exhaustive dialysis does not significantly remove
bound copper. On the other hand, a freshly prepared sample of
Mn(II)Az spontaneously releases Mn(II) as can be seen in Figure 3.

In the case of cobalt(II) epr studies have yielded more information. The epr spectra of Co(II)Az and the analagous derivative
of spinach plastocyanin (Co(II)Pl) are presented in Figure 4 and
Figure 5, respectively. In both cases the signals, particularly
that of Co(II)Pl, are seen to be very broad even at 4.2K. Two g
values are apparent in the case of Co(II)Az: $g_1 \approx 5.2$ and $g_2 \approx 3.8$.
The shape of the Co(II)Pl spectrum is quite different and probably
reflects 3 g values: $g_1 \approx 5.8$, $g_2 \approx 4.2$ and $g_3 \approx 2.6$. The broad
signals covering a wide range of g values and the large g values
observed demonstrate that both cobalt(II) centers are high spin.
In Figure 6 the temperature dependence of the signal from Co(II)Az

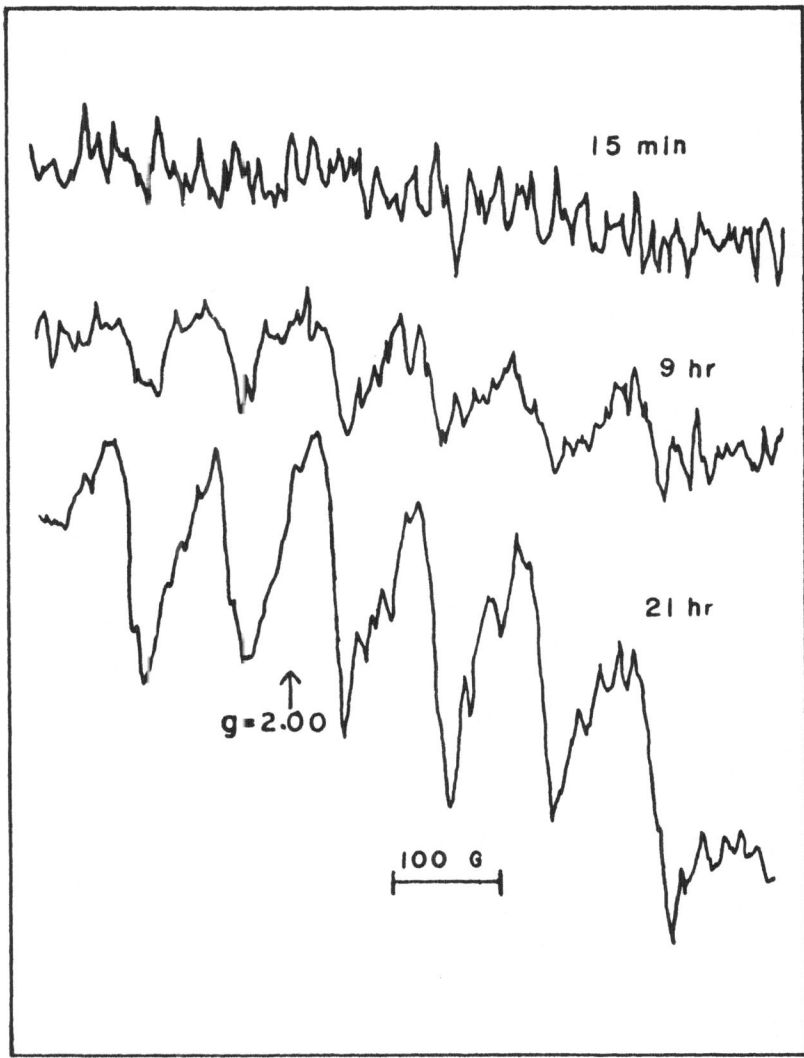

Figure 3.   X-band ESR spectrum of a Mn(II)Az sample at
25°C after gel filtrateion as a function of
time. The signal growing in is the same as
that of Mn(II) in buffer.

is presented. The signal broadens significantly as the temperature
is increased and it is apparent that very low temperatures are
required to observe signals from these high spin Co(II) centers.
This characteristic distinguishes high spin Co(II) centers from
low spin species.

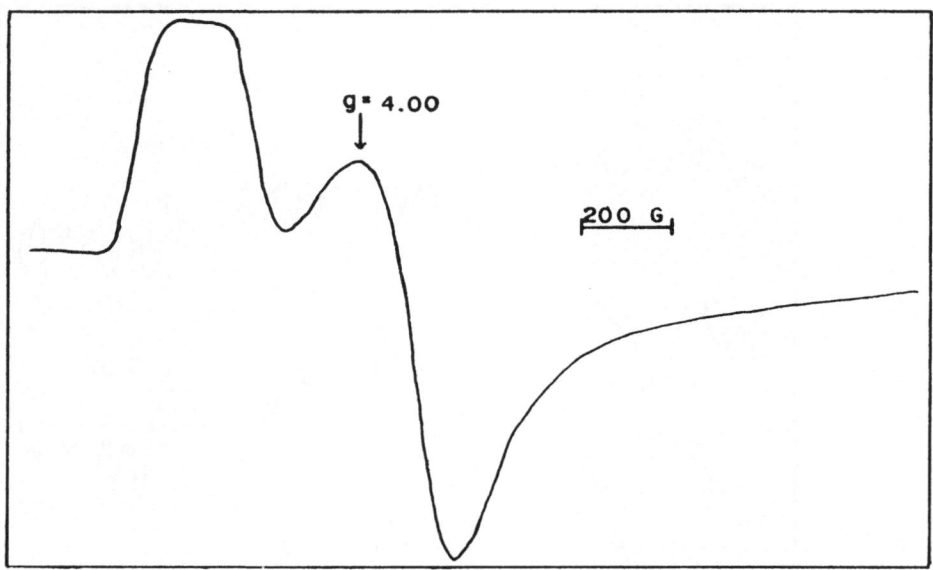

Figure 4.   X-band ESR spectrum of Co(II)Az in frozen buffer
           solution at 4K.

The spectrum of Co(II)Pl is very similar to spectra that have
been observed for a number of small molecule complexes of Co(II)
believed to involve a pseudotetrahedral coordination environment.
(16)   It also resembles the spectra observed for some cobalt(II)-
substituted proteins that are thought to involve distorted tetrahedral binding sites. (16-18)   The spectrum of Co(II)Az is not so
readily classified, however.   In both cases hyperfine coupling to
[59]Co appears to be absent.

In the case of Ni(II)Az a d^8 metal is involved. Assuming a
distorted tetrahedral configuration, the Ni(II) center would be
expected to be paramagnetic and to exhibit rapid spin relaxation.
If so, the Ni(II) center may be expected to induce large shifts in
the NMR resonances of nearby nuclei via Fermi contact and dipolar
interactions.   In Figure 7 the ^{1}H-NMR spectrum of Ni(II)Az is
presented, and signals are resolved which have been shifted many
ppm downfield from their positions in the spectrum of Cu(I)Az.
Other signals have been shifted upfield.   The detailed assignments
of these resonances remain to be made; however, it is clear that
the NMR spectrum of Ni(II)Az offers information which is not
readily obtainable from the corresponding Cu(II)Az spectra wherein
the unfavorable spin relaxation time of Cu(II) results in severe
broadening effects. (7)

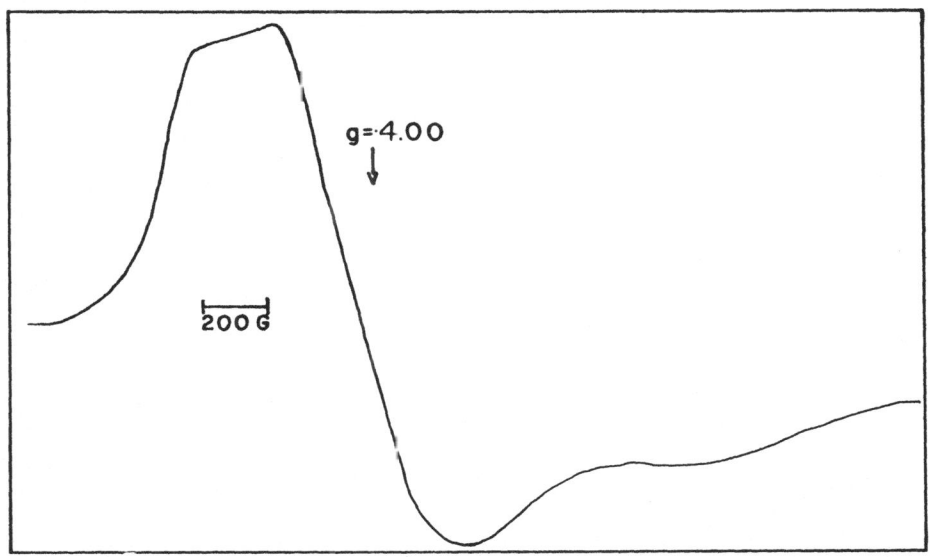

Figure 5. X-band ESR spectrum of Co(II)Pl in frozen buffer
           solution at 4K.

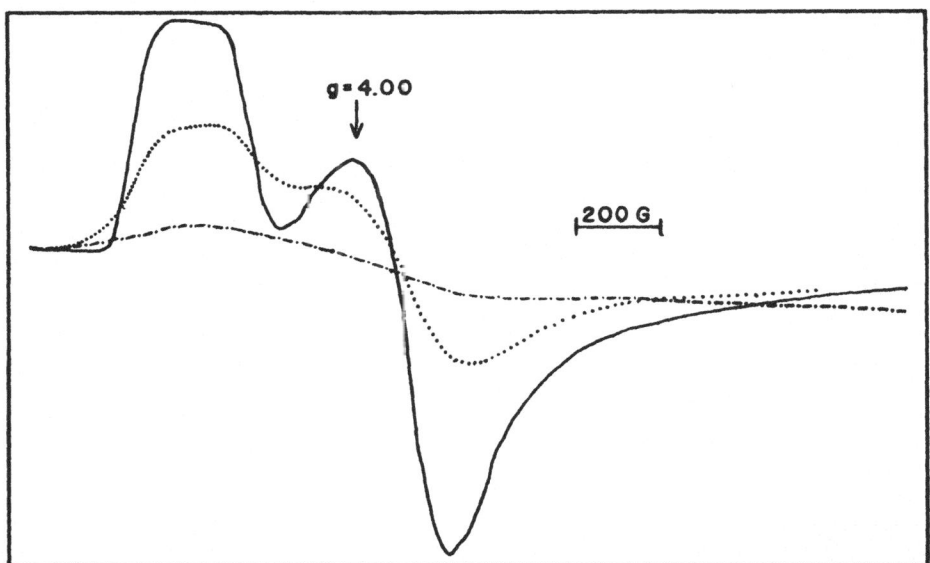

Figure 6. X-band ESR spectra of Co(II)Az as a function of temperature
           with a power of 5 milliwatts and a fixed gain setting.
           ————— 4K, · · · · · · 28K, — · — · — 42K.

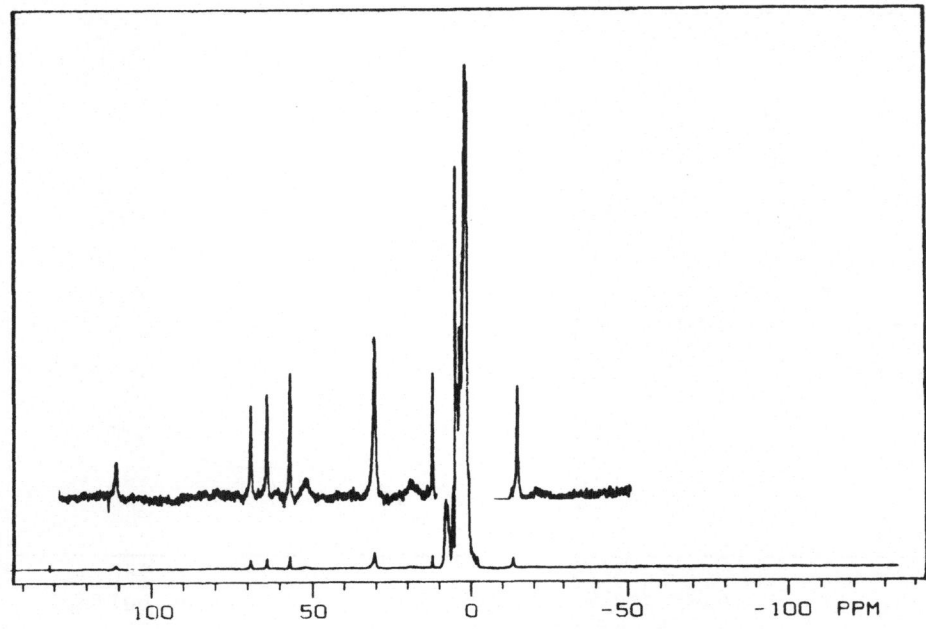

Figure 7.   360 MHz ^{1}H-NMR spectrum of a Ni(II)Az in D_2O buffer
             at 50°C.

3. SUMMARY

The spectral results obtained for Cu(II)Az and several metal-substituted derivatives of azurin have been shown to be very informative regarding the geometric and electronic structure of the blue copper site. The spectral interpretations are now in good accord with the low resolution crystal structure findings that have been recently reported. (19) It is apparent that the collaborative use of several physical methods is required to gain a reliable and detailed picture of the metal binding site of a protein. Where possible, spectral studies should be carried out in conjunction with chemical modification experiments, e.g. those involving metal replacement. Clearly, X-ray diffraction methods provide the most detailed insight into the structure of proteins, but spectral methods will remain valuable in cases where the diffraction methods are not readily applicable.

Acknowledgment.- The authors thank J. Blaszak, E. Ulrich, and J.L. Markley for measuring the ^{1}H-NMR of Ni(II)Az. This work has been supported by a grant from the National Institutes of Health Number GM 22764. The purchase of the EPR Spectrometer used was supported in part by a grant from the National Science Foundation, grant number PCM 75-19127.

REFERENCES

1. Fee, J.A.: 1975, Structure and Bonding 23, p. 1.

2. Brill, A.S., Bryce, G.F.: 1968, J. Chem. Phys. 48, p. 4398.

3. McGarvey, B.R.: 1956, J. Phys. Chem. 60, p. 71.

4. Malmström, B.G., Vänngård, T.: 1960, J. Molec. Biol. 2, p. 118.

5. Brill, A.S., Martin, R.B., Williams, R.J.P. in "Electronic Aspects of Biochemistry", ed. Pullman, B., Academic Press, New York: 1964, pp. 519-558.

6. Solomon, E.I., Hare, J.W., Gray, H.B.: 1976, Proc. Natl. Acad. Sci., U.S.A. 73, p. 1389.

7. Hill, H.A.O., Leer, J.C., Smith, B.E., Storm, C.B., Ambler, R.P.: 1976, Biochem. Biophys. Res. Commun. 70, p. 331.

8. Ugurbil, K., Norton, R.S., Allerhand, A., Bersohn, R.: 1977, Biochemistry 16, p. 886.

9. Hill, H.A.O., Smith, E.E.: 1978, Biochem. Biophys. Res. Commun. 81, p. 1201.

10. Ulrich, E.L., Markley, J.L.: 1978, Coord. Chem. Rev. 27, p. 109.

11. Tennent, D.L., McMillin, D.R.: 1979, J. Am. Chem. Soc. 101, p. 2307.

12. Jørgensen, C.K.: 1970, Prog. Inorg. Chem. 12, p. 101.

13. McMillin, D.R.: 1978, Bioinorg. Chem. 8, p. 179.

14. Reed, G.H., Ray, W.J., Jr.: 1971, Biochemistry 10, p. 3190.

15. Cohn, M., Townsend, J.: 1954, Nature (London) 173, p. 1090.

16. Kennedy, F.S., Hill, H.A.O., Kaden, T.A., Vallee, B.L.: 1972, Biochem. Biophys. Res. Commun. 48, p. 1533.

17. Grell, E., Bray, R.C.: 1971, Biochem. Biophys. Acta 236, p.503.

18. Desideri, A., Morpurgo, L., Raynor, J. B., Rotilio, G.: 1978, Biophys. Chem. 8, p. 267.

19. Adman, E.T., Stenkamp, R.E., Seiker, L.C., Jensen, L.H.: 1978, J. Mol. Biol. 123, p. 35.

PHYSICAL ASPECTS OF THE SPIN LABELLING TECHNIQUE

Philippe F. DEVAUX and Jean DAVOUST

Institut de Biologie Physico-Chimique
13, rue Pierre et Marie Curie, 75005 PARIS, France.

A brief review of the physical aspects of the spin labeling technique is given. Emphasis is put on the qualitative description of the phenomena which enable us to obtain information on molecular motion in biological and related systems. The conventional use of nitroxides, saturation transfer spectroscopy and electron spin-electron spin interactions are the three topics discussed.

I - Nitroxide radicals can give information about motion

Nitroxide radicals were introduced by Mc Connell in 1965 as paramagnetic probes for the study of molecular motions in biological
systems. This technique proved to be particularly useful in the
field of membrane biology, perhaps because of the great variety
of motions involved. Spin-labels are indeed very versatile.They
can be used to explore a wide range of correlation times and
diffusion constants. In that respect, spin-labeling is a unique
method in the mean time. It requires of the user a great adapta-
bility to various technical problems. Some people use nitroxides
according to magic recipes; others are so sophisticated that they
can only think about nitroxides in terms of spin hamiltonian and
only believe in computer description of the observed phenomena.
In between, we think there is place for physics.
The present article is an attempt to survey the physical basis
of the different ways of taking advantage of nitroxide radicals
paramagnetic properties. Our aim is not so much to give rigorous
and mathematical formulations of the phenomena but rather t. describe to non-physicists or to unspecialized physicists how nitroxide radicals can provide information about motion. Those inte-

*I. Bertini and R.S. Drago (eds.), ESR and NMR of Paramagnetic Species in Biological and
Related Systems. 419–422.*

rested in complete description of the method and an extended
list of references can find such information in several excellent
reviews. We strongly recommend the two volumes of the book edited
by L. Berliner in 1976 and 1979 : "Spin-labeling technique" (1).
They give a full depth view of the question. Each volume however
represents about 500 pages ; the goal is therefore quite different from the following presentation which should be considered
as an "essay" on the spin-labeling technique.

The spin-labeling method takes advantage of the paramagnetic properties of nitroxide radicals. They are derived from 5- or 6-
membered rings :

The radicals are relatively stable and correspond in most cases
to a moderate steric perturbation. However, this type of chemical
labeling has a limit. It is useless to search for too detailed
informations concerning the unlabeled system when using spin-
labels. First-order type of information and relative studies
should be favored. The great advantage of using nitroxide radicals is the high sensitivity of EPR : 1 µM of spin-label in 50 µl
can be detected by conventional EPR, 50 µM in the same volume is
required for the saturation transfer experiments. Furthermore,
spectral interpretation is generally unambiguous. As we shall see,
the spectral features are essentially influenced by the probe mobility. Three main approaches have been pursued. We will try to
make them understandable in simple terms.

II - Conventional EPR : $10^{-10} s \leq \tau \leq 10^{-7} s$

Oriented nitroxides paramagnetic properties.
If nitroxides are oriented with respect to the magnetic field,
three lines are visible around the resonance position corresponding to g = 2. This triplet is due to the interaction between
the unpaired electron on the nitrogen and the magnetic moment of
the nitrogen nucleus. ^{14}N has a nuclear spin 1, it gives rise to
3 possible values of the local magnetic field due to the nucleus,
and hence to 3 resonance positions. The distance between the lines is the hyperfine splitting. Because the unpaired electron has
axial symmetry, the interaction with the nitrogen is anisotropic
with approximately axial symmetry also. As shown in figure 1
a and b, if the magnetic field is parallel to the π orbital the
splitting (with X band, 9.35 GHz) is $A_{//}$ = 32,9 Gauss ; if the
field is 90° from the π orbital axis, the splitting is $A_{\perp}$ = 6 G.

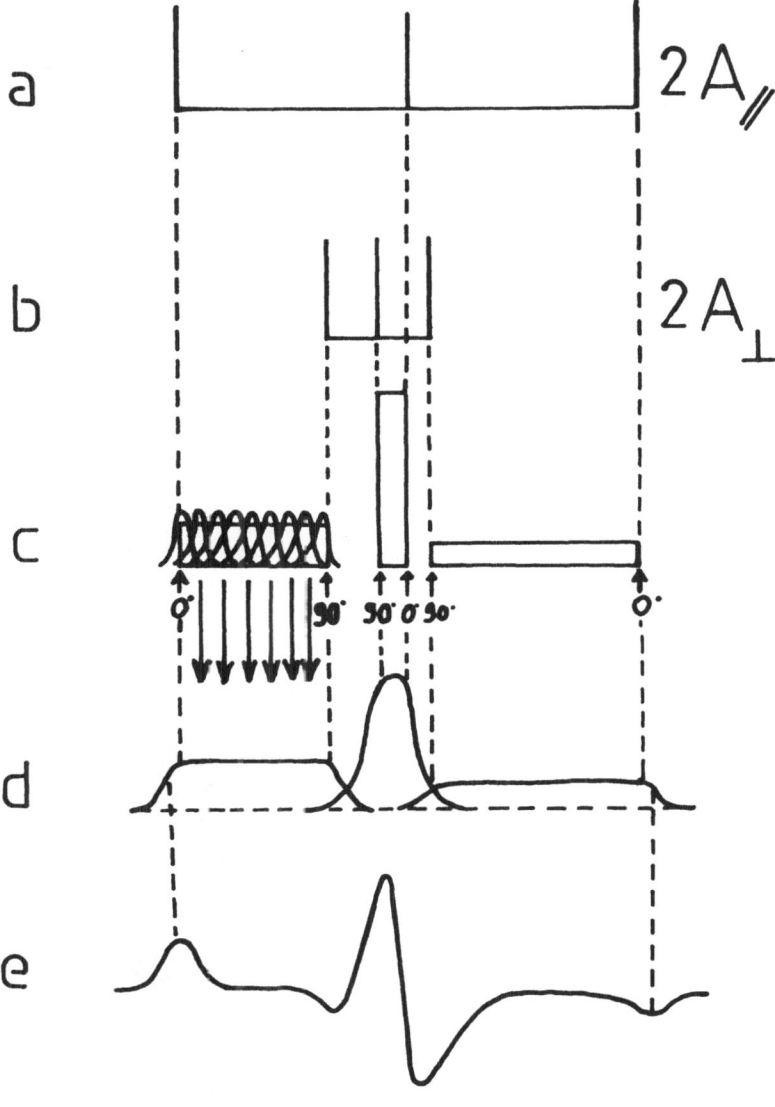

Figure 1 : Conventional ESR ; slow motion, $\tau_c \geq 10^{-7}$s.
Schematic construction of a "powder spectrum".

In addition to the difference in splitting, a small shift in the
resonance positions (Δg) is observed. It corresponds to a shift
of the central line of a few Gauss only. The positions of the
lines indicated in figure 1 a and b can be observed if the nitroxide is diluted in a diamagnetic crystal, and oriented (2).

Powder spectrum :
If the sample contains nitroxides with all possible orientations,
the spectrum is more complicated. The contribution of all orientations between 0° and 90° has to be added. The resonance positions H_{res} are distributed into 3 domains of the magnetic field
corresponding to the 3 nuclear spins (m = +1, m = 0, m = -1) as
schematized in figure 1-c. In each domain a magnetic field value
corresponds to the resonance field for a given orientation of the
spin-labels : H_{res} = f (θ). The low and high field lines (m = $\pm$ 1)
are spread over a larger field range than the central lines (m = 0).
As a result the ESR line density function (fig. 1-c), is more
intense in the central region than in the wings (3). Figure 1-d
is the envelope one can obtain by adding lines of finite lines
width in each domain and multiplying by the proper factor density.
The observed spectrum, due to modulation effects, is in fact the
1rst derivative of the absorption spectrum (figure 1-e). It is
important to note that the extreme splitting between low and high
field peaks of figure 1-e is a measure of 2A . Spectrum 1-e
represents a powder spectrum ; it is assumed that no motion
exists.

Fast isotropic motion :
If the nitroxide is tumbling rapidly in solution, all orientations
are present but a resonance exchange process takes place (figure 2).

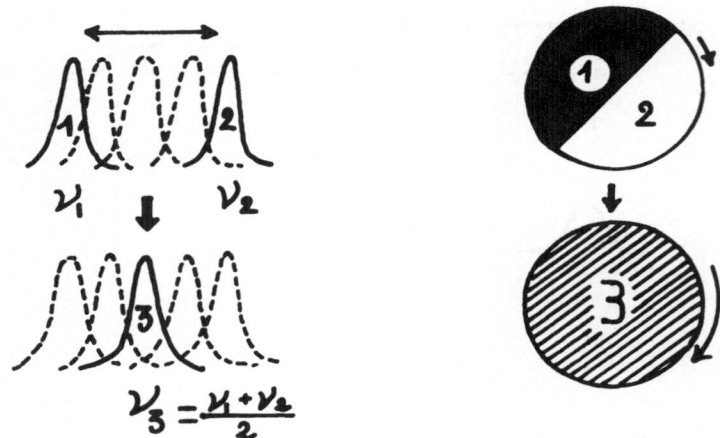

Figure 2 : Conventional ESR ; fast motion, $\tau_c \leq 10^{-10}$s.
              Shematic construction of "liquid spectrum".

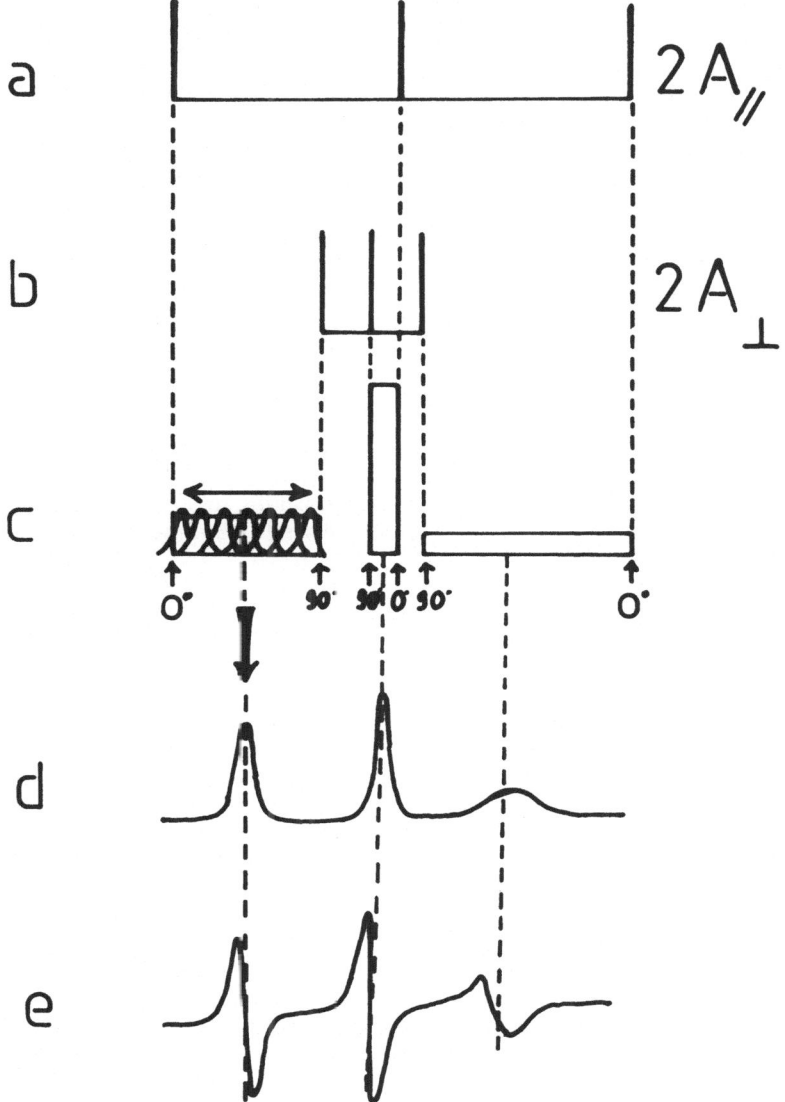

Figure 3 : Fast exchange between 2 states generates a fictious intermediate state.

Only one intermediate average state is defined for each nuclear
spin value. This "average state" is very precise for the central
peak$_+$ (m=0) ; it gives rise to a single narrow and intense line. For
m = $-$ 1, the average state is less accurately defined because the
averaging has to be done on larger field ranges. The resulting
peaks are slightly broader particularly for m = -1 (high field).
(See figure 3-e).

Intermediate cases : what is the time scale ?

All intermediate spectra between spectrum 1-e and spectrum 3-e
can be obtained experimentally by changing the spin-label tumbling rate, i.e. by changing the viscosity of the medium. Each
line shape can therefore be associated with a rotational correlation time. There is no simple feature of the spectra which can
be measured easily over the entire range of correlation times
explored and which could be used conveniently to parameterize the
spectra. At the fast motion limit, it is customary to measure
peak ratios,while at the slow motion limit, the extreme splitting
is the usual parameter.

From our point of view, the important question is : what determines the range of correlation times ? This can be very easily understood if figure 1 and figure 3 are thought of as being repre-
sentations of resonance positions in terms of frequencies (with
a fixed magnetic field). Each orientation is therefore assigned
a certain resonance frequency. When the molecules are tumbling,
the resonance position hops from one position (ν_1) to another
(ν_2). The hopping frequency is $\nu_c = 1/\tau_c$. Clearly if :

$$\nu_c \gg \Delta\nu = \nu_2 - \nu_1$$

there is no time for resonance to take place at each position.
The effective resonance frequency of the system is then :

$$(\nu_2 + \nu_1)/2 \text{ or } \nu_1 + \Delta\nu/2 \text{ (see figure 2 again).}$$

For nitroxides $\Delta\nu$ (which corresponds to the anisotropy of the hyperfine splitting expressed in frequency units) is of the order of
$10^8 sec^{-1}$. Therefore the time scale is given by $1/\Delta\nu \approx 10^{-8} sec$.
Fast motions have much shorter correlation times (10^{-10}s), slow
motions have much longer correlation times (10^{-6}s). This explains
the limits of applicability of conventional EPR when nitroxides
are used. It should be pointed our that the paramagnetic species
present in biological system (metal ions usually) have larger
hyperfine splittings or much larger values of Δg. As a result,
the corresponding $\Delta\nu$ would be larger and the reference time scale
much shorter. One is always at the slow motion limit with metal
ions. They are not suitable for probing biologically relevant
motions. Furthermore, if one finds a radical with a smaller hyperfine anisotropy than nitroxides, it is not obvious that it will
allow to explore the domain of slower motions, because one will
be faced with the problem of intrinsic line width of each single
line. Therefore nitroxides represent a good compromise. The range
of correlation times explored corresponds however to relatively

fast motions and is of little interest for big molecules such as
proteins. We shall indicate in section III how the spin-label
technique can be adapted (saturation transfer technique).
Fast anisotropic motion :
In the previous discussion all angular orientations are considered equally. However one often encounters situations where a spin
label can wobble in a restricted area, i.e. only within certain
directional limits. For example, if a spin-label oxazolidine is
attached to a fatty acid chain :

$$CH_3 - (CH_2)_n - \underset{\underset{\text{L}}{O}}{\overset{C}{\diagdown}}_{N-\bar{O}} (CH_2)_m - COOH$$

the π orbital of the unpaired electron is aligned in the direc
tion of the extended chain. When the fatty acid is incorporated
into a macroscopicaly oriented membrane, the average direction of
the π orbital is perpendicular to the place of the membrane. However restricted wobbling is permitted within a cone, whose aper-
ture depends on the exact position along the chain. We will consider only the case where the motion of the probe, within this
permited area, is <u>fast</u> (characteristic time smaller than 10^{-8}s).
We are dealing therefore with a <u>fast anisotropic motion</u>. This
situation can be accounted for rather easily if we consider again
the distribution function used on figures 1 and 2. The top of
figure 4 shows again the rectangular approximation. Each part of
the 3 blocks is associated with particular spin orientation (ranging from 0° to 90°). Let us suppose firstly that the average o-
rientation of the spins is parallel to the magnetic field. Because
the fast motion is restricted to a certain angular zone (around
0°), one obtains 3 narrow lines but centered differently than in
the case of fast isotropic motion. The splitting between the lines
is : $A'_{/\!/}$ with : $A'_{/\!/} < A_{/\!/}$.
If the membrane is tilted by 90°, the average line positions are
shifted towards a different field region. The splitting is now :
$A'_{\perp}$ with (see figure 4) : $A_{\perp} < A'_{\perp}$.
Intermediate orientations of the membrane correspond to 3 lines
situated in the 3 domains of figure 4-f. These domains represent
distribution function of the resonance lines for an isotropic
sample containing spin-labels undergoing fast anisotropic motion
(for example fatty acid spin-labels in a cell suspension). One
can now construct the absorption spectra and finally the observed
1rst derivative spectrum (figure 4-g). This latter spectrum is
obtained just as a powder spectrum, the values of the hyperfine splittings being $A_{/\!/}$ and $A_{\perp}$.The distance between the low field peak and
high field peak is a direct measure of $2 A_{/\!/}$ (see figure 4-g). In
some cases, $2 A'_{\perp}$ can also be measured directly from the spectrum
of the isotropic sample but not always.
Measure of the degree of anisotropy of the motion or order parameter :
If the spin label is allowed to reorient rapidly in a larger domain, $A_{/\!/}$ decreases whilst $A'_{\perp}$ increases. Therefore the quantity:

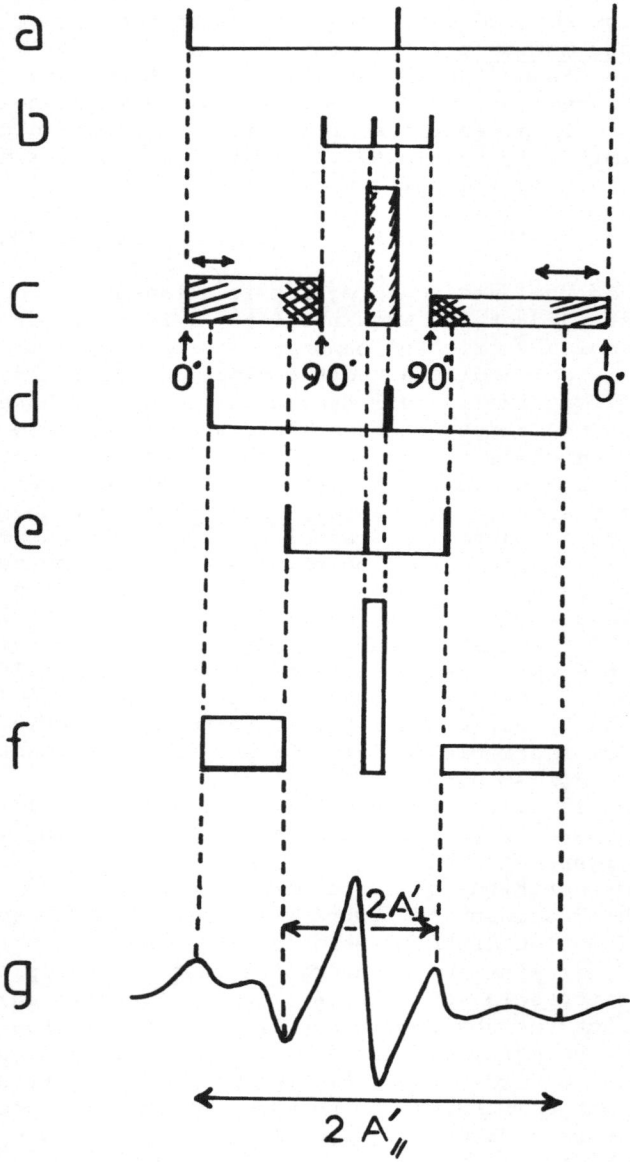

Figure 4 : Construction of the ESR spectrum corresponding to a
          fast anisotropic motion.

$$A'_\parallel - A'_\perp$$

reflects the degree of order of the spin-label. The best parameter to introduce is in fact the dimentionless quantity :

$$S = \frac{A'_\parallel - A'_\perp}{A_\parallel - A_\perp}$$

If $A'_\parallel = A'_\perp$, the spin-label undergoes fast isotropic motion :
the order is zero. If absolutely no wobbling is allowed (no fast
motion) then : $A'_\parallel = A_\parallel$ and $A'_\perp = A_\perp$ so that S=1 : the order
is 1.
Therefore S is a measure of the extension of the fast motion. It
should be pointed out however that for very special orientations
of the spins (namely around 45°), it is possible to obtain :
$A'_\parallel \sim A'_\perp$ and hence S = 0   even without any fast motion (4). This
points out the ambiguity of the commonly used parameter S which,
as can be found elsewhere, can be introduced with a more mathematical formulation.

In summary, conventional EPR can distinguish 2 different domains
when spin-labels are used :
a) Fast isotropic motion : The spectrum shows 3 well-defined and
relatively narrow lines. The motion is faster than 10^{-9}s.
b)   Slow or fast, but anisotropic motion : the spectrum is
broad ; the extreme splitting is close to 60 G (see figure 1 or
4). In spite of the ambiguity in the interpretation of the spectrum, one can definitely conclude that the motion is restricted
(either in rates and in orientations). Motions wich characteristic times smaller than 10^{-7} s cannot be measured.

III - Saturation transfer EPR spectroscopy : 10^{-7} s $\lesssim \tau \lesssim 10^{-3}$ s .
Saturation.
Saturation transfer EPR was introduced in 1972 by J. Hyde and
L. Dalton as an extension of the spin-labeling technique. It
became a technique applicable to biological samples in 1976,
when Thomas, Dalton and Hyde (5) showed how the 2nd harmonic,
out of phase display can provide enough information with a reasonable sensitivity : a concentration of about 5.10^{-5} M is required.
The reference time scale is now given by the relaxation time T_{1e}
of nitroxides. T_{1e} for nitroxides is of the order of 10^{-5}s :
it does not seem to vary very much with temperature nor with the
chemical nature of the nitroxide ring. In order to take advantage
of saturation effects, a high microwave power is used, typically
30 to 60 mW (6).
Detection out of phase of the modulation :
Let us suppose first that the magnetic field is fixed at a value
corresponding to H_{res} for a certain population of the nitroxides.
If the radio frequency power (rf) is modulated, but kept at a low
intensity (no saturation), the signal detected follows the modulation. There is no "out of phase" signal (figure 5-a). If the
power is increased in such a way that saturation takes place,

part of the signal "lacks behind" (figure 5-b). This out of phase
signal is entirely governed by saturation effects. In a real experiment, the modulation is obtained by a field modulation and not
by a rf power modulation. In order to gather the maximum information on the saturation behavior with a detection out of phase
of the modulation, it is convenient to have

$$T_{1e} \sim 1/\nu$$

where ν is the modulation frequency. This means $\nu \sim 10^5$ hertz,
which is precisely the standard modulation frequency used in most
EPR machines.

Effect of motion : adiabatic rapid passage :
How does motion interfer with the signal ? The motion does not
influence the relaxation time T_{1e}. However, since each orientation corresponds to a specific value of H_{res}, the tumbling of the
nitroxides means that the relaxation starts before the field is
off resonance. In fact, the tumbling produces a "transfer of saturation" to other part of the spectrum. The average effect is to
decrease the signal intensity. Figure 6 is an attempt to explain
such phenomenon. This figure at least explains why the original
name of this technique was adiabatic rapid passage. It is clear
in this figure that heat is accumulated when :

$$\tau_c \lesssim T_{1e}$$

In summary the following conditions must be fullfilled to reach
the best sensitivity :

$$\tau_c \sim T_{1e} \sim 1/\nu$$

Line shape :
The main effect of motion is to decrease the signal intensity of
the saturation transfer spectra. However the averaging is not
uniform on the whole spectrum. This is due to the fact that the
resonance position H_{res} are not linearly correlated to θ (θ is
the orientation of the nitroxide π orbital, with respect to the
magnetic field). It can be shown that a rotation of 10° around
90° or 0° produces a very small shift in H_{res}. On the contrary
around 45°, a 10° reorientation produces a large change in H_{res}.
As a consequence, the signal is more effectively decreased in
that part of the spectrum.
Two different artifices are used simultaneously to amplify the
spectral distorsions produced by motion. The first artifice is
to display the 2nd derivative of the absorption spectrum (instead
of the 1rst derivative). This procedure is a classical way of am-
plifying mild modifications of a spectral line shape. Technically
this means detecting the 2nd harmonic (out of phase of course).
The second artifice is to exagerate the modulation. This not only
increases the signal intensity, but it also distorts favorably
the signal. Typically a modulation of 5G is being used.
Figure 7 shows typical saturation transfer spectra (2nd harmonic,
out of phase) for various correlation times. When $\tau_c \ll 10^{-6}$s,
the saturation transfer spectrum is identical to the 2nd deriva-

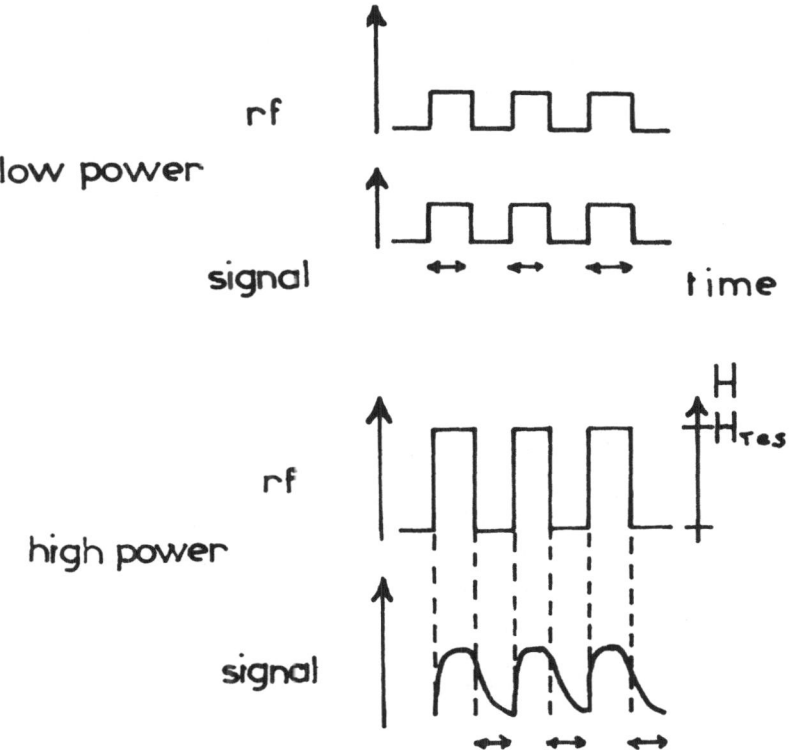

Figure 5 : At low microwave power, the signal follows the modulation. Detection is in phase. At high microwave
power, part of the signal lacks behind. The detection
of the saturation effect is best seen out of phase.

Figure 6 : Schematic representation of an experiment corresponding to an adiabatic rapid passage.

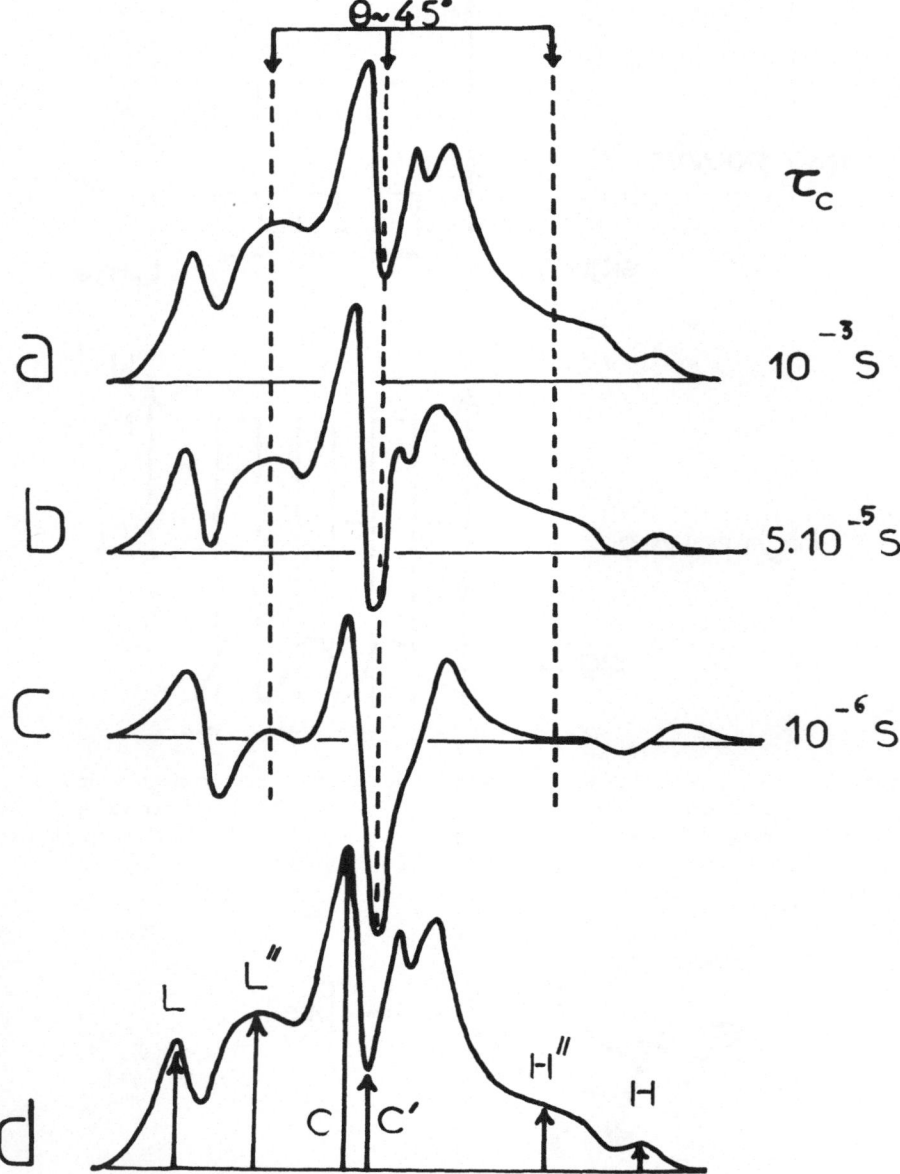

Figure 7 : Saturation transfer spectra (2nd harmonic ; out of
          phase ; modulation 5 G, powder 60 mW). Spectrum d
          indicates how the spectra can be parametrized. The
          ratios : H"/H, L"/L and C'C vary almost linearly
          with τ_c in the range 10^{-3} - 10^{-6}s.

tive in phase. Region indicated by the dotted lines spectra a-b-c
are the most affected by the motion (they correspond to $\theta \sim 45°$).
It is therefore quite reasonable to estimate the rotational correlation time from signal ratios of the type indicated on figure
5-d : H"/H, L"/L and C"/C. These parameters were introduced by
Thomas et al (5). They have an almost linear dependance with τ_c
in the range $10^{-6} - 10^3$s.

Anisotropic slow motions :
The above discussions is simplified by the implicite assumption
that the motion is isotropic. This is unrealistic for membrane
proteins for example. In the case of a cylindrical motion, the
orientation of the π orbital with respect to the axis of rotation
is crucial. If the two directions are parallel, there can be no
sensitivity at all to the motion. The part of the saturation
transfer spectrum most sensitive to the anisotropy of the motion
is the central part. In fact the central part of the spectrum is
sensitive to many things (anisotropy of the motion, non axial
symmetry of the hyperfine tensor, etc...). As a result, large variation in the central part should not be taken systematically as
a definite indication of a change in the rotational correlation
time. The consequence of this ambiguity brought about by anisotropic motion is that it is safer to have probes distributed over
different sites of a protein, or even probes with a residual motion allowing different orientations of the spin-labels : in such
cases the anisotropy is averaged out.

IV - The use of electron spin-spin interactions.

Whenever two nitroxides come into close contact, there is a certain probability of electron spin-exchange. As a result an electric spin undergoing a resonance effect in the vicinity of a nuclear spin +1 can suddenly be in the vicinity of a different nuclear spin, for example 0 or -1. In practice, this means that the
low-field line can exchange with the mid-field line or high-field
line and vice-versa. The appearance on the spectrum depends very
much on the exchange frequency. For the same reason as explained
in section II if the exchange frequency is of the order of the
distance (expressed in frequencies) between two exchanging lines,
a distortion appears. If the exchange rate is very high, the 2
(or 3) lines collapse (see figure 8). Eventually the exchange
rate narrows the unique line.
This phenomenon can be accounted for very well by a theory proposed in 1958 by McConnell to explain chemical exchange in NMR
(modified Bloch equations). Spectra of figure 7 were computed by
this method. The display in figure 7-B corresponds to the usual
1rst derivative.
Spin exchange frequencies are indicated in the figure. It is not
straightforward to relate the collision frequency and the exchange frequency. But these values are proportional (see reference 7).
There are two ways of increasing the collision frequency. One is

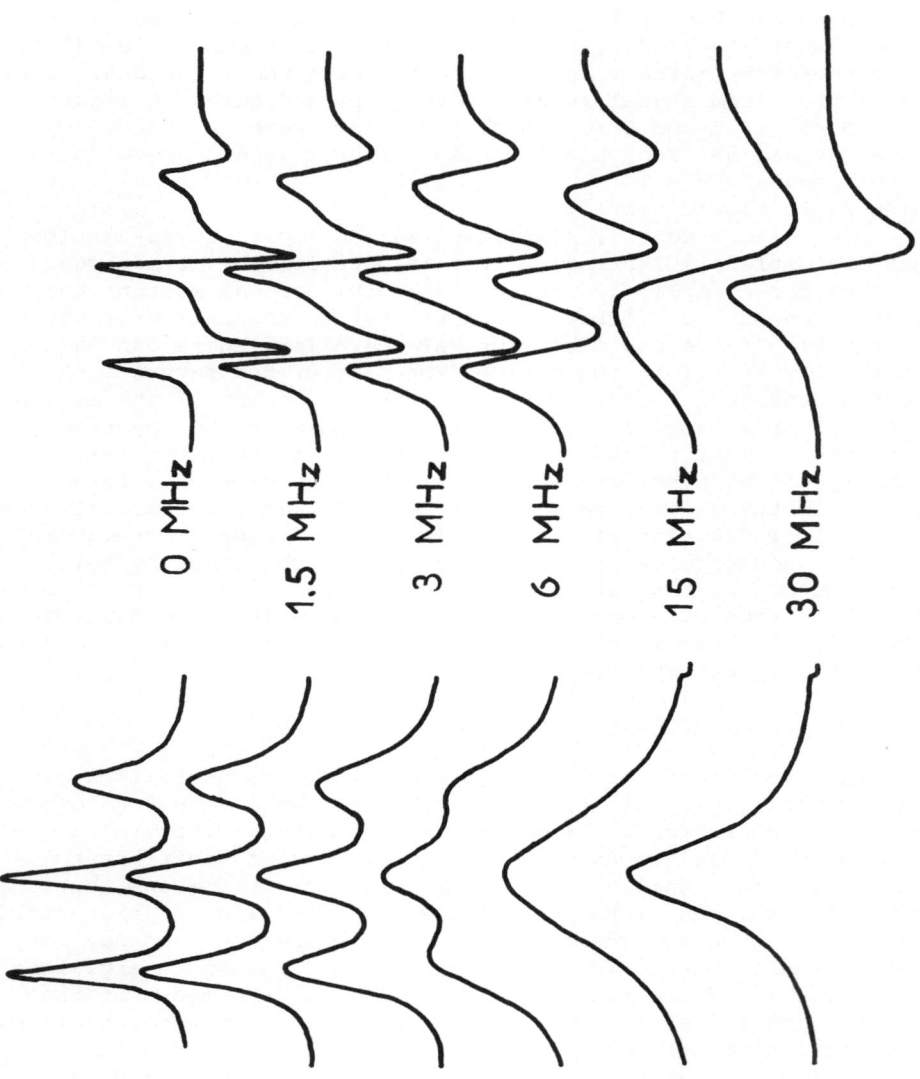

Figure 8 : Absorption spectra and 1rst derivative of ^{14}N nitroxides exchanging at different frequencies by spin
exchange. The exchange frequency is indicated in the
figure in MHz.

to increase the concentration of labels in a diffusing system,
the second is to increase the diffusion constant, for example, by
increasing the temperature. The top spectra of figure 8 correspond
to spin-labeled phospholipids (16-doxyl stearic derivatives) diluted in a membrane with a ratio of spin-labeled lipids to total
phospholipids equal to 1 : 100 (mole to mole) ; the interaction
can be seen when approximately 2% of the phospholipids are labeled. Bottom spectra correspond to pure phospholipid spin label.
Unfortunately, the broadening of the lines at low temperature is
governed by a second process : the dipole-dipole interaction.
This is a long distance interaction which depends on the relative
orientation of the 2 spins. For molecules tumbing rapidly, the
dipole-dipole interaction is averaged out. But at low temperatures
it dominates. The dipole-dipole interaction in principle gives
information on the distance between interacting species. However
it is difficult to account for rigorously. As a result it is less
usefull than the spin exchange process which gives rather direct
information on the collision frequencies. In the intermediate cases, the mixture of the 2 phenomenon makes the interpretation of
the spectra more complicated and hence less reliable.
Spin exchange was used some years ago to study rates of lateral
diffusion of phospholipids in membranes (7). Collision rates between spin-labeled phospholipids and spin-labeled proteins can be
mesured also by spin-spin interaction. The fundamental idea is to
label the proteins with ^{15}N nitroxide whilst the lipids are labeled with ordinary ^{14}N. The particular property of ^{15}N is to give
rise to two lines (nuclear spin = 1/2). In an isotropic sample,
with all labels tumbling rapidly only two narrow lines are seen
with ^{15}N spin-labels. This holds true for a nitroxide on a fatty
acid covalently bound to a protein such as rhodopsin (8). For
this purpose the following molecule is being used :

$$CH_3 - CH_2 - \underset{O \quad N-O}{C_{15}} - (CH_2)_{14} - N = C = O$$

We have shown that the viscosity is still low at the boundary
layer of such intrinsic protein. Therefore if the proteins are
labeled with ^{15}N derivatives whilst the lipids are labeled with
^{14}N, a total of 5 lines can be seen. In the absence of interactions, the lines are narrow (at 37° C), but if interaction is
allowed a spectacular decrease of some of the lines takes place.
Figure 9 shows an exemple of such phenomenon simulated by a computer. Experiments on model systems have given very similar results (9). On the top spectra phospholipids interact among themselves but do not perturb the protein labels. On the following
spectra, the collision rate between the labeled proteins and
labeled phospholipids is increased ; the lines corresponding to
the proteins are progressively erased.
This type of experiment allows to study the collision rates between proteins and phospholipids in a membrane. If the collision
rates however are smaller than 10^6 s^{-1}, conventional ESR only

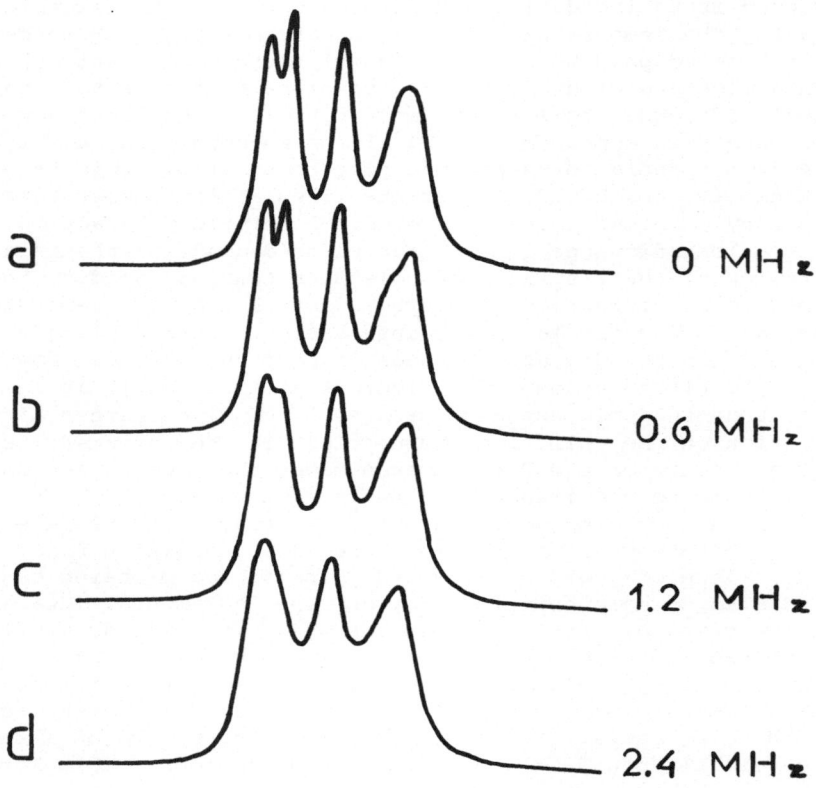

Figure 9 : Calculated absorption ESR spectra of a mixture of
^{14}N and ^{15}N spin labels. The ^{14}N represents spin
labeled phospholipids diluted into a membrane, with
the label near the W-2 carbon of the acyl-chain.
A small interaction is allowed between ^{14}N labeled
chains (1 MHz) ; this account for the lateral diffusion of the phospholipids. The ^{15}N represents proteins
labeled covalently with the same acyl chain positioned
at the lipid-protein interface by a isocyanate reacting
group. All spectra contain 3 time more ^{14}N spin labels.
The allowed exchange frequency between ^{15}N and ^{14}N is
indicated on the figure. High frequencies result in the
disappearance of the ^{15}N lines.

permits to give a upper limit.
The reader can find an other discussion about saturation transfer
ESR spectroscopy in the chapter by Dr. Ehrenberg.

ACKNOWLEDGEMENTS :

This work was supported by grant from the "Centre National de la
Recherche Scientifique, E.R.A. 690", the "Délégation à la Recherche Scientifique et Technique, Commission Membranes biologiques"
and the Université Paris VII.

REFERENCES AND NOTES :

(1) Spin-labeling, theory and applications. Acad. Press, Ed. L.
    J. Berliner, Vol. I, 1976, Vol.II in the press 1979.

(2) A pure spin-label crystal would give rise to a single very
    broad line due to spin-spin interactions, as indicated in
    section IV.

(3) In figure 1-c, the distribution function for the 3 lines is
    represented very crudely by 3 rectangles. The actual distri-
    bution is more complicated but this figure at least shows
    that the 3 nuclear spins play unequivalent roles. This disym-
    metry is due to the g value.

(4) The exact critical value is 54°7, this is because the distri-
    bution functions are not simple rectangles as indicated in
    figure 3.

(5) Thomas D., Dalton L.R. and Hyde J.S. (1976). J. Chem. Phys.,
    65, 3006-3024.

(6) The actual power on the sample is one of the crucial parame-
    ter to determine ; it requires a good calibration of the
    system.

(7) Devaux P., Scandella C.J. and Mc Connell H.M. (1973), J. Magn.
    Res., 9, 474-485.

(8) Favre E., Baroin A., Bienvenue A. and Devaux P.F. (1979),
    Biochem., 18, 1156-1162.

(9) Bienvenue A., Hervé P. and Devaux P.F. (1978), C.R. Acad. Sci.
    Paris, 287D, 247-1250.

ELECTRON SPIN DOUBLE RESONANCE TECHNIQUES AND SATURATION TRANSFER EPR

Anders Ehrenberg

Department of Biophysics, University of Stockholm, Arrhenius Laboratory, S-106 91 Stockholm, Sweden.

INTRODUCTION

This chapter will deal with three EPR techniques: Electron nuclear double resonance (ENDOR), Electron-electron double resonance (ELDOR), and saturation transfer EPR (ST-EPR). It will briefly outline the principles of each technique in a phenomenological way. Some particular instrumental points will be discussed. Emphasis will be placed on important applications in biological systems carried out in the past, and future possibilities will be outlined.

In all three techniques the EPR signal is partly saturated and effects of manipulation of the relaxation or of spin diffusion are observed. In common EPR one generally selects to work in a regime of low microwave power where the response observed is proportional to the applied microwave field B_1. For ENDOR, ELDOR and ST-EPR this is not the case since the EPR signal is partly saturated. These techniques are therefore often collectively referred to as non-linear techniques.

BACKGROUND

Before taking the trouble to learn about these somewhat more sophisticated techniques, as compared with EPR, one may ask what advantages they have to offer that are not obtained with normal EPR.

Already from our first work on EPR of flavin free radicals, free in solution and bound to a protein, it was apparent that the

399

I. Bertini and R.S. Drago (eds.), ESR and NMR of Paramagnetic Species in Biological and Related Systems, 419–422.
Copyright © 1979 by D. Reidel Publishing Company

slow tumbling motion of the protein was insufficient to average
out the anisotropy of the hyperfine couplings in the flavin radical (1,2). Hence, no resolved EPR spectra could be obtained for
the radicals of flavin enzymes and no hyperfine couplings could
be determined. It was suggested that ENDOR could be helpful to
extract such information in a case like this (2). A few years
later this was indeed realized in collaboration with Hyde (3,4)
who had constructed a suitable ENDOR instrument.

A similar situation is met when studying the EPR of frozen
solutions of biological transition metal ion complexes or models
for them. Because of large g-anisotropy and unresolved weak
proton hyperfine couplings it is rarely possible to determine
any hyperfine couplings directly from the powder EPR spectra.
Exceptions are of course the $A_{||}$ of Cu in copper proteins and in
case of heme proteins the strong coupling in the parallel direction to ^{19}F in the high spin fluoride compounds (5). ENDOR has
been particularly successful in case of heme (6,7) and iron
sulfur proteins (8,9). Hyperfine couplings to the iron, to
ligands and to nearby protons have been determined.

With spin labels normal EPR is capable to give information
about molecular rotation with correlation times in the region
10^{-11} to 10^{-7} s (10-13). Many interesting biomolecules and bio-
molecular aggregates move much slower. ST-EPR was successfully
developed with the aim to extend this time window up to 10^{-3} s
(14,15). ELDOR might turn out to be of particular importance for
investigation of slow anisotropic motion in this time domain (16).

ENDOR and ELDOR have recently been treated extensively in
two books (17,18) and ST-EPR applications in biochemistry has
been reviewed (19).

EPR DOUBLE RESONANCE EXPERIMENTS

In Fig. 1 the energy levels are shown schematically for a
spin system with S = 1/2, I = 1/2. The spin functions of the
four levels have been included. The spin lattice relaxation
mechanisms at thermal equilibrium leads to relative population
differences as indicated, with $\delta = \varepsilon = 1 - \exp(-g\beta B/kT)$, which
for kT >> $g\beta B$ gives $\varepsilon \approx g\beta B/kT$.

In a normal EPR experiment a microwave frequency ν_e is
applied. The microwave power is low, so that at resonance no
change in population difference occurs. The field B is swept
so that the two hyperfine transitions 1 - 4 and 2 - 3 are hit
in succession.

When a double resonance experiment is performed the field
B is adjusted so that ν_e (now called the observe frequency ν_{eo})

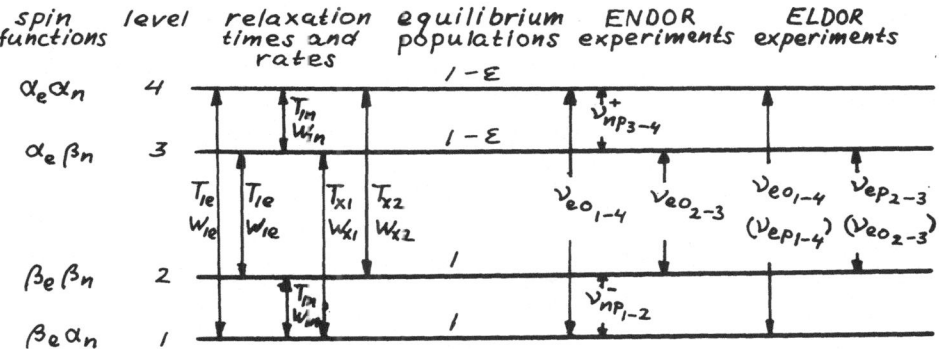

Fig. 1. Energy levels for a spin system with S = 1/2, I = 1/2. Relevant relaxation pathways are shown with their relaxation times T and rates W. Relative populations at thermal equilibrium are indicated. Observe and pump frequencies are shown for possible ENDOR and ELDOR experiments.

coincides with one of those hyperfine transitions, say the transition 1 – 4. A high microwave power has to be used so that partial saturation of the transition is obtained, i.e. the population difference between levels 1 and 4 is decreased, i.e. $\delta < \varepsilon$. This saturation is established against the relaxation rates W_e, W_n, W_x, and W_{x2} operating between the levels. In a double resonance experiment a second alternating field is applied over the sample.

In an ENDOR experiment this second field is a radio frequency field which is swept so that it can coincide with the frequencies of the nuclear transitions 1 – 2 or 3 – 4. Also this radiation must be of high power so that these transitions at resonance are saturated, at least partly saturated. This field is pumping between the two nuclear levels and is called the pumping field. Its frequency is the nuclear resonance pumping frequency ν_{np}. When ν_{np} operates between levels 3 and 4 it means that the transition rate between these two levels is enhanced. The result is that the total relaxation rate between levels 1 and 4 is enhanced. The observe frequency was partly saturating this transition. As a consequence of the increased total relaxation rate the saturation will be less marked, there will be a partial desaturation, i.e. the observed EPR signal will increase.

A similar but in general not identical result (since $T_{x1} \neq T_{x2}$) is obtained with the same observe transition, 1 – 4, but with ν_{np} operating between levels 1 and 2. For the observe transition between levels 2 and 3 the same two pump transitions, 3 – 4 and 1 – 2, will in principle lead to ENDOR response.

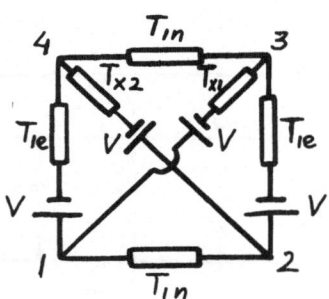

Fig. 2. Equivalent electrical circuit for the experiments depicted in Fig. 1.

In an ELDOR experiment we again saturate with the observe frequency between levels 1 and 4. The pumping field is in this case a second microwave frequency which is swept so that it can hit the other EPR transition, 2 - 3. The microwave pumping power must be large enough to cause saturation of this transition. This saturation enhances relaxation transitions between levels 1 and 4 and counteracts the observe saturation. Thus the ELDOR response is detected as an increase of the EPR signal at the observe frequency. An analogous but in general not identical response is obtained for the observe transition between levels 2 and 3 and the pump transition between levels 1 and 4.

It is worth noting that in ENDOR the observe and pump transitions always have one level in common whereas in ELDOR there is no common level for the two transitions.

From the previous it should be clear that saturation at both the observe transition and the pump transition is a requirement for a successful ENDOR or ELDOR experiment. In order to achieve these saturations the applied alternating fields must be strong enough to excite transitions more rapidly than the inherent relaxation mechanisms. It is important to determine the maximum enhancement factor for any given system. It turns out to be convenient for this purpose to use an equivalent electrical circuit as outlined in Fig. 2 for the same system as shown in Fig. 1. Each relaxation time is equivalent to a resistance. The relative population differences at thermal equilibrium are represented as voltages. Saturation between two levels of Fig. 1 means application of a low resistance or a short circuit between the corresponding two points in Fig. 2.

ENDOR SPECTRA

In order to interpret ENDOR spectra it is important to know

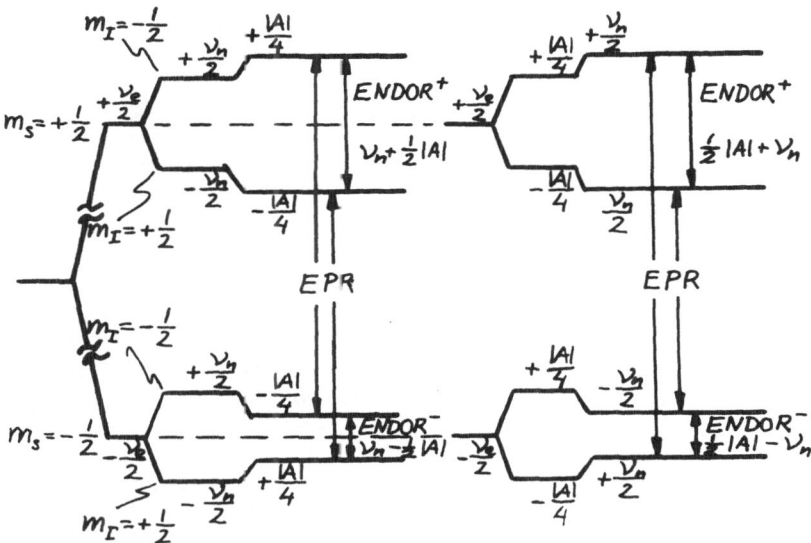

Fig. 3. Energy levels (in Hz) for a spin system with S = 1/2,
I = 1/2.

how an ENDOR spectrum is built up and what factors influence the
line intensities.

The energy levels for our spin system S = 1/2, I = 1/2 are
shown in Fig. 3. When the nuclear Zeeman frequency is larger than
half the hyperfine coupling, $\nu_n > \frac{1}{2}|A|$, the two ENDOR lines are
centered at ν_n and separated by $|A|$. When we change the magnetic
field, i.e. move from one EPR line to the other, ν_n, the center
of the lines, will change, whereas the spacing will remain exact-
ly $|A|$. This is depicted in Fig. 4a.

For $\nu_n < \frac{1}{2}|A|$, see Fig. 3, the two ENDOR lines will be cen-
tered at $\frac{1}{2}|A|$ and spaced by $2\nu_n$. When the magnetic field is
changed the center of the lines will remain unchanged but the
spacing $2\nu_n$ will change as is shown in Fig. 4b.

In either case the positions of the lines are described by
the equation

$$\nu_\pm = |\nu_n \pm \frac{A}{2}| \qquad (1)$$

By changing the magnetic field it is possible from the ENDOR
spectrum to determine the hyperfine coupling and the nuclear
Zeeman frequency, i.e. the kind of nucleus involved.

The scheme outlined is applicable to nuclei with I = 1/2,

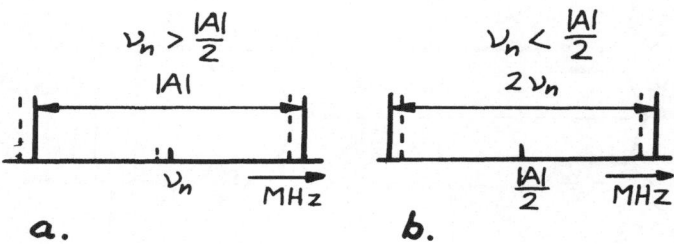

Fig. 4. ENDOR spectra for the system S = 1/2, I = 1/2 and the two cases $\nu_n > \frac{1}{2}|A|$ and $\nu_n < \frac{1}{2}|A|$. The solid lines are for EPR observation on the high field hyperfine line and the dashed lines for EPR observation on the low field hyperfine line.

i.e. protons and ^{57}Fe. Thus each set of equivalent protons will give a pair of lines as described. The scheme is easily extended to other nuclei, for instance ^{14}N in a solid sample. In this case an extra term, $\pm P_z$, describing the quadrupolar coupling in the direction observed, has to be included in Eqn. (1). This means that each set of equivalent nitrogen nuclei will give four lines. When looking for nitrogen ENDOR lines one has to examine the spectrum for lines centered around the nitrogen Zeeman frequency or spaced with twice that frequency.

Relative line intensities will depend on relationships between the relaxation times of the system. A few examples from our system S = 1/2 will make this clear.

Let us first consider the case $T_{1e} \approx T_{1n} \ll T_{x1}$, T_{x2}. This approximates the situation for radicals in solution. Fig. 5 shows the equivalent electrical circuits for the ENDOR experiments. In Fig. 5a the saturation between levels 1 and 4 by the observe microwave frequency is represented by the small resistance R_0. When the rf saturates between levels 4 and 3 a short circuit connects these two points. It is easy to show that if $R_0 \ll T_{1e}$, T_{1n} (i.e. efficient saturation) the voltage ratio V_b/V_a (the maximum enhancement factor) is approximated by the ratio of the conductances (the relaxation rates) between points (levels) 4 and 1 for the two cases.

$$\frac{V_b}{V_a} \approx \frac{T_{1e}^{-1} + (T_{1e} + T_{1n})^{-1}}{T_{1e}^{-1} + (T_{1n} + T_{1e} + T_{1n})^{-1}} = \frac{1.5}{1.33} = 1.12$$

The two ENDOR transitions are in this case predicted to give lines of equal intensity.

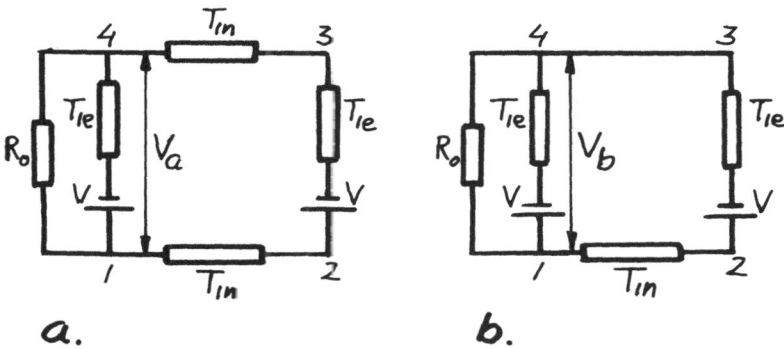

Fig. 5. Equivalent electrical circuits for the case $T_{1e} \approx T_{1n}$ $\ll T_{x1}$, T_{x2}. (a) With only observe microwave saturation between levels 1 and 4. (b) With additional ENDOR rf saturation between levels 3 and 4.

As long as we are dealing with isotropic hyperfine couplings we may neglect T_{x2}. Hence let us consider the case $T_{1e} \approx T_{1n} \approx T_{x1} \ll T_{x2}$. The equivalent electrical circuits are given in Fig. 6. Again the enhancement factor V_b/V_a may be approximated by the ratio between the total conductances between points 4 and 1.

For the ENDOR transition 4 – 3 we obtain

$$\left(\frac{V_b}{V_a}\right)_{4-3} \approx \frac{T_{1e} + T_{x1}^{-1} + \left(T_{1e} + T_{1n}\right)^{-1}}{T_{1e}^{-1} + T_{1n} + \left(\dfrac{1}{T_{x1}^{-1} + \left(T_{1e}+T_{1n}\right)^{-1}}\right)^{-1}} = \frac{2.5}{1.6} = 1.56$$

and for the other ENDOR transition

$$\left(\frac{V_b}{V_a}\right)_{2-3} \approx \frac{T_{1e}^{-1} + \left(T_{1n} + \dfrac{1}{T_{x1}^{-1} + T_{1e}^{-1}}\right)^{-1}}{T_{1e}^{-1} + \left(T_{1n} + \dfrac{1}{T_{1e}^{-1} + \left(T_{1e}+T_{1n}\right)^{-1}}\right)^{-1}} = \frac{1.67}{1.6} = 1.05$$

For the case $T_{1e} \approx T_{x1} \ll T_{1n}$, T_{x2} the just given expressions give the enhancement factors

$$\left(\frac{V_b}{V_a}\right)_{4-3} \approx 2 \quad \text{and} \quad \left(\frac{V_b}{V_a}\right)_{2-3} \approx 1$$

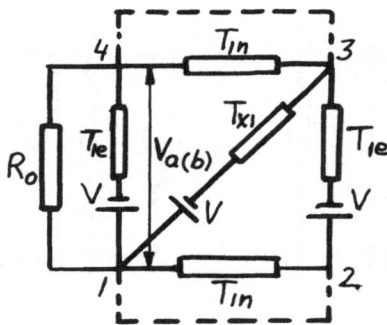

Fig. 6. Equivalent electrical circuits for the case $T_{1e} \approx T_{1n} \approx$ $T_{x1} \ll T_{x2}$. Observe microwave saturation is applied between levels 1 and 4. ENDOR rf saturation is applied between levels 4 and 3, or 1 and 2.

It is thus evident that as long as the cross relaxation pathway T_{x1} is rapid enough to give significant contribution to the total relaxation rate, the intensity of the two ENDOR lines will be un- equal. It is straightforward to include also the other cross re- laxation pathway T_{x2}. In general the relative intensity of ENDOR lines does not give a reliable value of the number of equivalent nuclei contributing to a transition.

The most striking feature of ENDOR when compared to EPR is the much improved resolution. The linewidths in the two types of spectra are roughly the same (100 to 500 kHz) but the spectral density differs. In an EPR spectrum the number of lines is built up in a multiplicative way from all sets of equivalent nuclei with hyperfine coupling, whereas in an ENDOR spectrum the number of lines is obtained as a sum over all the same sets of nuclei. In the ENDOR spectrum the different Zeeman frequencies of the different nuclei may give an additional spectral spreading not present in EPR.

ENDOR TECHNIQUE

In order to record ENDOR spectra one has to introduce a radio frequency field over the sample in the cavity of an EPR instrument. This rf field has to be swept over the widest possi- ble range from close to zero up to 40 MHz or higher. If only experiments at very low temperatures (liquid He temperatures) are intended the relaxation rates are low and the power of the rf field does not need to be particularly high. In this case it is not too difficult to obtain a wide sweep range without dis-

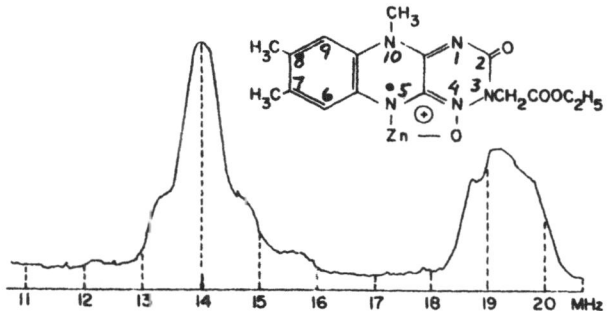

Fig. 7. Proton ENDOR spectrum of lumiflavin radical chelate with zinc in dimethylformamide at ca 113 K. The free proton frequency has been adjusted to 14 MHz. From ref. 4.

turbing power changes. Feher, who was the first to make ENDOR experiments, and his collaborators have used this technique (20-22). If one wants to make experiments at higher temperatures, where the relaxation rates are larger, the rf field must be of higher power. Hyde has described (23,24) a useful high power ENDOR instrument where the tuning of the rf output circuit is servo controlled.

It is important to include a possibility to vary the temperature of the sample over the largest possible range. In this way it is possible to find the best experimental conditions for balance between available microwave and rf power, speed of relaxation processes and internal molecular motion.

In general, a signal to noise ratio of the EPR spectrum of 100:1 or more is necessary in order to make it worthwhile to look for ENDOR signals.

ENDOR APPLICATIONS

As mentioned already in the introduction, the first extensive applications of ENDOR in biophysical chemistry were made on flavin radicals. In the early experiments the flavin radicals were stabilized as metal chelates (4), see Fig. 7. A helpful feature is that rapidly rotating methyl groups give strong ENDOR signals with a small axial anisotropy. The broad line of Fig. 7 between 18.3 and 20.3 MHz is composed of contributions from the methyl groups bound to C-8, peak at 19 MHz, and N-10, peak at 18.5 MHz. Definitive assignments of the various peaks were made with isotopic and chemical substitutions (4). The peak centered

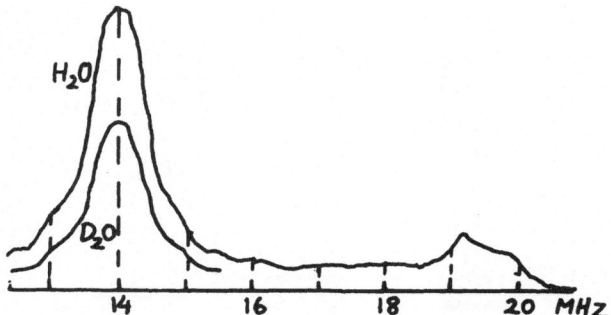

Fig. 8. Proton ENDOR spectra of NADPH dehydrogenase in H_2O and
D_2O at 153 K. The free proton frequency has been adjusted to
14 MHz. From ref. 26.

at the free proton frequency 14.0 MHz is due to weak coupling to
nearby protons of the solvent and has been called the matrix
ENDOR signal (25).

     Fig. 8 shows spectra from a partly reduced flavin enzyme in
H_2O and in D_2O (26). The line with peak at 19.2 MHz is the axial
signal from the methyl group at position C-8. This group in the
protein is hence still rotating rapidly at 153 K. The methyl
group at N-10 is now substituted by a ribityl group. It cannot
rotate and the two remaining methylen protons give different
couplings. The matrix ENDOR signal at 14 MHz is reduced by 40%
when the enzyme is dissolved in D_2O. This shows that the flavin
radical is only partly surrounded by water protons and/or exchangeable protein protons.

     The spin density on C-8 of the flavin radical depends on
the protonation state of the radical. This is then also reflected in the size of the coupling to the methyl protons, which is
in the range 9.5 to 11.4 MHz for anionic radicals and in the
range 6.7 to 8.4 for neutral radicals (4,27). Dependent on the
solvent pH and the effect of the nearby amino acid side chains
either of these two radicals could be found in a flavin enzyme
(27).
     The axial symmetry of the signal of the C-8 methyl group
may be used to further resolve the spectrum. This is illustrated
in Fig. 9. The axis of the methyl group is along the bond to C-8
(cf Fig. 7). The peak of the signal (smallest coupling) corresponds to the plane perpendicular to this bond. One direction in
this plane is also perpendicular to the benzenoid ring of the
flavin. The hyperfine couplings to N-5 and N-10 both have large
anisotropy with the axes of largest coupling along the p-orbitals

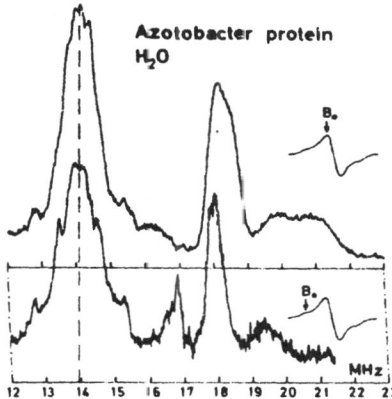

Fig. 9. Proton ENDOR spectra of the Azotobacter flavoprotein at
153 K. Recordings at two selected positions on the EPR spectrum
as indicated on insets. From ref. 27.

perpendicular to the flavin plane. If we sit on the peak of the
EPR derivative signal radicals with a multitude of orientations
contribute and a broad axial-like ENDOR signal is obtained. If
we sit far out on the wing of the EPR signal only molecules with
the N-5 and N-10 p-orbitals in the direction of the field are
selected and a sharp ENDOR signal is obtained corresponding to
the low frequency edge of the broad signal in the other spectrum.
This proves in fact that the flavin radical has a planar configuration in the protein. In Fig. 9 it is also seen that other signals change shape and position, both in the matrix portion and
elsewhere, showing that also these signals are anisotropic.

When these ENDOR experiments were carried out it had for
quite some time been a challenge to find out on which point of
the flavin molecule the covalent link to the protein was present
in enzymes such as succinate dehydrogenase, from which flavin
could not be dissociated in the usual way. It could be shown (28)
that the signal from the C-8 methyl group was absent in ENDOR
spectra of the radical of the flavin peptide obtained from succinate dehydrogenase. This demonstrated that the covalent linkage
must be to this point of the molecule, which has later been confirmed in other ways (29).

Another free radical problem that has been solved by ENDOR
concerns the nature of the chlorophyll radical formed in reaction
centers of photosynthetic bacteria as a result of electron donation to the primary acceptor. Feher et al. (30) and Norris et al.
(31) showed independently that the hyperfine coupling to methyl

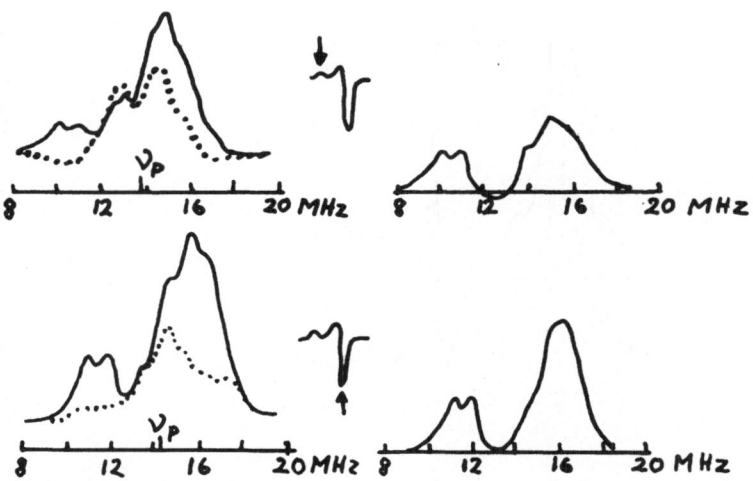

Fig. 10. ENDOR spectra of oxidized native (·····) and ^{57}Fe-
enriched (————) high-potential iron-sulfur protein with difference spectra (to the right). Positions on EPR spectrum are indicated on insets. Temperature 4.2 K. From ref. 9.

groups of the chlorophyll radicals obtained by chemical oxidation of pure monomeric chlorophyll are twice as large as the
corresponding hyperfine couplings in light-induced radicals in
the reaction centers. These findings prove that the entity of
the reaction center carrying the radical is a dimer of bacterial
chlorophyll.

In iron-sulfur proteins it is often easy to split off the
iron and labile sulfide ions and later reconstitute the active
protein. In this way ^{57}Fe may be introduced. Sands and his collaborators have made penetrating investigations of the iron ENDOR
of reduced two-iron two-sulfur ferredoxins with enriched ^{57}Fe (8).
Their results showed that the two-iron atoms of each center are
inequivalent, and that the total electronic spin is 1/2. A strong
support was obtained for the model containing a high-spin Fe(III)
atom and a high-spin Fe(II) atom in an antiferromagnetically
coupled pair forming the S = 1/2 system.

Also the more complicated systems containing centers consisting of iron atoms and four labile sulfur atoms have been
studied by ENDOR (9). From this work two spectra obtained with
the oxidized high-potential iron-sulfur protein from Chromatium
are shown in Fig. 10. The iron ENDOR spectra overlap with the
proton spectra so that it is necessary to calculate the difference spectra. The low frequency peak is a clear doublet with a

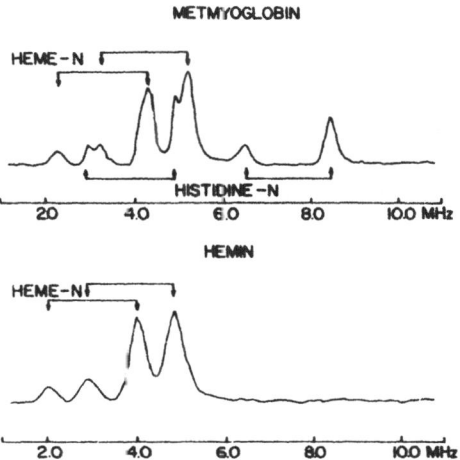

Fig. 11. Part of the ENDOR spectra of metmyoglobin and hemin showing the interaction of the Fe^{3+} spin with its ligand nitrogen nuclei. Temperature 2.1 K. A set of equivalent nitrogen nuclei gives two pairs of lines with a separation within each pair of twice the free nitrogen frequency. From ref. 6.

separation of twice the iron nuclear Zeeman frequency (0.9 MHz). The doublet of the peak at higher frequency is more difficult to resolve since it is obtained as a difference on the slope of the overlapping proton signals. The EPR spectrum, shown as inset, reveals that the system has nearly axial symmetry. Both ENDOR peaks move toward higher frequencies when the EPR observation field moves from a position selecting the parallel direction to a position selecting perpendicular directions. It is concluded that there are two types of iron with the approximate hyperfine components "$A_\parallel$" = 21.2 (29.8) MHz and "$A_\perp$" = 29.8 (32.0) MHz, respectively (9).

The investigation of heme proteins with ENDOR has been pioneered by Feher and his collaborators (6,7). In Fig. 11 the ^{14}N ENDOR signals are shown for metmyoglobin observed in the parallel direction, i.e. g = 2 (6). Pairs of peaks separated by twice the nitrogen nuclear Zeeman frequency are identified. Comparison with hemin shows which signals come from the heme-nitrogens and which from the histidin-nitrogens. Likewise it was shown (32) that with methemoglobin of a mutant hemoglobin of type M, $HbM_{Hyde\ Park}$, with the proximal histidine of the β chain replaced by tyrosine the ENDOR spectrum was similar to that of hemin, i.e. the signals of histidine were lacking. Also the hyperfine interaction with ^{57}Fe of metmyoglobin and methemoglobin has been

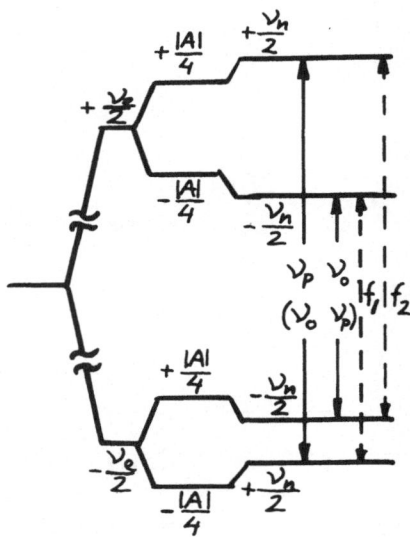

Fig. 12. Energy levels for a spin system with $S = 1/2$, $I = 1/2$, and $\nu_n < \frac{1}{2}|A|$. The two possibilities for allowed(pump)-allowed (observe) transitions in an ELDOR experiment are indicated. The two "forbidden" transitions ($\Delta m_I = \pm 1$) are shown by broken lines. When the pump frequency coincides with one of those transitions a forbidden-allowed ELDOR line may be obtained.

studied (33). Because of the very high precision of the ENDOR method definitive differences between the two proteins could be detected. Scholes and van Camp have extended this type of work and measured the ENDOR signals of protons and nitrogen ligands also in low spin forms of hemoproteins (34).

ELDOR SPECTRA

An example of an energy level diagram for a system with $S = 1/2$, $I = 1/2$ is shown in Fig. 12. The allowed-allowed transitions give a single line at $|\nu_p - \nu_o| = |A|$ MHz, which is independent on the magnetic field. Depending on the transition probabilities and the relaxation rates other lines may also be observed: forbidden-allowed lines at $|1/2 A \pm \nu_n|$ MHz and a forbidden-forbidden line at $2\nu_n$ MHz. The positions of these lines obviously depend on the magnetic field.

The equivalent electrical circuit diagrams described in connection with ENDOR may be used to calculate the maximum values of the enhancement factor which again depends on the relaxation rates of the system.

ELDOR TECHNIQUE

In order to make an ELDOR experiment, two microwave frequencies must be applied over the sample. Basically this means that the EPR microwave bridge must be supplemented with an extra microwave source and a cavity with facility to irradiate with the two frequencies, e.g. a bimodal cavity or a cavity combined with a helix. A system must also be provided for sweeping the pump klystron frequency and the cavity tuning combined with counting of the frequency difference for x-axis drive of the recording device. These solutions are discussed in detail in the book by Kevan and Kispert (17). In summary, the technical requirements for ELDOR experiments are rather complicated.

ELDOR APPLICATIONS

Not many biochemical applications have hitherto been made with the ELDOR technique. One interesting example is offered by xanthinoxidase, an enzyme containing iron-sulfur centers, molybdenum ions and a flavin coenzyme (FAD). All centers may obtain paramagnetic properties. During a visit in our laboratory Bray found by EPR that the signal from Mo-ions was split at low temperatures. This phenomenon was investigated further by Bray and collaborators (35) and it was suggested that the splitting was due to an isotropic coupling to one iron-sulfur center about 30 Å away. Lowe and Hyde made ELDOR and EPR saturation studies of this system (36). The ELDOR results showed that the coupling had more anisotropic dipolar character than previously believed, but still compatible with a distance of about 30 Å between the centers. The relaxation coupling between the centers, however, suggested that the distance would be of the order of 2 Å (36). This paradox remains to be solved.

ELDOR, SATURATION TRANSFER EPR AND MOLECULAR MOTION

Spin labels are paramagnetic reporter groups that can be included in a biological or biochemical system and give information about its motional and/or conformational state. Particularly stable nitroxide free radicals have found a wide use (37). The normal EPR spectra of these spin labels are sensitive for rotational motions with correlation times τ_c in the range 10^{-11} to 10^{-7} s (10-13). In this section the possibility will be discussed to extend this time window to 10^{-3} s by means of ELDOR (16) and ST-EPR (14,15).

In a nitroxide spin label the major hyperfine coupling is along the p-orbital (see Fig. 13). In the perpendicular plane the hyperfine coupling is approximately constant. For the present

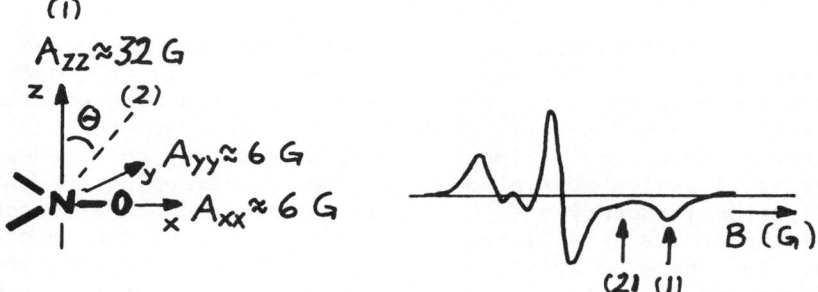

Fig. 13. Left: Schematic drawing of the nitroxide group with hyperfine coupling components. Right: EPR spectrum of a spin label in a high viscosity medium. Fields (1) and (2) correspond to the directions similarly labeled in the drawing at left.

discussion we will neglect the g-anisotropy. When $\tau_c > 10^{-7}$ s a powder-like EPR spectrum is obtained as depicted in Fig. 13. The wings, the turning points, are due to molecules oriented with A_{zz} parallel to the field. An intense microwave irradiation at for instance the high field turning point (point 1) in the time interval from t = 0 to t = Δt saturates the molecules (number of molecules δn_0) oriented at that moment in that direction. At time t the rotational diffusion of the spin label molecules has changed the orientation of these molecules, partly randomized their orientation. The resonant field of each molecule will correspond to another part of the spectrum. Some of them will give resonance at the field indicated by the arrow labeled 2. The portion of the molecules that are still saturated depends on T_{1e}:

$$\delta n_t = \delta n_0 \exp(-t/T_{1e})$$

With a second weak (non-saturating) observe frequency at position 2 the degree of saturation at this position of the spectrum could be determined. The property of saturation stored in the molecules has been transferred from one portion of the spectrum to another by spectral diffusion. The distance in the spectrum that can be influenced in this way depends on (a) the anisotropy of the magnetic interactions A and g, (b) the rotational correlation time τ_c, and (c) the saturation memory time T_{1e}.

For nitroxide spin labels in biochemical systems T_{1e} has been determined to be ca 10^{-5} s (38). Hence it will be possible to study the range $\tau_c \approx 10^{-7}$ to 10^{-3} s.

Such ELDOR experiments can be made in two modes: (I) Pulsed, time domain ELDOR: Pulsing pump power at position 1, and observing the appearance and disappearance of the response to the puls at position 2; and (II) Staticnary ELDOR: Continuous pumping at position a, and measuring the steady state EPR signal at position 2, with and without power at position 1. In either case one of the frequencies may be swept over the spectrum. The time domain ELDOR will be of particular value for investigation of the details of anisotropic motion which is prevailing in biological systems. Stationary ELDOR has been explored by Hyde (16) and it was demonstrated that very slow motions with correlation times of the order of 10^{-3} indeed could be studied by this method. The drawback of these two methods is of course that the complicated ELDOR instrumentation is needed.

An alternative and technically much simpler method is offered by saturation transfer EPR (ST-EPR). The usefulness and possibilities of this method have been fully demonstrated by Hyde, Thomas and Dalton and their collaborators (14,15,19).

In this method the EPR microwave power is of a strength sufficient to partly saturate the spin system. Because of the field modulation, usually 5 G ptp, the EPR irradiation and detection point is continuously moving over a portion of the spectrum. The saturation achieved at each point is simultaneously spreading over the spectrum by spectral diffusion. The two processes will interfere with each other in governing the passage of spins through the resonance condition. The result is dependent on (a) the sensitivity of the resonance field to an orientational change $\partial B_{res}/\partial\theta$, (b) the rotational correlation time τ_c, (c) the saturation memory time T_{1e}, and (d) the modulation angular frequency ω_m.

It turns out that the sensitivity to changes in rotational correlation time is optimum when $\omega_m^{-1} \approx \tau_c \approx T_{1e}$. The best sensitivity for long τ_c (slow motion) is obtained at spectral portions between the outer ard central turning points, where $\partial B_{res}/\partial\theta$ is largest.

Several different modes of detection are possible. It has been shown (15,19) that detection of the second harmonic absorption $90°$ out of phase is most suitable. Since $T_{1e} \approx 10^{-5}$ s, a field modulation of 50 kHz with detection at 100 kHz, the standard frequency of modern EPR instruments, is close to optimum. It is technically a fairly simple thing to add the 50 kHz modulation facility and the necessary phase control to an existing EPR spectrometer.

In principle the detection of the first harmonic dispersion $90°$ out of phase would be the best, but the klystron noise is coming through and gives poor sensitivity.

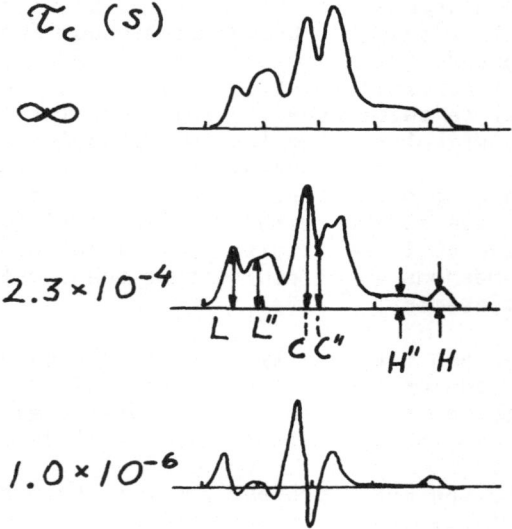

Fig. 14. Saturation-transfer EPR spectra from spin labeled hemoglobin (maleimide label) in glycerol-water mixtures of different
viscosities. Reference curves are obtained from amplitude ratios
measured at the positions indicated. From ref. 15.

Examples of spectra are shown in Fig. 14. These are for
hemoglobin with a maleimide spin label attached (15). The spin
label is firmly bound, the rotational motion of the molecule is
very nearly isotropic and τ_c may be calculated for various viscosities by means of the Stokes Einstein formula $\tau_c = 4\pi\eta r^3/3kT$.
Spectra of this kind are used as standard spectra and the parameters C'/C, H"/H and L"/L are measured and plotted versus τ_c to
give standard curves for evaluation of other experimental spectra
(15,19). The parameter H"/H is best for measurement of slow motion
since $\partial B_{res}/\partial\theta$ has its largest value at the field of H". The parameter C'/C is best for estimates of rapid motions since $\partial B_{res}/\partial\theta$
is small in this spectral region.

Extensive studies have so far been made on spin labeled
muscular myosin and its derivatives by Thomas et al. (39). The
usefulness of the method is emphasized by the excellent agreement with analogous measurements using fluorescence depolariza-
tion (19). Applications to membrane systems are dealt with by
Devaux in another chapter of this book.

It should be emphasized that when anisotropic motion is
investigated comparison with standard spectra and standard curves
obtained with hemoglobin, which has an isotropic motion, must be
made with great caution.

The effects observed by ST-EPR may also be calculated theoretically. This has been done by Dalton and coworkers (15,40). In order to obtain accurate theoretical simulations of ST-EPR spectra quite expensive computer calculations are however needed. Nevertheless, the results are very promising and the mechanisms are well understood. Also anisotropic motion may be dealt with in this way (40).

REFERENCES

1. Ehrenberg, A.: 1960, Acta Chem. Scand. 14, pp. 766-767.
2. Ehrenberg, A.: 1962, Arkiv Kemi 19, pp. 97-117.
3. Ehrenberg, A., Eriksson, L.E.G., and Hyde, J.S.: 1968, Biochim. Biophys. Acta 167, pp. 482-484.
4. Eriksson, L.E.G., Hyde, J.S., and Ehrenberg, A.: 1969, Biochim. Biophys. Acta 192, pp. 211-230.
5. Ehrenberg, A.: 1966, in "Instrumentation in Biochemistry" (ed. T.W. Goodwin), Academic Press, New York, Biochem. Soc. Symp. No. 26, pp. 41-51.
6. Scholes, C.P., Isaacssor, R.A., and Feher, G.: 1972, Biochim. Biophys. Acta 263, pp. 448-452.
7. Scholes, C.P.: 1979, in "Multiple Electronic Resonance Spectroscopy" (eds. M.M. Dorio and J.H. Freed), Plenum Press, New York.
8. Fritz, J., Anderson, R., Fee, J., Palmer, G., Sands, R.H., Tsibris, J.C.M., Gunsalus, I.C., Orme-Johnson, W.H., and Beinert, H.: 1971, Biochim. Biophys. Acta 253, pp. 110-133.
9. Anderson, R., Anger, G., Petersson, L., Ehrenberg, A., Cammack, R., Hall, D.O., Mullinger, R., and Rao, K.K.: 1975, Biochim. Biophys. Acta 376, pp. 63-71.
10. Stone, R., Buckman, T., Nordio, P., and McConnell, H.M.: 1965, Proc. Nat. Acad. Sci. U.S. 54, pp. 1010-1017.
11. McGalley, R.C., Shimshick, E.J., and McConnell, H.M.: 1972, Chem. Phys. Letters 13, pp. 115-119.
12. Goldman, S.A., Bruno, G.V., Polnaszek, C.F., and Freed, J.H.:1972, J. Chem. Phys. 56, pp. 716-735.
13. Israelachvili, J., Sjösten, J., Eriksson, L.E.G., Ehrström, M., Gräslund, A., and Ehrenberg, A.: 1975, Biochim. Biophys. Acta 382, pp. 125-141.
14. Hyde, J.S., and Thomas, D.D.: 1973, Ann. N.Y. Acad. Sci. 222, pp. 680-692.
15. Thomas, D.D., Dalton, L.R., and Hyde, J.S.: 1976, J. Chem. Phys. 65, pp. 3006-3024.
16. Hyde, J.S., Smigel, M.D., Dalton, L.R., and Dalton, L.A.: 1975, J. Chem. Phys. 62, pp. 1655-1667.
17. Kevan, L., and Kispert, L.D.: 1976, "Electron Spin Double Resonance Spectroscopy", Wiley.
18. Dorio, M.M., and Freed, J.H. (eds.): 1979, "Multiple Resonance Spectroscopy", Plenum Press.

19. Hyde, J.S.: 1978, in "Methods in Enzymology" 49, pp. 480-511.
20. Feher, G., and Gere, E.A.: 1956, Phys. Rev. 103, pp. 501-503.
21. Feher, G.: 1956, Phys. Rev. 103, pp. 834-835.
22. Feher, G.: 1959, Phys. Rev. 111, pp. 1219-1244.
23. Hyde, J.S.: 1965, J. Chem. Phys. 43, pp. 1806-1818.
24. Ranon, U., and Hyde, J.S.: 1966, Phys. Rev. 141, pp. 259-274.
25. Hyde, J.S., Rist, G.H., and Eriksson, L.E.G.: 1968, J. Phys. Chem. 72, pp. 4269-4276.
26. Eriksson, L.E.G., Ehrenberg, A., and Hyde, J.S.: 1970, Eur. J. Biochem. 17, pp. 539-543.
27. Eriksson, L.E.G., and Ehrenberg, A.: 1973, Biochim. Biophys. Acta 293, pp. 57-66.
28. Walker, W.H., Salach, J., Gutman, M., Singer, T.P., Hyde, J.S., and Ehrenberg, A.: 1969, FEBS Letters 5, pp. 237-240.
29. Salach, J., Walker, W., Singer, T.P., Ehrenberg, A., Hemmerich, P., Ghisla, S., and Hartmann, U.: 1972, Eur. J. Biochem. 26, pp. 267-278.
30. Feher, G., Hoff, A.J., Isaacsson, R.A., and McElroy, J.D.: 2973, Biophys. Soc. Abstr. VIIth Ann. Meeting, p. 61.
31. Norris, J.R., Druyan, M.E., and Katz, J.J.: 1973, J. Am. Chem. Soc. 95, pp. 1680-1682.
32. Feher, G., Isaacsson, R.A., Scholes, C.P., and Nagel, R.: 1973, Ann. N.Y. Acad. Sci. 222, pp. 86-101.
33. Scholes, C.P., Isaacsson, R.A., Yonetani, T., and Feher, G.: 1973, Biochim. Biophys. Acta 322, pp. 457-462.
34. Scholes, C.P., and van Camp, H.L.: 1976, Biochim. Biophys. Acta 434, pp. 290-296.
35. Lowe, D.J., Lynden-Bell, R.M., and Bray, R.C.: 1972, Biochem. J. 130, pp. 239-249.
36. Lowe, D.J., and Hyde, J.S.: 1975, Biochim. Biophys. Acta 377, pp. 205-210.
37. Berliner, L.J. (ed.): 1976, "Spin Labeling, Theory and Applications", Academic Press, New York.
38. Huisjen, M., and Hyde, J.S.: 1974, Rev. Sci. Instrum. 45, pp. 669-675.
39. Thomas, D.D., Seidel, J.C., Gergely, J., and Hyde, J.S.: 1975, J. Supramol. Struct. 3, pp. 376-390.
40. Dalton, L.R., Robinson, B.H., Dalton, L.A., and Coffey, P.: 1976, Adv. Magn. Res. 8, pp. 149-259.

RECENT INSTRUMENTAL DEVELOPMENTS ON ENDOR SPECTROSCOPY

Arturo Colligiani

Laboratorio di Chimica Quantistica ed Energetica Mole-
colare of the C.N.R. (National Research Council) and
Institute of Physical Chemistry, University of Pisa,
56100 PISA, Italy.

When someone wants to build a continuous-wave ENDOR spectrometer,
it is usually sufficient to add to a standard ESR spectrometer
some ancillary external electronic components. The most important
of these external components is the radio-frequency (r.f.) power
amplifier which must allow a high r.f. current to flow in the
"loop" irradiating the sample inside the microwave cavity. The microwave cavity is the main point in an ENDOR spectrometer. It must
be decided if a standard rectangular (TE_{102}) or cylindrical (TE_{011})
cavity is to be used or if one wants to develop a special cavity
working in a different mode (for instance TM_{110}). The choice between these two possibilities is conditioned by the conformation
of the loop which generates the r.f. magnetic field B_E on the sample and by the proper orientation of B_E with respect to the static
magnetic field B_0 and to the microwave field. In fact the B_E field
has to be orthogonal to B_0 and it can be indifferently orthogonal
or parallel to the microwave field in order to excite the allowed
ENDOR transitions ($\Delta m_I = \pm 1$). The loop generating the ENDOR field
B_E can be 1) a true multi-turn r.f. coil or 2) a single-turn r.f.
coil which can be built by metallic rods or strips. In both cases
all the metallic parts of the loop must be orthogonal to the electric component of the microwave field in order to reduce to a minimum the coupling between them, if the Q factor of the cavity is
to be saved. In the case 1), for instance, a multi-turn coil cannot be introduced into a standard cavity with its axis parallel
to the axis of the cavity itself. In fact in doing so, evidently
the electric coupling is at a maximum and the Q factor becomes
nearly zero also if the ENDOR field is properly oriented at 90°
with respect to B_0. On the contrary, a multi-turn coil can be introduced into a standard cavity through its collet, without a dras-

I. Bertini and R.S. Drago (eds.), ESR and NMR of Paramagnetic Species in Biological and
Related Systems, 419–422.

tic lowering of the Q factor, if the axis of the coil is held orthogonal to the axis of the cavity (1). It is clear that the axis
of the coil can be easily placed also orthogonal to B_0. In this
way ENDOR fields up to about 15 G can be obtained <u>at a given fre-
quency</u> but their values rapidly vary with the frequency as a consequence of the variation of the impedance of the coil. The constancy of the B_E value at different frequencies and during a sufficiently broad frequency sweep has not received due attention in
the past. The constancy of B_E can be particularly important when
studies on ENDOR lineshapes are to be carried out. The variation
of the impedance of the coil also gives origin to current reflections in the irradiating line with a consequent poor exploitation
of the r.f. power amplifier. Moreover, a given value of B_E can be
found only in a very small volume at the center of the coil, hence
limiting the use of this system to small samples. However, it must
be mentioned that the study of this system has suggested an interesting way of measuring the effective B_E value by taking into account the sidebands generated when the B_E field is used as a modulation of the B_0 field, placing the two fields parallel (2).
A more interesting ENDOR spectrometer still of the type 1) has
also been built (3); in such spectrometer the side-wall of the cavity, operating in the TE_{011} mode, is also the ENDOR multi-turn
r.f. coil. In this apparatus the irradiating coil is constantly
held at the resonance during a frequency sweep, by a discriminator
circuit and the impedance of the resonant circuit is continuosly
matched to the 50 Ω output impedance of the r.f. power amplifier.
In this way power reflections in the irradiating line are reduced
to a minimum and the B_E value is very constant also during very
broad frequency sweeps (of the order of 20 MHz). The obtained B_E
values (with a r.f. power amplifier of 1 KW) are of the order of
some tens of Gauss over a large volume allowing the study of large
solid or liquid samples. The ENDOR signals can be processed by an
on-line minicomputer which also controls many functions of the
spectrometer. Although the very good features of the above-de-
scribed instrument, it is to be stressed that its actual construction is rather difficult especially for people not so expert in
general electronics and who want to carry out ENDOR experiments
by the use of an ESR apparatus without important modifications.
A more simple ENDOR apparatus to build is the one using a special
TM_{110} cavity (4). It is a cavity of the cylindrical type but, differently to the standard TE_{011} mode, the electric component of the
microwave field is parallel to the axis of the cavity. This means
that a multi-turn r.f. coil can be introduced with its axis parallel to the axis of the cavity without influencing the Q factor.
The microwave magnetic field (laying on planes orthogonal to the
axis of the cavity) has a maximum all along the axis of the cavity
where the coil and the sample are placed. For this reason relatively large samples can be employed. As verified in our Laboratory, it is easy to suppress other spurious resonance modes which
tend to settle simply by a careful choice of the ratio between the

diameter and the length of the cavity (5). The only disadvantage of this cavity is its inherently low Q value. We have also verified that it is not necessary to use a coaxial cable in order to excite the TM_{110} mode, as indicated in ref.(4). The cavity can be driven by a standard rectangular waveguide *via* the usual circular iris open in the side-wall. The multi-turn r.f. coil has been included in a series resonant circuit which, having a very low Q value of about one, is characterized by a large bandwidth of some tens of MHz. This means that the r.f. current in the coil, and therefore the B_E value, varies only of few percent during a frequency sweep not wider than the bandwidth. In order to have such a low Q value, the 50 Ω output impedance of the r.f. power amplifier has been stepped up to 200 Ω by a broadband "coaxial transformer" I shall describe in the following of this lesson. The irradiation line is terminated on a 200 Ω dummy load, which is also the resistance of the resonant circuit, in order to reduce to a minimum the power reflections. In doing so, along the axis of the r.f. coil, B_E values of about 15 Gauss can be obtained by using a r.f. power amplifier of only 100 W.

The type 1) ENDOR spectrometers, that is those using a single-turn irradiation loop made of metallic rods or strips, are surely the easiest to build at least as far as the cavity is concerned. This can be a standard rectangular or cylindrical one into which the irradiating loop can be introduced either through the upper collet or it can be built steadily in its inside.

In an interesting case (6) the side-wall of the cylindrical cavity has been constructed by using a rectangular block of a material called "Wonderstone", which is a naturally occurring rock found in South Africa. Wonderstone is rather soft, it can be readily machined, after firing at 1090°C it becomes very hard and can be coated with a thin layer of conducting silver paint. The important advantages of Wonderstone are its extremely low thermal expansion coefficient ensuring a very good frequency stability and its permeability to 100 KHz modulation fields. The ENDOR irradiating rods, symmetric with respect to the center of the cavity and parallel to its axis, pass through Teflon sleeves fixed at the ending brass plates. The metallic rods are directly connected to the r.f. power amplifier and to a R = 50 Ω resistor terminating the irradiation line. In this way, however, from the relationship:

$$B_E = (4\mu_0/2\pi)(i/d) = 8 \cdot 10^{-3} \, i/d \text{ Gauss}$$
$$i = \text{r.f. current in Ampères.}$$
$$d = \text{distance between the rods in meters.}$$

which gives the ENDOR field value in the middle of the two rods where the sample is located, it is easy to verify that if R = 50 Ω, B_E = 1.13 G when using a r.f. power of 100 W, i = 1.41 A and d = 10^{-2} m. Therefore, the ENDOR field in this system is very small and can be considered useful only when the relaxation times of the ENDOR transitions are particularly long as, for instance, at very low temperatures. Therefore, in order to increase the r.f. current in the irradiating loop, and so the value of B_E, while using the same r.f. power of 100 W, we have recently built (7) in our Laboratory an ENDOR apparatus of the type 1) in which the usual 50 Ω output impedance of the power amplifier has been stepped down to 5.55 Ω, i.e. of a factor of 9. In doing so the r.f. current in the loop, actually made of a double silver strip introduced inside a standard rectangular cavity through the upper collet, increases from 1.41 A to 4.2 A. If a value of d = $6 \cdot 10^{-3}$ m is used, an ENDOR field of B_E = 5.64 G is obtained and it allows to observe good spectra from organic radicals also at room temperature. The stepping down of the impedance is easily accomplished by the "coaxial transformers" (8) already mentioned which are built by coiling a few turns of coaxial cable of proper characteristic impedance on suitable toroids of ferromagnetic material. When the inductance of the irradiating loop is introduced in a series resonant circuit, the system works well till 65 MHz and the B_E value is held constant within 4% during a frequency sweep of ±3 MHz around a given resonance frequency of the tuned circuit.

Up to now I have summarized the more relevant features of continuous-wave ENDOR spectrometers i.e. apparatus in which both the microwave and the r.f. fields are continuosly applied on the paramagnetic sample. However, it is not difficult to foresee that in the future pulsed experiments will become more and more widely exploited both from a theoretical point of view (9) and in order to increase the sensitivity of the spectrometers.

An interesting ENDOR cavity for spin-echo experiments has been recently published (10); in this cavity, by using a half wave stripline resonator in the X-band, both the microwave and the r.f. fields are applied as short pulses with high repetition rate.

REFERENCES

(1) Miyagawa, I., Davidson, R.B., Helms, H.A.Jr., and Wilkinson, B.A.Jr.: 1973, J.Magn.Resonance 10, pp. 156-161.
(2) Miyagawa, I., Hayashi, Y., and Koiaki, Y.: 1977, J.Magn.Resonance 25, pp. 183-195.
(3) Gruber, K., Forrer, J., Schweiger, A., and Günthard, H.Hs.: 1974, J.Phys.E: Sci.Instrum. 6, pp. 569-574.
(4) Biehl, R., Plato, M., and Möbius, K.: 1975, J.Chem.Phys. 63, pp. 3515-3522.
(5) Poole, C.P.Jr.: Electron Spin Resonance, Interscience Publishers, John Wiley & Sons, Inc., New York, 1967.
(6) Chacko, V.P.: 1978, Rev.Sci.Instrum. 49, pp. 1012-1014.
(7) Colligiani, A., Pinzino, C., and Bertolini, M.: 1979, J.Magn. Resonance 33, pp. 511-518.
(8) Ruthroff, C.L.: August 1959, Proc.IRE, p. 1337.
(9) Stillman, A.E., and Schwartz, R.N.: 1978, Molec.Phys. 35, pp. 301-313.
(10) Davis, J.L., and Mims, W.B.: 1978, Rev.Sci.Instrum. 49, pp. 1095-1097.

SUBJECT INDEX

423